Medical: M
40 -
CCO 618

AF681206

Muscle and Motility

Volume 2

Muscle and Motility Volume 2

Proceedings of the XIXth European Conference in Brussels

Edited by

G. Maréchal and U. Carraro

Intercept
Andover, Hampshire

British Library CIP data available

ISBN 0 946707 34 0

Copyright

© Intercept Ltd 1990

All rights reserved. No part of this publication may be reproduced (including photocopying), stored in a retrieval system of any kind, or transmitted by any means without the written permission of the Publishers. Permission to copy is granted to libraries and other users on the condition that the appropriate fee is paid directly to the Copyright Clearance Center, 21 Congress Street, Salem, MA01970, USA. For *Muscle and Motility* the copying fee per chapter is $20.00.

Published in December 1990 by Intercept Limited, P.O. Box 716, Andover, Hants SP10 1YG, England.

Laserset in 'Times' by
Reachword Limited, Sudbrooke, Lincoln.

Printed in Great Britain by
Athenaeum Press, Newcastle upon Tyne.

Contributors

ACHAZI R. K., *Institut für Tierphysiologie, Freie Universität Berlin, Grunewaldstrasse 34, D-1000 Berlin 41, Germany.*

ARPESELLA G., *Chirurgia del Cuore e dei Grossi Vasi, University of Bologna, Italy.*

ASMUSSEN G., *Carl-Ludwig-Institute of Physiology, University of Leipzig, Liebigstrasse 27, 7010 Leipzig, Germany.*

AUGIER N., *Pathologie Moléculaire du muscle, INSERM U300, Faculté de Pharmacie, Avenue Charles Flahault, 34060, Montpellier, Cedex, France.*

AZAN G., *Department of Neurology, University of Pavia, Forlanini 16, 27100 Pavia, Italy.*

BAKSH S., *Department of Anatomy, University of Toronto, Toronto, Ontario, Canada M5S 1A8.*

BALLOCCHI E., *Istituto di Anatomia degli Animali Domestici con Istologia ed Embriologia, Via Celoria, 10-20133 Milano, Italy.*

BECKERS-BLEUKX G., *Department of Physiology, Catholic University of Louvain, 5540, 1200 Brussels, Belgium.*

BEINBRECH G., *University of Münster, Zoological Institute, Hindenburgplatz 55, D-4400 Münster, Germany.*

BELLUCO S., *CNR Unit for Muscle Biology and Physiopathology, Institute of General Pathology, University of Padova.*

BELOUSOVA L. V., *A.V. Shubnikov Institute of Crystallography, U.S.S.R. Academy of Sciences, Moscow, U.S.S.R.*

BENNETT P., *MRC Cell Biophysics Unit, 26-29 Drury Lane, London WC2B 5RC, U.K.*

BENYAMIN Y., *UPR 8402 (CNRS)-U 249 (INSERM)-EPHE, Montpellier Université I CRBM (CNRS), BP 5051-34033 Montpellier Cedex, France.*

BERQUIN A., *Department of Physiology, UCL 5528, 1200 Brussels, Belgium.*

BERSHITSKY S. Y., *Department of Biophysics, Institute of Physiology, Urals Branch of Academy of Sciences of the U.S.S.R., Popova 30 Sverdlovsk 620014, U.S.S.R.*

BLANGÉ T., *Department of Physiology, University of Amsterdam, Meibergdreef 15, 1105 AZ Amsterdam, The Netherlands.*

BLESSE N., *Kerckhoff-Clinic, Department of Thoracic and Cardiovascular Surgery, Benekestrasse 2, 6350 Bad Nauheim, Germany.*

BLYAKHMAN A., *Department of Biophysics, Institute of Physiology, Ural's Branch of Academy of Sciences of the U.S.S.R., Sverdlovsk, U.S.S.R.*

BOROVIKOV YU. S., *Group of Cell Motility, Institute of Cytology, Academy of Sciences of the U.S.S.R., 194064 Leningrad, Tikhoretskii str.4, U.S.S.R.*

BRAND T., *Kerckhoff-Clinic, Department of Thoracic and Cardiovascular Surgery, Benekestrasse 2, 6350 Bad Nauheim, Germany.*

BREDMAN J. J., *Department of Anatomy and Embryology, University of Amsterdam, Meibergdreef 15, 1105 AZ, Amsterdam, The Netherlands.*

BREUER E. M., *Institute for Veterinary Pathology Freie Universität Berlin, Strasse 518, Nx15, 1000 Berlin 37, Germany.*

BUKATINA A. E., *Institute of Biological Physics, U.S.S.R. Academy of Sciences, Pushchino, Moscow region 142292, U.S.S.R.*

BURNS K., *Department of Anatomy, University of Toronto, Toronto, Ontario, Canada M5S 1A8.*

BUTLER-BROWNE G. S., *INSERM U262, Clinique Universitaire Baudelocque, 123 Bd Port-Royal, 75674 Paris, Cedex 14, France.*

CARRARO U., *CNR Unit for Muscle Biology and Physiopathology, Institute of General Pathology, University of Padova, Italy.*

CARTMILL T., *Department of Paediatric Surgery, Royal Alexandra Hospital for Children, Camperdown, U.K.*

CATANI C., *Institute of General Pathology, Via Trieste 75, University of Padova, 1-35100 Padova, Italy.*

CAVAGNA G. A., *Istituto di Fisiologia Umana I, Università degli studi di Milano, Via Mangiagalli 32, 20133, Milano, Italy.*

CHANG V., *National Heart Transplantation Program, St. Vincent's Hospital, Darlinghurst, Australia.*

CHASE P. B., *Department of Radiology, University of Washington, Seattle, Washington 98195, U.S.A.*

CHINET A., *Faculty of Medicine CMU, 1 rue St Michel-Servet, 1211 Geneva 4, Switzerland.*

CIRILLO M., *Chirurgia del Cuore e dei Grossi Vasi, University of Bologna, Italy.*

COULIC V. P., *P. Lumumba, Peoples' Friendship University, Moscow 117198, U.S.S.R.*

CUMMING D. V. E., *Department of Cardiac Surgery, National Heart and Lung Institute and the Royal Veterinary College, London, U.K.*

DAHMS S., *Molecular Biology Institute and Rees-Stealy Research Foundation, San Diego State University, San Diego, CA 92182, U.S.A.*

DALLA LIBERA L., *CNR Unit for Muscle biology and Physiopathology, Institute of General Pathology, University of Padova, Italy.*

DANG D. C., *INSERM U262, Clinique Universitaire Baudelocque, 123 Bd Port-Royal, 75674 Paris, Cedex 14, France.*

DANIELI-BETTO D., *Institute of General Pathology, Via Trieste 75, University of Padova, 1-35100 Padova, Italy.*

DE BEER E., *Department of Medical Physiology, University of Utrecht, Vondellaan 24, NL 3521 GG, Utrecht, The Netherlands.*

DECROUY A., *Faculty of Medicine CMU, 1 rue St Michel-Servet, 1211 Geneva 4, Switzerland.*

DHOOT G. K., *The Royal Veterinary and University College, University of London, U.K.*

DOS REMEDIOS C., *Muscle Research Unit, Department of Anatomy, The University of Sydney, Australia.*

DUNN M. J., *Department of Cardiac Surgery, National Heart and Lung Institute and the Royal Veterinary College, London, U.K.*

ELZINGA G., *Laboratory for Physiology, Free University, van der Boechorststraat 7, 1081 BT Amsterdam, The Netherlands.*

FEDOSOV S. N., *A.V. Shubnikov Institute of Crystallography, U.S.S.R. Academy of Sciences, Moscow, U.S.S.R.*

FUCHI T., *Departments of Physiology III and Pharmacology, Karolinska Institutet, Box 5626, S-114 86, Stockholm, Sweden.*

GAUNITZ U., *Carl-Ludwig-Institute of Physiology, University of Leipzig, Liebigstrasse 27, 7010 Leipzig, Germany.*

GERLACH G-F., *Unit of Molecular and Cellular Biology, The Royal Veterinary College, London University, London NW1 OTU, U.K.*

GOLDSPINK G., *Unit of Molecular and Cellular Biology, The Royal Veterinary College, London University, London NW1 OTU, U.K.*

GROUNDS M. D., *University Department of Pathology, Queen Elizabeth II Medical Centre, Nedlands, Western Australia, 6009.*

GRUNDEMAN R., *Department of Medical Physiology, University of Utrecht, Vondellaan 24, NL 3521 GG, Utrecht, The Netherlands.*

GUBA F., *Department of Biochemistry, A. Szent-Györgyi Medical University, 6701 Szeged, P.O. Box 415, Hungary.*

GUNNING P., *Muscle Genetics Unit, Children's Medical Research Foundation, University of Sydney, Australia.*

HARRY J. D., *Istituto di Fisiologia Umana I, Università degli studi di Milano, Via Mangiagalli 32, 20133, Milano, Italy.*

HASSELBAINK H. D. J., *Department of Anatomy and Embryology, University of Amsterdam, The Netherlands.*

HAUCK R., *Institut für Tierphysiologie, Freie universität Berlin, Grunewaldstrasse 34, D-1000 Berlin 41, Germany.*

HEGLUND N. C., *Istituto di Fisiologia Umana I, Università degli studi di Milano, Via Mangiagalli 32, 20133, Milano, Italy.*

HEILIG R., *Laboratoire de Génétique Moléculaire des Eucaryotes du CNRS, INSERM U184, 11 rue Humann, 67085, Strasbourg, Cedex, France.*

HELLWIG G., *Institut für Tierphysiologie, Freie universität Berlin, Grunewaldstrasse 34, D-1000 Berlin 41, Germany.*

HENRIKSSON J., *Departments of Physiology III and Pharmacology, Karolinska Institutet, Box 5626, S-114 86, Stockholm, Sweden.*

HOUMEIDA A., *UPR 8402 (CNRS)-U 249 (INSERM)-EPHE, Montpellier Université I CRBM (CNRS), BP 5051-34033 Montpellier Cedex, France.*

HUBER P. A. J., *Institute of Pharmacology, University of Zürich, Gloriastrasse, 32, CH-8006 Zürich, Switzerland.*

IZAKOV Ya. V., *Department of Biophysics, Institute of Physiology, Ural's Branch of Academy of Sciences of the U.S.S.R., Sverdlovsk, U.S.S.R.*

JACOBER A., *Institute of Pharmacology, University of Zürich, Gloriastrasse, 32, CH-8006 Zürich, Switzerland.*

JAENICKE, TH., *Unit of Molecular and Cellular Biology, The Royal Veterinary College, London University, London NW1 OTU, U.K.*

JAKUBIEC-PUKA A., *Laboratory of Protein Metabolism, Nencki Institute of Experimental Biology, 02-093 Warszawa, ul. Pasteura 3, Poland.*

JUNG D. W. G., *Department of Physiology, University of Amsterdam, Meibergdreef 15, 1105 AZ Amsterdam, The Netherlands.*

KATSNELSON L. B., *Department of Biophysics, Institute of Physiology, Ural's Branch of Academy of Sciences of the U.S.S.R., Sverdlovsk, U.S.S.R.*

KHVOROV N. V., *A.N. Bach Institute of Biochemistry, U.S.S.R. Academy of Sciences, Leninsky prospect 33, Moscow 117071, U.S.S.R.*

KOHAMA K., *Department of Pharmacology, Gunma University School of Medicine, Maebashi, Gunma 371, Japan.*

KÖSSLER F., *Central Institute of Occupational Medicine, Department of Work Physiology, Berlin 1134, Germany.*

KÜCHLER G., *Central Institute of Occupational Medecine, Department of Work Physiology, Berlin 1134, Germany.*

KUSHMERICK M. J., *Department of Physiology and Biophysic, University of Washington, Seattle, Washington 98195, U.S.A.*

LAMERS W. H., *Department of Anatomy and Embryology, University of Amsterdam, Meibergdreef 15, 1105 AZ Amsterdam, The Netherlands.*

LANGE F., *Central Institute of Occupational Medicine, Department of Work Physiology, Berlin 1134, Germany.*

LEBACQ J., *Department of Physiology, UCL 5528, 1200 Brussels, Belgium.*

LÉGÉR J-J., *Pathologie Moléculaire du muscle, INSERM U300, Faculté de Pharmacie, Avenue Charles Flahault, 34060, Montpellier, Cedex, France.*

LÉGÉR JOC., *Pathologie Moléculaire du muscle, INSERM U300, Faculté de Pharmacie, Avenue Charles Flahault, 34060, Montpellier, Cedex, France.*

LEVITSKY D. I., *A.N. Bach Institute of Biochemistry, U.S.S.R. Academy of Sciences, Leninsky prospect 33, Moscow 117071, U.S.S.R.*

LINSEL G., *Central Institute of Occupational Medecine, Department of Work Physiology, Berlin 1134, Germany.*

LOUGHNA P., *Unit of Molecular and Cellular Biology, The Royal Veterinary College, London University, London NW1 OTU, U.K.*

LU D.-X., *Institute of Anatomy, University of Zürich-Irchel, Winterthurerstrasse 190, CH-8057 Zürich, Switzerland.*

MANUSSEN T., *Department of Biomedical Pharmacology, University of Utrecht, Vondellaan 24, NL 3521 GG, Utrecht, The Netherlands.*

MARÉCHAL G., *Department of Physiology, Catholic University of Louvain 5540, 1200 Brussels, Belgium.*

MARKHASIN V. S., *Department of Biophysics, Institute of Physiology, Ural's Branch of Academy of Sciences of the U.S.S.R., Sverdlovsk, U.S.S.R.*

MARSTON S. B., *Department of Cardiac Medicine, National Heart and Lung Institute, Devehouse Street, London SW3 6LY, U.K.*

MARTINDALE J., *Unit of Molecular and Cellular Biology, The Royal Veterinary College, London University, London NW1 OTU, U.K.*

MASCARELLO F., *Istituto di Anatomia degli Animali Domestici con Istologia ed Embriologia, Via Celoria, 10-20133 Milano, Italy.*

MEDBØ J. I., *Department of Physiology, National Institute of Occupational Health, PO Box 8149 dep., N-0033-8149 Oslo 1, Norway.*

MEGIGHIAN A., *Institute of General Pathology, Via Trieste 75, University of Padova, 1-35100 Padova, Italy.*

MEYER H. E., *University of Bochum, Institute of Physiology and Chemistry I, Univ.-Str.150, D-463 Bochum, Germany.*

MICHALAK M., *Department of Anatomy, University of Toronto, Toronto, Ontario, Canada M5S 1A8.*

MIDRIO M., *Institute of General Pathology, Via Trieste 75, University of Padova, 1-35100 Padova, Italy.*

MIJNDERS T. A., *Department of Anatomy and Embryology, University of Amsterdam, Meibergdreef 15, 1105 AZ, Amsterdam, The Netherlands.*

MIKUS P., *Chirurgia del Cuore e dei Grossi Vasi, University of Bologna, Italy.*

MILNER R. E., *Department of Anatomy, University of Toronto, Toronto, Ontario, Canada M5S 1A8.*

MIRACOLI G., *Sorin Biomedica, Cardiovascular Division, Saluggia, Italy.*

MOENS P. D., *Department of Physiology, Catholic University of Louvain, 5540 B-1200 Brussels, Belgium.*

MOMMAERTS W. F., H. M. *Max Planck Institute for Medical Research, Heidelberg, Germany.*

MOORMAN A. F. M., *Department of Anatomy and Embryology, University of Amsterdam, The Netherlands.*

MORNET D., *Pathologie Moléculaire du muscle, INSERM U300, Faculté de Pharmacie, Avenue Charles Flahault, 34060, Montpellier, Cedex, France.*

MUNKEL B., *Kerckhoff-Clinic, Department of Thoracic and Cardiovascular Surgery, Benekestrasse 2, 6350 bad Nauheim, Germany.*

MÜNTENER M., *Institute of Anatomy, University of Zürich-Irchel, Winterthurerstrasse 190, CH-8057 Zürich, Switzerland.*

NAKAMURA T., *Department of Othopedic Surgery, University of Occupational and Environmental Health, 1-1 Logaoka, Yawatanishi-ku, Kitakyusyu City, 807 Japan.*

O'BRIEN G. A., *Department of Cardiac Surgery, National Heart and Lung Institute and the Royal Veterinary College, London, U.K.*

OOSTHOEK P. W., *Department of Anatomy and Embryology, University of Amsterdam, Meibergdreef 15, 1105 AZ Amsterdam, The Netherlands.*

OPAS M., *Department of Anatomy, University of Toronto, Toronto, Ontario, Canada M5S 1A8.*

ORLOVA E. V., *A.V. Shubnikov Institute of Crystallography, U.S.S.R. Academy of Sciences, Moscow, U.S.S.R.*

OSHIDA Y., *Departments of Physiology III and Pharmacology, Karolinska Institutet, Box 5626, S-114 86, Stockholm, Sweden.*

PAPP Z., *Department of Physiology, University, Medical School of Debrecen, PO Box 22, Debrecen, H4012 Hungary.*

PARLAPIANO M., *Chirurgia del Cuore e dei Grossi Vasi, University of Bologna, Italy.*

PASSERINI L., *Istituto di Fisiologia Umana, Università degli Studi di Milano, Via Mangiogalli 32, 20133 Milano, Italy.*

PATTISON C. W., *Department of Cardiac Surgery, National Heart and Lung Institute and the Royal Veterinary College, London, U.K.*

PEDROSA F., *Department of Anatomy, University of Umea, 5-901 87 Umea, Sweden.*

PERMYAKOV E. A., *Institute of Biological Physics, U.S.S.R. Academy of Sciences, Pushchino, Moscow region 142292, U.S.S.R.*

PIERANGELI A., *Chirurgia del Cuore e dei Grossi Vasi, University of Bologna, Italy.*

POGGI P., *Institute of Histology, University of Pavia, Via Forlanini 16, 27100 Pavia, Italy.*

POGLAZOV B. F., *A.N. Bach Institute of Biochemistry, U.S.S.R. Academy of Sciences, Leninsky prospect 33, Moscow 117071, U.S.S.R.*

PONS F., *Pathologie Générale, Faculté de Médecine, Place Henri IV, 34000, Montpellier, France.*

PRITCHARD K., *Cardiac Medicine Department, National Heart and Lung Institute, Dovehouse Street, London SW3 6LY, U.K.*

RANATUNGA K. W., *Department of Physiology, Medical School, Bristol BS81TD, U.K.*

REDWOOD C., *Department of Cardiac Medicine, National Heart and Lung Institute, Devehouse Street, London SW3 6LY, U.K.*

RINDERKNECHT D., *Institute of Anatomy, University of Zürich-Irchel, Winterthurerstrasse 190, CH-8057 Zürich, Switzerland.*

RIZZI C., *CNR Unit for Muscle biology and Physiopathology, Institute of General Pathology, University of Padova.*

ROHÁN J., *Department of Biochemistry, A. Szent-Györgyi Medical University, 6701 Szeged, P.O. Box 415, Hungary.*

ROMANENKO A., *A.V. Palladin Institute of Biochemistry Academy of Sciences of the Ukrainian SSR, 252030, Kiev, Leontovich str.9, U.S.S.R.*

ROSDAHL H., *Departments of Physiology III and Pharmacology, Karolinska Institutet, Box 5626, S-114 86, Stockholm, Sweden.*

ROUSTANT C., *UPR 8402 (CNRS)-U 249 (INSERM)-EPHE, Montpellier Université I CRBM (CNRS), BP 5051-34033 Montpellier Cedex, France.*

ROWLERSON A., *Department of Physiology, UMDS, St. Thoma's Hospital, Lambeth Palace Road, SE1 7EH London, U.K.*

RUPP-DAHLEM M., *Institut für Tierphysiologie, Freie Universität Berlin, Grunewaldstrasse 34, D-1000 Berlin 41, Germany.*

SAILER M., *Institut für Tierphysiologie, Freie Universität Berlin, Grunewaldstrasse 34, D-1000 Berlin 41, Germany.*

SALVATORI S., *C.N.R. - Centre per lo Studio della Biologia e Fisiopatologia Muscolare, cio Istituto di Patologia generale, Università di Padova, 35121-Padova, Italy.*

SCELSI R., *Department of Human Pathology, Institute of Histology, University of Pavia, Via Forlanini 16, 27100 Pavia, Italy.*

SCHAPER W., *Max-Planck Institute, Department of Experimental Cardiology, 6350 Bad Nauheim, Germany.*

SCHAPER J., *Kerckhoff-Clinic, Department of Thoracic and Cardiovascular Surgery, Benekestrasse 2, 6350 Bad Nauheim, Germany.*

SCHAUB M. C., *Institute of Pharmacology, University of Zürich, Gloriastrasse, 32, CH-8006 Zürich, Switzerland.*

SCUTT A., *Unit of Molecular and Cellular Biology, The Royal Veterinary College, London University, London NW1 OTU, U.K.*

SENNI M., *Chirurgia del Cuore e dei Grossi Vasi, University of Bologna, Italy.*

SHARMA H. S., *Kerckhoff-Clinic, Department of Thoracic and Cardiovascular Surgery, Benekestrasse 2, 6350 Bad Nauheim, Germany.*

SHIERECK P., *Biomedical Pharmacology, University of Utrecht, Vondellaan 24 Nl3521GG Utrecht, The Netherlands.*

SHNYROV V. L., *Institute of Biological Physics, U.S.S.R. Academy of Sciences, Pushchino, Moscow region 142292, U.S.S.R.*

SOFFERS C. A. S., *Department of Anatomy and Embryology, University of Amsterdam, Meibergdreef 15, 1105 AZ, Amsterdam, The Netherlands.*

SOHÁR I., *Department of Biochemistry, A. Szent-Györgyi Medical University, 6701 Szeged, P.O. Box 415, Hungary.*

SOUKUP T., *Institute of Physiology, Czechoslovak Academy of Sciences 142 20 Prague, Czechoslovakia.*

SOUSSI-YANICOSTAS N., *INSERM U262, Clinique Universitaire Baudelocque, 123 Bd Port-Royal, 75674 Paris, Cedex 14, France.*

SPURWAY N. C., *Department of Physiology, University of Glasgow, Glasgow WG 128QQ, U.K.*

STELMASCHUK V. J., *A.V. Shubnikov Institute of Crystallography, U.S.S.R. Academy of Sciences, Moscow, U.S.S.R.*

STIENEN G. J. M., *Laboratory for Physiology, Free University, van der Boechorststraat 7, 1081 BT Amsterdam, The Netherlands.*

SZCZEPANOWSKA J., *Laboratory of Protein Metabolism, Nencki Institute of Experimental Biology, 02-093 Warsaw, Pasteur str.3, Poland.*

TAKAGI M., *Department of Orthopedic Surgery, Saitama Medical Centre, 1981 Tsujido-cho, Kamodo, Kawagoe City, 350 Japan.*

TAKANO-OHMURO H., *The Tokyo Metropolitan Institute of Medical Science, Honkomagome, 3-18-22, Bunkyo-Ku, Tokyo 113, Japan.*

THORNELL L-E., *Department of Anatomy, University of Umea, 5-901 87 Umea, Sweden.*

TRAHAIR T., *Muscle Research Unit, Department of Anatomy, The University of Sydney, Australia.*

TSATURYAN A. K., *Institute of Mechanics.M.V. Lomonosov State University, Leninsky Gory, Moscow 117234, U.S.S.R.*

TSUZUKI N., *Department of Pharmacology, Gunma University, School of Medecine, 3-39-22, Showa-machi, Maebashi City, 371 Japan.*

UNGERSTEDT U., *Departments of Physiology III and Pharmacology, Karolinska Institutet, Box 5626, S-114 86, Stockholm, Sweden.*

van der BENT V., *University Departments of Physiology, Faculty of Sciences, 1 rue St Michel-Servet, 1211 Geneva 4, Switzerland.*

VAN BUUREN K., *Biomedical Pharmacology, University of Utrecht, Vondellaan 24 Nl3521GG Utrecht, The Netherlands.*

VAN KEMPEN M. J. A., *Department of Anatomy and Embryology, University of Amsterdam, Meibergdreef 15, 1105 AZ Amsterdam, The Netherlands.*

VEDENKINA N. S., *Institute of Biological Physics, U.S.S.R. Academy of Sciences, Pushchino, Moscow region 142292, U.S.S.R.*

VERMEULEN J. L. M., *Department of Anatomy and Embryology, University of Amsterdam, Meibergdreef 15, 1105 AZ, Amsterdam. The Netherlands.*

VERSTEEG P. G. A., *Laboratory for Physiology, Free University, van der Boechorststraat 7, 1081 BT Amsterdam, The Netherlands.*

VOGTMANN M., *Carl-Ludwig-Institute of Physiology, University of Leipzig, Liebigstrasse 27, 7010 Leipzig, Germany.*

VOSBERG H. P., *Max Planck Institute for Medical Research, Heidelberg, Germany.*

VRBOVÁ G., *The Royal Veterinary and University College, University of London, U.K.*

WEIJS, *University of Amsterdam, Meiberdreef 15; 1105 AZ Amsterdam, The Netherlands.*

WENINK A. C. G., *Department of Anatomy and Embryology, University of Leiden, The Netherlands.*

WESSELS A., *Department of Anatomy and Embryology, University of Amsterdam, Meibergdreef 15, 1105 AZ Amsterdam, The Netherlands.*

WILLIAMSON H. A., *Department of Cardiac Surgery, National Heart and Lung Institute and the Royal Veterinary College, London, U.K.*

WOUTER H., *Department of Anatomy and Embryology, University of Amsterdam, Meibergdreef 15, 1105 AZ Amsterdam, The Netherlands.*

YACOUB M. H., *Department of Cardiac Surgery, National Heart and Lung Institute and the Royal Veterinary College, London, U.K.*

YAZAKI P., *Molecular Biology Institute and Rees-Stealy Research Foundation, San Diego State University, San Diego, CA 92182, U.S.A.*

YEOH T., *Muscle Research Unit, Department of Anatomy, The University of Sydney, Australia.*

ZAWADOWSKA B., *Institute of Anatomy, University of Zürich-Irchel, Winterthurerstrasse 190, 8057 Zürich, Switzerland.*

ZIEGLER C., *University of Bochum, Institute of Physiology and Chemistry I, Univ.-Str.150, D-463 Bochum, Germany.*

Preface

Twenty years have passed since Prof. Hamoir organized the first Meeting of European Muscle scientists in Liège (Belgium); there the European Muscle Club was informally founded as a genuinely friendly and interactive research association. The Club has since grown into the European Conference on Muscle and Motility, which, over the years, has covered all major achievements in research on striated muscle, smooth muscle and cell motility.

This year the XIXth Conference was held in Brussels from September 10 to 13th. The conference programme was mainly devoted to muscle physiology, although cell biology of myogenesis was well represented, with many outstanding contributions to the field of muscle development and regeneration; equally largely present were molecular approaches, as in the sessions on 'New muscle proteins and their possible function' or 'Genetics and structure of myosins and actins'. As usual, under the light hand of Marcus Schaub, the background of the Local Organizer appeared, without overwhelming the interests of other Muscle Club members.

The excellence of the speakers made the conference up to date and interesting even to non-subspecialists, a rare event in these days of femto-specialization. A pinch of perspectives in Applied Myology attracted physicians and surgeons to the conference. It is a really exciting fact that less than ten years after Dirk Pette edited *Muscle Plasticity* (one of the best editorial products of European muscle research), this basic concept is close to being implemented as essential in biochemical devices, inspiring major breakthroughs in the clinical management of heart diseases, just as successful as the cardiac pacemaker. The rapid development of clinical applications and the accumulation of new results in experimental research were reasons for inviting outstanding specialists in basic research in this field and in that of cellular therapies of muscle dystrophies.

The success of the conference was guaranteed by the contribution of all participants. Thanks are due to all who came to Brussels and contributed to such a friendly and constructive atmosphere.

The XVIIth Conference held in Padova in 1988 also met with great success and the book published then by U. Cararro was named '*Sarcomeric and Non-sarcomeric Muscles: Basic and Applied Research Prospects for the 90s*' (Unipress, Padova). The success of this book prompted us to propose to our colleagues a successor, which, accordingly, has been referred to as 'Volume 2'. Some of the papers presented at the conference and some additional relevant material is published in these Proceedings. This was made possible by funds granted to cover the publishing expenses of the papers presented by workers from Eastern Europe. We thank here the Belgian Fonds de la Recherche Scientifique and Unipress Padova. It has been the aim of the editors to avoid long delays and to publish the proceedings within six months of the conference. All authors, who handed in manuscripts in time and made the plan feasible, are warmly acknowledged. We are most indebted to all the contributors who represent those in the front line of muscle research in both the most traditional and the very young but exciting new fields. The papers have been reviewed by several of our colleagues during and after the Conference. We want to pay special homage to our colleague, Dr Anne Godfraind-De Becker who checked all the manuscripts and made many important suggestions and improvements. However good the help of these colleagues, the final acceptance and editing of the papers published in this book remain our responsibility.

Dr A. Berquin, Mrs G. Beckers-Bleukx and all the collaborators of the Local Organizer are to be acknowledged by all participants for the valuable help they gave in organizing the XIXth European Conference on Muscle and Motility, which provided the foundation for compiling this book. Thanks are due for her very efficient secretarial assistance to Mrs A. Mollet.

It is our hope that this book will summarize and further stimulate research in free, basic biological science and in object-oriented, applied medical knowledge, so that both of them ultimately will bring benefit to many of the patients with disorders of sarcomeric and non-sarcomeric muscles.

Georges Maréchal
Local Organizer and Editor

Ugo Carraro
Co-editor

Contents

SECTION THREE: MUSCLE PLASTICITY

SECTION FIVE; INTRACELLULAR REGULATIONS

SECTION SIX: MECHANICS AND CROSSBRIDGES

SECTION SEVEN: CONTRACTILE PROTEINS

Section One
New Proteins

1
Isolation, Visualization and Substructural Comparison of Chicken Skeletal and Smooth muscle Dystrophins

N. AUGIER*, F. PONS**, Joc. LÉGER*, R. HEILIG[†], J-J. LÉGER* AND D. MORNET*

*Pathologie Moléculaire du muscle, INSERM U300, Faculté de Pharmacie, Av. Charles Flahault, 34060, Montpellier, Cedex, France.
**Pathologie Générale, Faculté de Médecine, Place Henri IV, 34000, Montpellier, France.
[†]Laboratoire de Génétique Moléculaire des Eucaryotes du CNRS, INSERM U184, 11 rue Humann, 67085, Strasbourg, Cedex, France

From the Gene to the Protein

Duchenne muscular dystrophy (DMD) is the most common and severe of the human muscular dystrophies. The gene responsible for DMD and BMD has been assigned to band Xp21 of the human X chromosome (1, 2). Cloning of the DMD cDNA by Kunkel's group (3) was a major step for identifying the gene and its protein product. From a 14 Kb total cDNA, assumed to be the full coding sequence, the amino acid sequence was deduced. It has been predicted that this new molecule was constituted of 3685 residues in human (4) but only 3660 residues in chicken skeletal muscles (5).

The simple examination of its primary structure showed that it shared an actin-functional N-terminal domain homologous with α-actinin, and a rod-shaped central part similar to the spectrin structural arrangement (6–8), while no predictive similarities were found within the C-terminal portion (4). Even now, the use of specific antibodies against fusion protein (9) or synthetic oligopeptides (10–11) has allowed for the visualization of dystrophin by both cryostat trans-

Muscle and Motility Vol. 2. Proceedings of XIXth European Conference in Brussels
© Intercept Ltd, PO Box 716, Andover, Hampshire, SP10 1YG, UK

verse section of muscle fibre and immunofluorescence detection (12) or by the Western immunoblot procedure (9). This protein, which is lacking in DMD patients, is localized to the cytoplasmic face of the muscle fibre, labelling its periphery (13–14).

However, dystrophin is not a protein restricted to the muscular tissue, it is also expressed in the brain tissue (15) and its gene expression involves a tissue-specific promoter (16–17). The alternative transcript splicing, using the mRNA/PCR technique, has also demonstrated the existence of several dystrophin patterns, particularly different on the N-terminal smooth muscle dystrophin (18).

Although the DMD gene product, due to its rather large sequence and low concentration in all tissues, has not been isolated, tentatives by others have obtained the molecule either denaturated or associated with a membrane glycoprotein (19–20). Our aim in this report is to describe the strategy used to prepare isolated dystrophin. Using electron microscopy we investigated its general shape and by the limited proteolysis approach we show the evidence of a tissue-specific structure for this molecule.

The Protein and its Isolation

The first step was to produce antibodies and characterize their relative affinity. The fusion protein expressed by chicken segments of cDNA cloned in expression vectors span three different regions of the molecule: 43 to 760, A; 1173 to 1728, C, and 3357 to 3660, H. The respective polyclonal antibodies are later referred to as A-DYS, C-DYS and H-DYS. Their affinity was tested by ELISA and the most affined was the C-DYS (21). All of them recognize both skeletal and smooth muscle dystrophin and because of an easier purification we started from chicken gizzard smooth muscle. After a buffered rapid homogenization of the muscle in presence of a cocktail of inhibitors and 0.5% Triton X100 with 10 mM iodoacetamide, we obtained a supernatant which was treated by 20% ammonium sulphate. The pellet was removed with a 0.1 M Tris buffer pH 9 without Triton. The final supernatant was then concentrated and submitted to a final step of purification. An affinity column was made with C-DYS antibody and the dystrophin solution was eluted through it. The chromatography revealed that dystrophin is selectively retained, and that elution by acidic pH specifically removed dystrophin and a 43 kDa protein together. With C-DYS antibody the Western blot pattern of this purification detected only one protein of about 400 kDa. This result indicates that this procedure does not destroy dystrophin which, extracted from smooth muscle, corresponds to a protein of 400 kDa as expected and found for skeletal muscle (22).

Electron Microscopic Images of Dystrophin

The protein does not present a simple form but rather a large variety of conformation arrangements which, according to the known structure already described, have been interpreted as α-actinin and spectrin (23). The major species allows us to propose that dystrophin gives a rotary-shadowing picture which corresponds to an elongated and asymmetrical molecule of 175 nm (24). This assumed monomer can be twice as long, representing a dimer formation. Even larger so-called polymeric structures have been found. However, some filaments do not exactly fit into such a classification, even if an end-to-end or end-to-tail elongation pattern exists, but this molecule is also able to overlap in part giving rise to images about 200 nm long. From all these findings, dystrophin is thought to have the properties of an elastic ribbon which could envelop each muscular fibre with significant flexibility.

Specific Proteolysis of the Molecule

In an attempt to learn more about the dystrophin substructure, we used the limited proteolysis approach which allows for easy detection of the main domain of a protein; that is the case, for example, with myosin (25,26). A large battery of enzymes, with different specificities, was used to comparatively split the dystrophin preparation obtained from smooth muscle, skeletal muscle and cardiac muscle.

The proteolytic pattern was followed by electrophoresis on gradient acrylamide slab gel (2.5–7%) in denaturing condition and then transferred onto a nitrocelluslose sheet for revelation with A-DYS, C-DYS or H-DYS antibodies. The protein bands detected, when present, revealed that the proteolytic cleavage was tissue-dependent. The major proteolytic products can be identified and classified as follows: limited proteolysis with trypsin gives rise to a C-terminal fragment of about 320–350 kDa, chymotrypsin produces a N-terminal fragment of about 210–230 kDa, V8-protease and thermolysin make a shortened central fragment from 170 to 90 kDa, depending on the enzyme used.

Questions for the Origin of the Difference between each Tissue-dependent Dystrophin

It has been reported, by analysis of the amino acid sequence of dystrophin, that the molecule contained different proline-rich regions (27–28). Such fragments may correspond to an exposed structure, easily targeted by proteolytic enzymes. In fact, the natural intrinsic proteolytic susceptibility of dystrophin was rapidly detected (29,30) and extensively demonstrated in recent reports by Nicholson (31). By the proteolytic approach, dystrophin was selectively cleaved under controlled conditions from its N-terminal portion by trypsin, releasing a 320–350 kDa major C-terminal fragment, or from its C-terminal end by chymotrypsin, giving rise to a 210–230 kDa major N-terminal segment. The limited proteolytic method was found both tissue-specific and usefully efficient in providing shorter dystrophin pieces on which functional properties can be investigated. Proteolytic products corresponding to resistant central portions of the dystrophin were also obtained using proteases as V8 or thermolysin and also provide new materials to understand the functional role of this molecule.

Aside from the fact of their difference in size, due to the origin of the tissue from which dystrophin was prepared, the fragment obtained which revealed a difference in the proteolytical susceptibility of dystrophin calls for an explanation. This could correspond to only an alternative splicing as described at the mRNA level (18) and, although a different promoter seems to be a specific feature for glial and neuronal cells (17), the data presented here could also arise from three other different reasons: the possibility of an autosomal transcript as described by Love *et al.* (32), the putative presence of another dystrophin-like gene, or a tissue-specific post-transcriptional methylation or phosphorylation. Work is now underway to interpret the dystrophin rotary-shadowing electron microscopic image in more detail, to explore the antibody mapping on smooth muscle dystrophin but also on skeletal muscle dystrophin, and to find out whether or not a dystrophin family exists either as isoenzymes from a single gene or as dystrophin-like products.

References

1. Murray, J. M., Davies, K. E., Harper, P. S., Meredith, L., Mueller, C. R., Williamson, R., 1982, *Nature*, **300**, 69–71.
2. Davies, K. E., Pearson, P. L., Harper, P. S., Murray, J. M., O'Brien, T., Sarfarazi, M., Williamson, R., 1983, *Nucleic Acids Res.*, **11**, 2303–2312.
3. Monaco, A. P., Bertelson, C. J., Middlesworth, W., Colletti, C. A., Aldridge, J., Fischbeck, K. H., Bartlett, R., Pericak-Vance, Roses, A. D., Kunkel, L. M., 1985, *Nature*, **316**, 842–845.
4. Koenig, M., Monaco, A. P., Kunkel, L. M., 1988, *Cell*, **53**, 219–228.
5. Lemaire, C., Heilig, R., Mandel, J. L., 1988, *EMBO J.*, **7**, 4157–4162.
6. Hammonds, R. G. Jr. 1987, *Cell*, **51**, 1.
7. Davidson, M. D., Critchley, D. R., 1988, *Cell*, **52**, 159–160.
8. Byers, T. J., Husainchist, A., Dubreuil, R. R., Branton, D., Goldstein, L. S. B., 1989, *J. Cell. Biol.*, **109**, 1633–1641.
9. Hoffman, E. P., Brown, R. H., Kunkel, L. M., 1987, *Cell*, **51**, 919–928.
10. Kao, L., Drstenansky, J., Mendel, L. J., Rammohan, K. W., Gruenstein, E., 1988, *Proc. Natl. Acad. Sci. USA*, **85**, 4491–4495.
11. Sugita, H., Arahata, K., Ishiguro, T., Suhara, Y., Ishiura, S., Eguchi, C., Nonaka, I., Zawa, E., 1988, *Proc. Jpn. Acad.*, **64**, 37–39.
12. Bonilla, E., Samitt, C. E., Miranda, A. F., Mays, A. P., Salviati, G., Dimauro, S., Kunkel, L. M., Hoffman, E. P., Rowland, C. P., 1988, *Cell*, **54**, 447–452.
13. Zubrycka-Gaarn, E., Bulman, D. E., Karpati, G., Burghes, A. H. M., Belfall, B., Klamut, H. J., Talbot, J., Hodges, R. S., Ray, P. N., Worton, R. G., 1988, *Nature*, **333**, 466–469.
14. Watkins, S. C., Hoffman, E. P., Slayter, A. S., Kunkel, L. M., 1988, *Nature*, **333**, 863–866.
15. Nudel, U., Zuk, D., Zeelon, E., Levy, Z., Neuman, S., Yaffe, D., 1989, *Nature*, **337**, 76–78.
16. Nudel, U., Robzik, K., Yaffe, D., 1988, *Nature*, **331**, 635–638.

17. Chelly, J., Hamard, G., Koulakoff, A., Kaplan, J. C., Kahn, A., Berwald-Better, J., 1990, *Nature*, **344**, 64–65.
18. Feener, C., Koenig, M., Kunkel, L. M., 1989, *Nature*, **338**, 509–511.
19. Matsumara, K., Shimizu, T., Mannen, T., 1989, *Biomed. Res.*, **10**, 325–328.
20. Campbell, K. P., Kahl, S. D., 1989, *Nature*, **338**, 259–262.
21. Augier, N., Heilig, R., Boucraut, J., Léger, Joc., Lemaire, C., Georgesco, M., Eschenne, B., Mandel, J. L., Pons, F., Pellissier, J. F., Léger, J., 1990, in press.
22. Knudson, C. M., Hoffman, E. P., Kahl, S. D., Kunkel, L. M., Campbell, K. P., 1988, *J. Biol. Chem.*, **263**, 8480–8484.
23. Whiting, A., Wardale, J., Trinick, J., 1989. *J. Mol. Biol.*, **205**, 263–268.
24. Pons, F., Augier, N., Heilig, R., Léger, Joc., Mornet, D., Léger, J. J., 1990, *Proc. Natl. Acad. Sci. USA*, **87**, 7851–7855.
25. Offer, G., Moos, C., Starr, R., 1973, *J. Mol. Biol.*, **74**, 653–676.
26. Mornet, D., Ue, K., Morales, M. F., 1984, *Proc. Natl. Acad. Sci. USA*, **81**, 736–739.
27. Koenig, M., Kunkel, L. M., 1990, *J. Biol. Chem.*, **265**, 4560–4566.
28. Mandel, J. L., 1989, *Nature*, **339**, 584–586.
29. Arahata, K., Ishura, S., Ishiguro, T., Tsukahara, T., Suhara, Y., Eguchi, C., Ishihara, T., Nonaka, I., Ozawa, E., Sugita, H., 1988, *Nature*, **333**, 861–863.
30. Matsumara, K., Shimizu, T., Mannen, T., 1989, *Biomed. Res.*, **10**, 325–328.
31. Nicholson, L. V. B., Davison, K., Johnson, M. A., Slater, C. R., Young, C., Bhattacharya, S., Gardnermedwin, D., Harris, J. B., 1989, *J. Neurol. Sci.*, **94**, 1–13.
32. Love, R. D.,.Hill, D. F., Dickson, G., Spurr, N. K., Byth, B. C., Marsden, R. F., Walsh, F. S., Edwards, Y. H., Davies, K. E., 1989, *Nature*, **339**, 55–58.

2
Identification and Immunological Localization of a High Molecular Weight Protein in Arthropod Muscles

C. ZIEGLER, H.E. MEYER* AND G. BEINBRECH

University of Münster, Zoological Institute, Hindenburgplatz 55, D-4400 Münster, Germany
**University of Bochum, Institute of Physiology and Chemistry I, Univ.-Str.150, D-463 Bochum, Germany*

Introduction

Asynchronous flight muscles of insects are able to contract with frequencies of up to 1000 Hz. The ultrastructural background of this unusual capability is a highly ordered arrangement of actin and myosin filaments. This arrangement is obviously maintained by connections of the myosin filaments to the Z-line (1, 2, 3) reducing thereby the width of the I-bands of asynchronous flight muscles to narrow stripes barely recognizable on electron micrographs. The molecular basis of these connections seem to be proteins like connecting protein (4) or projectin (5). The description of a protein (titin) anchoring the myosin filaments in the Z-line of vertebrate skeletal muscles (6) suggested a search for a protein with a similar function also in synchronous muscles of invertebrates.

Methods

Myofibrils of the flight muscles of the locust, *Locusta migratoria*, have been extracted with high ionic strength solutions (40 mM $Na_4P_2O_7$, 5 mM $MgCl_2$, 5 mM ATP, 1 mM EGTA, 10 mM Tris/HCl, pH 7.5 + proteinase inhibitors). After centrifugation at 200 000 × *g* the supernatant was applied to Mono Q-FPLC.

Muscle and Motility Vol. 2. Proceedings of XIXth European Conference in Brussels
© Intercept Ltd, PO Box 716, Andover, Hampshire, SP10 1YG, UK

The unbound components, mainly paramyosin and a high molecular weight protein (HMWP), were further separated by isoelectric precipitation and centrifugation (30 min at 50 000 × *g*) of paramyosin and minor impurities. The supernatant was eventually concentrated by lyophilization. The M_r of the protein was estimated using 7.5 % SDS-PAGE (7) and 3.2 % SDS-PAGE (8).

Antibodies against the HMWP were raised in rabbits. POD-,FITC- and 5 nm gold-labelled goat anti-rabbit IgG (Sigma) were used for Western blots, immunofluorescence and negative staining (Zeiss EM 9) studies respectively.

Results and discussion.

PTC-amino acid analysis (Table 1) of the purified HMWP shows a similar composition as projectin (5) and twitchin (9). Only the serine and glycine contents seem to be slightly higher. Estimations of its M_r by SDS-PAGE yielded results obviously depending on the acrylamide concentration of the gels. On 7.5 % SDS-gels a M_r of about 350 kD (Fig. 1) comparable to projectin (5) could be calculated. Using 3.2% SDS-gels a M_r between crosslinked rabbit myosin (440 kD) and nebulin (600–800 kD, see 10) has been found (not shown). A similar migration distance of the HMWP band in Fig. 1, lane a (locust myofibrillar proteins), and Fig. 1, lane c (purified protein), suggests that proteolysis did not occur during the preparation procedure.

Table 1 PTC-amino acid analysis of the HMWP from locust flight muscle (percentages of total amount). Cysteine and tryptophane contents have not been determined.

Asp/Asn	8.2	Thr	4.6	Arg	4.7	Ile	3.7
Glu/Gln	12.9	Ala	8.2	Tyr	2.6	Leu	5.7
Ser	11.1	His	2.7	Val	5.4	Phe	2.8
Gly	14.9	Pro	6.1	Met	1.2	Lys	5.3

Shape and size of the molecules have been investigated by rotary shadowing. The molecules are flexible in shape with a diameter of 3–4 nm (Fig. 2). Measurements of the lengths of the molecules gave pronounced peaks at 270 nm (fig. 3) and 220 nm (not shown). The average length was 282 nm if molecules smaller than 240 nm were considered to be fragments caused by the spraying procedure and were not included into the calculation.

To localize the HMWP in the sarcomere antibodies were added to myofibrils of different muscles and visualized by FITC-labelled goat anti-rabbit IgG. In myofibrils of locust flight muscles prepared at resting length, the fluorescence could be observed forming two narrow stripes at either side of the Z-line (Fig. 4a). This can also be seen in the sarcomeres of the flight muscles of the fleshfly and crayfish abdominal muscles (Fig. 4d and e respectively). The width of one

Fig. 1 a–e

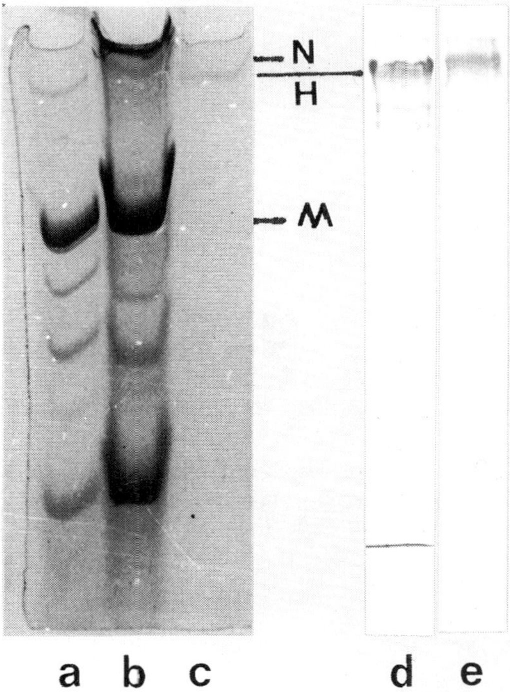

Fig. 2

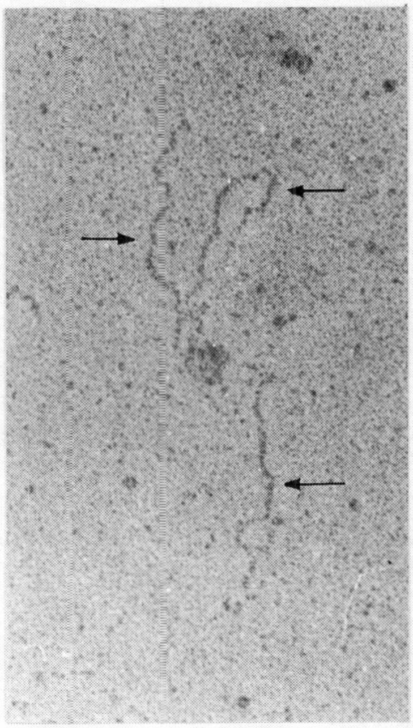

Fig. 1 Electrophoretic analyses of myofibrillar protein components and purification steps of the HMWP. Lanes a–c 3.2% gels with myofibrillar proteins of locust flight muscle (a), rabbit skeletal muscle (b) and purified HMWP (c). Lane d shows a 7.5% gel loaded with impure HMWP. Lane e is the corresponding Western blot. N = rabbit nebulin, H = HMWP, M = myosin. All gels were stained with Coomassie blue.

Fig. 2 HMWP molecules (arrows) after rotary shadowing with platinum. The curved shape of the molecules demonstrates a high flexibility. Magnification 142 000 ×.

fluorescent stripe in Fig. 4a and d was about 0.3 µm, corresponding to the average length of one HMWP molecule (Fig. 3). The Z-line is usually free of fluorescence. This can be seen best if the sarcomeres are slightly stretched as shown in the crayfish sarcomeres (Fig. 4e). The antibody labelling was different in locust leg muscles. Fig. 4c shows that the fluorescence in myofibrils of leg muscles is restricted to the ends of the A-bands.

If the sarcomeres were extensively stretched from a resting length of 3 µm to a length of about 8 mm as in the case of the myofibrils of locust flight muscles shown in Fig. 4b the whole I-band (width 4.8 µm) seemed to be labelled. This suggests that the HMWP has been stretched from 0.3 to 2.4 µm. It suggests also that one end of the protein inserts in the Z-line and the other end is connected to the myosin filaments. A maximum length of 2.4 mm is well compatible with a

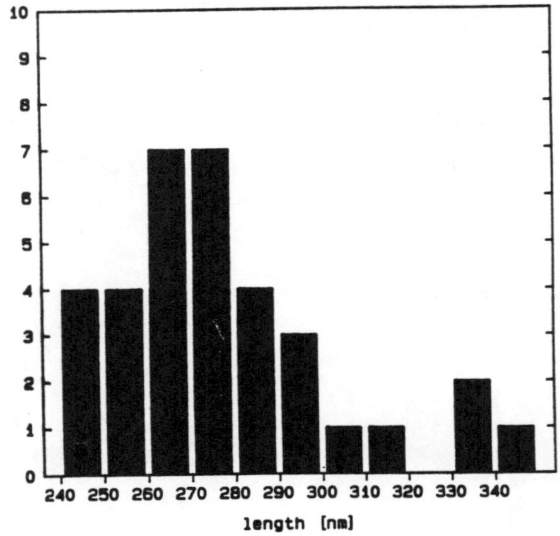

Fig. 3 Diagram of the length distribution of rotary shadowed HMWP molecules (number plotted versus length). Molecules shorter than 240 nm have not been included.

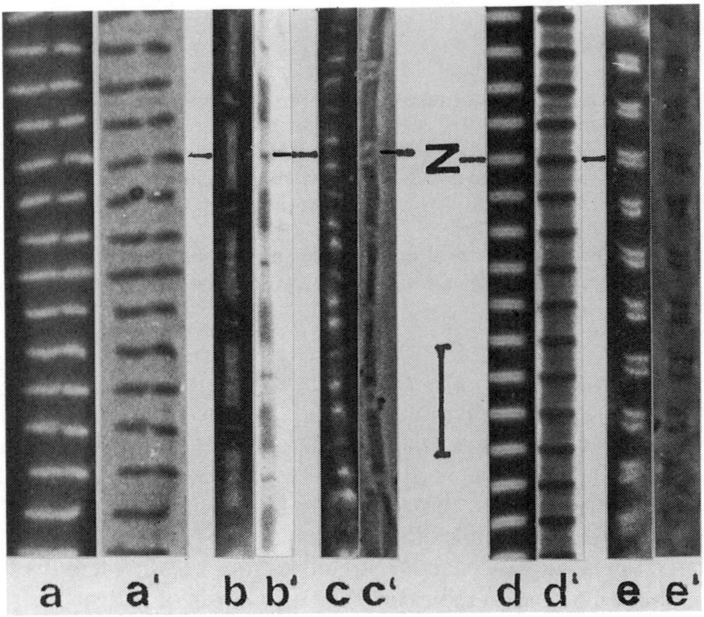

Fig. 4 Immunofluorescent localization of HMWP in sarcomeres of locust flight muscles at resting length (a) and stretched (b), locust leg muscle (c), fleshfly (Phormia) flight muscle (d) and crayfish (Orconectes) fast abdominal muscle (e), respectively. The images in a′-e′ are the corresponding phase contrast micrographs. Z = Z-line. Bar 10 mm.

M_r of 500–700 kD if we assume that the molecules have only small a-helical and large elastic b-sheet portions.

The connection of the HMWP to the myosin filaments could also be demonstrated by antibody labelling and negative staining of native myosin filaments on carbon coated grids. The localization of the immunogold particles is restricted to epitopes at the ends of the myosin filaments (Fig. 5a). A positive control of synthetic filaments formed by coaggregation of myosin, paramyosin and HMWP (over night dialysis) showed gold particles all over the length of the filaments (Fig. 5b). This localization of the HMWP seems to inhibit the activity of the actomyosin-ATPase of these coaggregates for about 90% compared to the ATPase activity of coaggregates of myosin and paramyosin without HMWP.

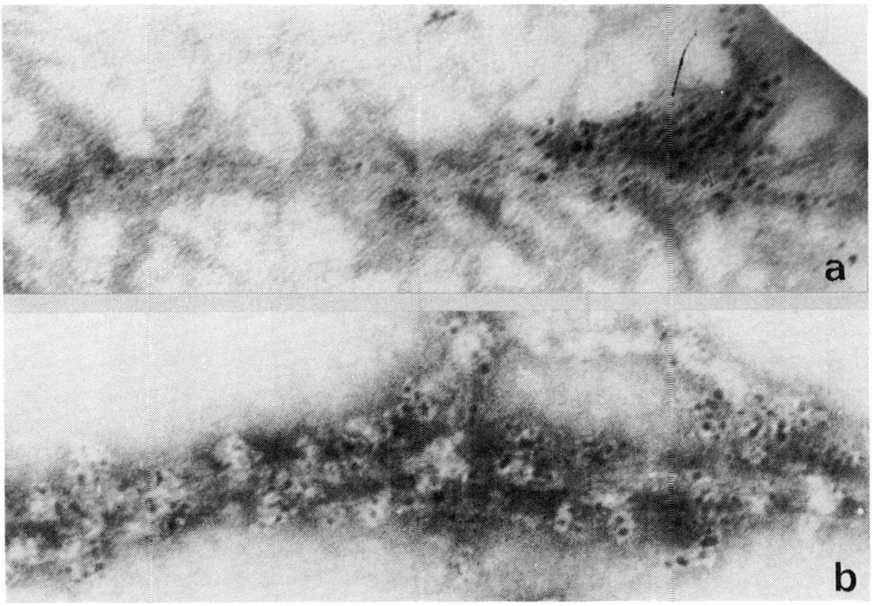

Fig. 5 Immunogold (5 nm) labelled and negatively stained native myosin filament of locust flight muscles (a) and (b) coaggregates of locust flight muscle myosin, Mytilus paramyosin and locust flight muscle HMWP. In (a) gold particles are mainly localized at the end of the filament half. In (b) the particles are distributed all over the length of the coaggregates. Magnification 155 000 ×.

Our findings suggest that the HMWP in locust flight muscles is similar in M_r, amino acid composition and localization to other myofibrillar proteins of invertebrates like connecting protein (4), projectin (5),twitchin (9) or mini-titin (11). Unlike titin of vertebrate skeletal muscles it could not be localized at the surface of native thick filaments but seems to be inserted at their ends.

Acknowledgements

We thank Gereon Ader and David Troyer for rotary shadowing, Kirsten Kielbassa for the immunogold labeling, Bernd Kölsch for the ATPase measurements, Thomas Pforr and Rotraud Schustereit for excellent technical assistance and the Deutsche Forschungs-gemeinschaft for financial support (Be 347/8-1).

References

1. Auber, J., Couteaux, R., 1962, *C. R. Acad. Sci. Paris*, **254**, 3425–3426.
2. Garamvölgyi, N., 1965, *J. Ultrastruct. Res.*, **13**, 409–424.
3. Zebe, E., Meinrenken, W., Rüegg, J. C., 1968, *Z. Zellforsch.*, **87**, 603–621.
4. Bullard, B., Hammond, K. S., Luke, B. M., 1977, *J. Mol. Biol.*, **115**, 417–440.
5. Laemmli, U. K., 1970, *Nature*, **227**, 680–685.
6. Wang, K., McClure, J., Tu, A., 1979, *Proc. Natl. Acad. Sci.*, **76**, 3698–3702.
7. Saide, J. D., 1981, *J. Mol. Biol.*, **153**, 661–679.
8. Wang, K., 1982, *Meth. Enzymol.*, **85**, 264–274.
9. Benian, G. M., Kiff, J. E., Neckelmann, N., Moermann, D. G., Waterston, R. H., 1989, *Nature*, **324**, 45–50.
10. Wang, K., Wright, J., 1988, *J. Cell Biol.*, **107**, 2199–2212
11. Nave, R., Weber, K., 1990, *J. Cell Sci.*, **95**, 535–544.

3
Calreticulin, a Ca^{2+}-binding Protein of Sarcoplasmic and Endoplasmic Reticulum Membranes

M. MICHALAK, S. BAKSH, K. BURNS, R. E. MILNER AND M. OPAS*

Cardiovascular Disease Research Group, Departments of Pediatrics and Biochemistry, University of Alberta, Edmonton, Alberta, Canada T6G 2S2
**Department of Anatomy, University of Toronto, Toronto, Ontario, Canada M5S 1A8*

Introduction

In many different cell types intracellular membrane systems are responsible for the regulation of Ca^{2+} levels in the cytoplasm, by which means they control a variety of cellular functions (1). For example, in muscle cells Ca^{2+} storage, release and uptake are regulated by the proteins of the sarcoplasmic reticulum (SR)(2). Calreticulin is one of several Ca^{2+}-binding proteins found in skeletal muscle SR membranes, which was initially identified as the 'high affinity Ca^{2+}-binding protein' (3). It is localized within the lumen of the SR (4) and it is known to bind Ca^{2+} with both high and low affinity and capacity (5). Recently, we have demonstrated that this protein is also present in a variety of non-muscle tissues (6). We found (6) that calreticulin is identical to calregulin, a Ca^{2+}-binding protein, which was recently identified in bovine liver (7), and that it is a major Ca^{2+}-binding protein of both liver endoplasmic reticulum (ER) membranes and smooth muscle SR (6, 8). Calsequestrin, another SR Ca^{2+}-

Muscle and Motility Vol. 2. Proceedings of XIXth European Conference in Brussels
© Intercept Ltd, PO Box 716, Andover, Hampshire, SP10 1YG, UK

binding protein, has been localized specifically to the lumen of the skeletal and cardiac muscle SR (9). Only low amounts of calsequestrin (cardiac form) have been found in smooth muscle cells (8, 10), and we have not detected any calsequestrin in the non-muscle tissues (8). Because of the opposing distribution of calreticulin and calsequestrin in different tissues we propose that calreticulin may play an important role in Ca^{2+} sequestration by smooth muscle SR and non-muscle ER membrane systems, and that it may be a functional analogue of calsequestrin.

We have recently cloned a cDNA, which encodes rabbit skeletal muscle calreticulin and have deduced the amino acid sequence of the protein from the nucleotide sequence of this cDNA (11). Using the amino acid sequence information obtained, we conclude that calreticulin is identical (over 92% amino acid sequence identity) to CRP55 (mouse calreticulin), which was identified recently as a major Ca^{2+}-binding protein in ER membranes (12), and to the Ro/SS-A autoantigen (human calreticulin), which is specific to patients with systemic lupus erythematosus and Sjögren syndrome (13). Other molecules that are highly homologous to calreticulin have also recently been identified in *Onchocerca volvulus, Drosophila melanogaster* and *Aplysia californica* (14).

Here we present a folding model for calreticulin, which is based on the deduced amino acid sequence, and describe initial results of studies, which aim to determine the functional significance of each of the four distinct structural domains.

Experimental Procedures

ISOLATION OF CALRETICULIN

Calreticulin was isolated by a selective ammonium sulphate precipitation procedure in the presence of a mixture of protease inhibitors followed by DEAE (6) and hydroxylapatite chromatography, as described earlier (8).

SDS-POLYACRYLAMIDE GEL ELECTROPHORESIS AND ANALYSIS OF $^{45}Ca^{2+}$ BINDING

SDS-polyacrylamide gel electrophoresis (SDS-PAGE) was carried out using 10% polyacrylamide gels (15). After electrophoresis, gels were either stained with Coomassie blue or the proteins were transferred to nitrocellulose membranes and tested for $^{45}Ca^{2+}$ with a $^{45}Ca^{2+}$ overlay technique (16).

CONSTRUCTION AND EXPRESSION OF A GLUTATHIONE S-TRANSFERASE FUSION PROTEIN

All recombinant DNA techniques were conducted according to standard protocols (17). The BamH1-BamH1 restriction fragment (944 bp) of calreticulin cDNA (for restriction map of calreticulin, see 11) was subcloned into a unique BamH1 restriction site of the pGEX-3X (glutathione S-transferase) expression vector in order to construct the glutathione S-transferase-calreticulin fusion protein. The fusion protein used in this study encoded the last 129 C-terminal amino acids of calreticulin. A protease deficient BNN103 strain of *E. coli* was used for the transformation and expression experiments. Expression and purification of the fusion protein was carried out as described earlier (18).

Results and Discussion

THE PREDICTED MODEL OF FOLDING OF CALRETICULIN

A cDNA encoding calreticulin was isolated in our laboratory and the amino acid sequence of the protein was deduced from the nucleotide sequence of this cDNA (11). Fig. 1 shows the model of folding of calreticulin, which we propose, based on the amino acid sequence information and on the secondary structure predictions for this sequence. Secondary structure predictions for the protein suggest that it might have a 'lollipop' shape (Fig. 1). Apparently, there are no long hydrophobic segments in the sequence, which would be capable of spanning the membrane bilayer confirming that calreticulin is a peripheral membrane protein (4).

Mature calreticulin comprises 401 amino acids with a M_r of 46 567 (11). The protein sequence begins with a helix-turn-helix motif (residues 1-74) (Fig. 1, helix-turn-helix domain). This is followed by a sequence predicted to form eight anti-parallel β-strands connected by protein loops (Fig. 1, β-sheets domain). Two regions of short α-helices are predicted at residues 98–103 and 149–154. The sequence from residue 187–317 contains an abundance of prolines, some of which are spaced regularly every four or five amino acids (Fig. 1, P-rich domain). This sequence also contains a repeating, highly charged, rigid turn structure, which separates the globular head of the protein from the acidic tail (Fig. 1, acidic tail domain). The sequence from residue 187–317 also contains a proline-, serine- and threonine-rich sequence and the sequence KPEDWD is repeated three times (Fig. 1, KPEDWD repeat). The C-terminal 20% of the protein is highly acidic. In the last 56 residues, 37 are acidic while 10 are basic. Although predicted to form an α-helical structure this region is more likely to be in an extended conformation, due to a charge repulsion, and, it may therefore provide a site of interaction with other protein components.

One of the interesting features of the amino acid sequence of calreticulin is the presence of the KDEL sequence at the C-terminal end (2). A number of other peripheral membrane proteins in the ER are known to contain the KDEL sequence at the C-terminus (19), including some heat shock proteins and some glucose regulated proteins, both of which also contain a relatively large number of acidic amino acids residues within their C-termini (19). The function of these acidic sequences is not known, but they could be involved in both Ca^{2+}-binding and in protein retention in the ER. The KDEL sequence is known to be responsible for retention of newly synthesized proteins within the lumen of the ER (20, 21). Since calreticulin contains this sequence it is likely salvaged as an ER/SR protein (21). This is in agreement with our earlier observation that calreticulin is localized to the lumen of the SR/ER (6, 4).

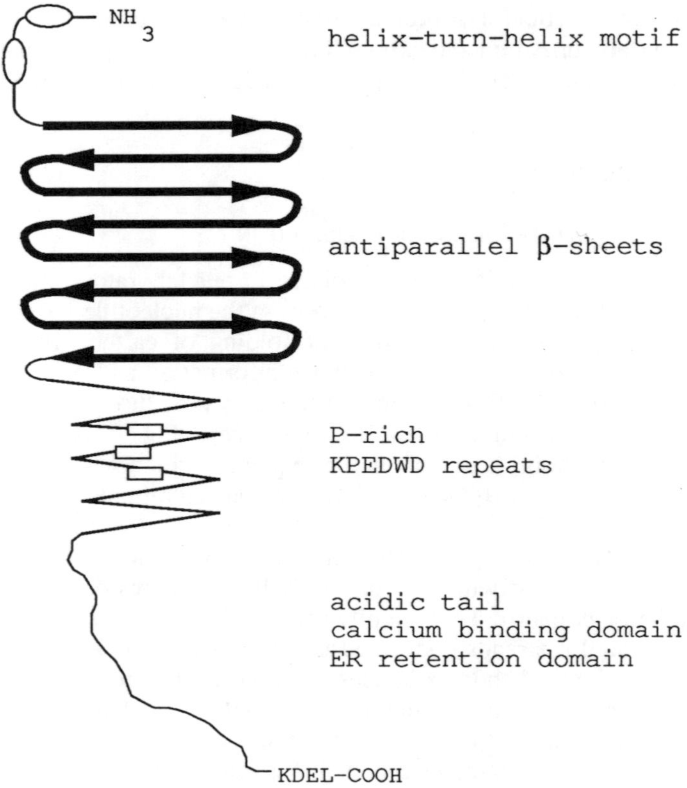

Fig. 1 Calreticulin. This model is based on the amino acid sequence deduced from the nucleotide sequence of a cDNA encoding the skeletal muscle calreticulin (11). Four distinct domains are indicated.

IDENTIFICATION OF A Ca^{2+} BINDING DOMAIN

Along with a number of low affinity Ca^{2+} binding sites (K_d = ~250 μM, B_{max} = 25 moles of Ca^{2+}/mole of protein) calreticulin also has a single high affinity site (K_d = ~1 μM, B_{max} = 1 mole of Ca^{2+}/mole of protein) (5). As a consequence, we anticipated that an 'EF hand' type of Ca^{2+} binding site would be detected in the primary sequence of calreticulin. No evidence of this type of sequence was observed, however (11), and, therefore, the location of the high affinity Ca^{2+}-binding site is, as yet unknown. The helix-turn-helix motif present at the extreme NH_2-terminus might be a candidate.

Thirty-two out of 40 residues in the sequence between residues 341 and 390 are acidic (11) (Fig. 1, 'acidic tail' domain). This sequence is similar to the acidic structure found in calsequestrin, which binds up to 40 moles of Ca^{2+}/mole of protein (9, 19). Since calreticulin has been shown to bind about 25 moles of Ca^{2+} per mole with a low affinity (18), it is probable that this highly acidic sequence is responsible for the low-affinity Ca^{2+} binding.

In order to determine if the 'acidic tail' domain (Fig. 1) of calreticulin is responsible for Ca^{2+} binding by the protein we subcloned and expressed, as a fusion protein, the last 129 C-terminal amino acids of the protein, which encode the 'acidic tail' domain. As an expression vector we used a plasmid encoding glutathione S-transferase (~26 kDa) (18). The glutathione S-transferase-calreticulin fusion protein (~40 kDa) was isolated using glutathione-Sepharose affinity chromatography and assayed for Ca^{2+} binding by a $^{45}Ca^{2+}$ overlay technique. Fig. 2A shows Coomassie blue staining of mature calreticulin (~60 kDa) and the glutathione S-transferase-calreticulin fusion protein (~40 kDa) after SDS-PAGE. Both mature calreticulin and the recombinant protein bound $^{45}Ca^{2+}$ (Fig. 2B, lanes 2 and 3), whereas the glutathione S-transferase control did not (Fig. 2B, lane 1). This shows that the 'acidic tail' domain of calreticulin does contain a Ca^{2+} binding site(s) of the protein.

In summary, we have presented a folding model for calreticulin, which shows that the protein can be structurally and functionally divided into four distinct domains. The C-terminal 'acidic tail' domain is involved in Ca^{2+} binding by calreticulin.

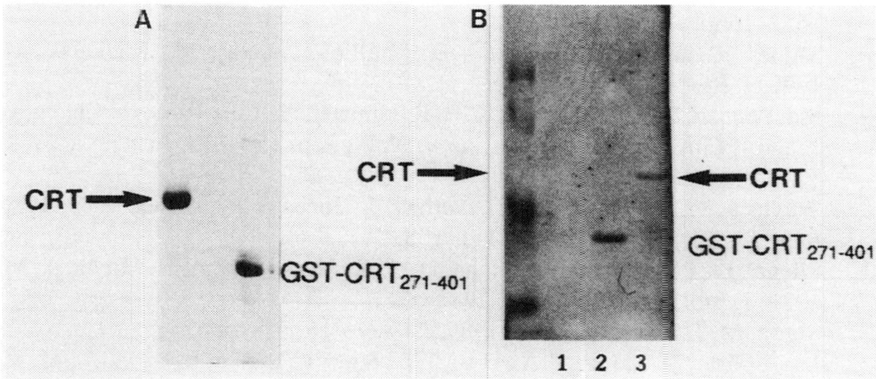

Fig. 2. Ca^{2+} binding to calreticulin and the glutathione S-transferase-calreticulin fusion protein. Glutathione S-transferase-calreticulin fusion protein was expressed in *E. coli*, isolated by affinity chromatography techniques, separated by SDS-PAGE, stained with Coomassie blue (A), transferred to nitrocellulose filters and incubated with $^{45}Ca^{2+}$ (B). The arrows indicate the position of calreticulin. CRT, calreticulin (~60 kDa). GST-$CRT_{271-401}$, glutathione S-transferase-calreticulin fusion protein (~40kDa). The Bio-Rad prestained molecular weight marker proteins are shown in *B*.

Acknowledgements

We would like to thank Dr J. Weiner for providing BNN103 strain of *E. coli* and his help with the fusion protein experiments. This work was supported by the Medical Research Council of Canada, the Alberta Heart and Stroke Foundation and the Alberta Heritage Foundation for Medical Research. M. M. is a Scholar of the Alberta Heritage Foundation for Medical Research. R. E. M. is an Alberta Heritage Foundation for Medical Research Postdoctoral Fellow. M. O. is a Scholar of the Medical Research Council of Canada.

References

1. Walz, B., Baumann, O., 1989, *Topics Histochem. Cytochem.*, **20**, 1–45.
2. Michalak, M., 1985, In: Martonosi, A.N. (ed.) T*he Enzymes of Biological Membranes*, , vol. 3, pp. 115–155. Plenum Press, New York and London.
3. Ostwald, T. J., MacLennan, D. H., 1974, *J. Biol. Chem.*, **249**, 974–979.
4. Michalak, M., Campbell, K. P., MacLennan, D. H., 1980, *J. Biol. Chem.*, **255**, 1317–1326.
5. MacLennan, D. H., Yip, C. C., Iles, G. H., Seeman, P., 1972, *Cold Spring Harbor Symp. Quant. Biol.*, **37**, 469–477.
6. Fliegel, L., Burns, K., Opas, M., Michalak, M., 1989, Biochim. Biophys. Acta, 982, 1-8.
7. Waisman, D .M., Salimath, B. P., Anderson, M. J., 1985, *J. Biol. Chem.*, **260**, 1652–1660.
8. Milner, R., Baksh, S., Vance, J. E., Smillie, L., Opas, M., Michalak, M., 1990, *J. Biol. Chem.*, submitted
9. MacLennan, D. H., Campbell, K. P., Reithmeier, A. A. F., 1983. In: Cheng, W. Y. (ed.), *Calcium and Cell Function*, vol. 4, pp. 151–173. Academic Press, New York.
10. Wuytack, F., Raeymaekers, L., Verbist, J., Jones, L.R., Casteels, R., 1987, *Biochim. Biophys. Acta*, **899**, 151–158.
11. Fliegel, L., Burns, K., MacLennan, D. H., Reithmeier, R. A. F., Michalak, M., 1989, *J. Biol. Chem.*, **264**, 21522–21528.
12. Smith, M. J., Koch, G. L. E., 1989, *EMBO J.*, **8**, 3581–3586.
13. McCauliffe, D. P., Lux, F. A., Lieu, T. S., Sanz, I., Hanke, J., Newkirk, M. M., Bachinski, L. L., Itoh, Y., Siciliano, M. J., Reichlin, M., Sontheimer, R.D., Capra, J.D., 1990, *J. Clin. Invest.*, **85**, 1379–1391.
14. McCauliffe, D. P., Zappi, E., Lieu, T. S., Michalak, M., Sontheimer, R. D., Capra, J. D., 1990, J. *Clin. Invest.*, **86**, 332–335.
15. Laemmli, U. K., 1970, *Nature*, **227**, 680–685.
16. Michalak, M., Fliegel, L.,Wlasichuk, K., 1990, *J. Biol. Chem.*, **265**, 5869–5874.
17. Maniatis, T., Fritsch, E. F., Sambrook, J., 1982, *Molecular Cloning. A Laboratory Manual.* Cold Spring Harbor Laboratory, Cold Spring Harbor Laboratory, New York.
18. Smith, D. B., Johnson, K. S., 1988, *Gene*, **67**, 31–40.

19. Fliegel, L., Burns, K., Wlasichuk, K., Michalak, M., 1989, *Biochem. Cell Biol.*, **67**, 696–702.
20. Andres, D. A., Dickerson, I. M., Dixon, J. E., 1990, *J. Biol. Chem.*, **265**, 5952–5955.
21. Pelham, H. R. B., 1989, *Annu. Rev. Cell Biol.*, **5**, 1–23.

4
Structural Characteristics of Chicken Calsequestrin

S. SALVATORI*, P. YAZAKI** AND S. DAHMS**

**C.N.R.-Centro per lo Studio della Biologia e Fisiopatologia muscolare, c/o Istituto di Patologia generale, Universita' di Padova, 35121-Padova, Italy*
***Molecular Biology Institute and Rees-Stealy Research Foundation, San Diego State University, San Diego, CA 92182, USA*

Introduction

Calsequestrin (CS) is a high-capacity, moderate affinity Ca^{2+}-binding protein localized within the lumen of terminal cisternae of the sarcoplasmic reticulum (SR) of muscle cells. Depending on the source, CS binds 35–50 mol of Ca^{2+} per mole of protein, with a K_d of about 1 mM, at physiological concentration of salt.

Since the first report on this protein by MacLennan and Wong (1) and Ikemoto *et al.* (2) it has been suggested that its main function is buffering calcium inside the SR. Morphological studies (3, 4) and X-ray microprobe analysis (5) indicated that CS is localized in the terminal cisternae specifically (6, 7), at the junctional face of SR, that is the specialized area where calcium release occur. Recent evidence (8, 9) suggests that calsequestrin could play an important role during excitation–contraction coupling, since the presence of CS-binding proteins in the junctional face membrane allows the calcium concentration to increase where Ca^{2+} movements occur.

Different isoforms of CS have been described (8, 10, 11, 12) and a few of them have been sequenced (11, 12). The comparison of amino acid sequence of CS of different tissues and species could be of help in understanding both the existing variability of Ca^{2+}-binding levels (8, 13) and the relevant homology of the Ca^{2+}-sensitive hydrophobic region, which is responsible for binding CS to

Muscle and Motility Vol. 2. Proceedings of XIXth European Conference in Brussels
© Intercept Ltd, PO Box 716, Andover, Hampshire, SP10 1YG, UK

phenyl-sepharose (14) and, presumably, of binding to CS-binding proteins (8). Chicken CS exhibits unusual properties such as reduced maximal Ca^{2+}-binding levels and pH-dependent shift in mobility in SDS-PAGE, compared to CS purified from other species or tissues, as well as the metachromatic staining properties with the cationic carbocyanine dye Stains-AII (15). Based on the NH_2-terminal sequence, we recently proposed (16, 17) that chicken CS is identical with the previous described laminin-binding protein (18), called 'aspartactin'. In this study, we report the nucleotide sequence of chicken calsequestrin (see also ref. 17) and some evidence for functional domains of the protein.

Methods

CS was purified from total muscle homogenates, according to the procedure described by Slupsky *et al.* (19). Proteolysis of purified chicken CS was carried out essentially as described by Mitchell *et al.* (8). PAGE was carried out according to Laemmli (20) both, in the presence or in the absence of SDS. Electrophoretic transfer of CS and tryptic fragments onto Immobilon-P membrane and automated microsequence analysis were carried out as described by Matsudaira (21) and Yazaki *et al* (16, 17). Stains-AII staining and ^{45}Ca-autoradiography were performed as previously described (15).
^{3}H-Trifluoperazine (TFP) binding and autoradiography were done as described by Dabbeni-Sala *et al* (22).

Results

Fig. 1 shows the characterization by different staining procedures of chicken CS before and after limited tryptic cleavage. Coomassie blue staining (panel a) indicates that after digestion by trypsin, one prominent peptide of about 30 kDa is formed, which is pink-stained with Stains-AII (panel b) and contains the immunoreactive site toward anti-chicken CS antibody (panel c). As expected from the Stains-AII staining, this 30 kDa fragment does not bind $^{45}Ca^{2+}$ (panel d) and corresponds to the amino terminus of the protein, as stated by microsequence analysis (see also Fig. 2). On the other hand, Stains-AII staining makes evident that other two peptides are formed after tryptic cleavage, which are heavily labelled with $^{45}Ca^{2+}$ (panel d) and hence belong to the C-terminus of the protein (23). Fig. 1 also shows (panel e) the comparison between Coomassie blue-stained electrophoretic pattern of peptide fragments obtained by tryptic cleavage at very low (lane 1) and at high (lane 2) ionic strength. Lane 1 displays a main proteolytic fragment of about 45 kDa, whose N-terminus (VSQKQFQMTEMVLELAAQVLEP) is different from that of the 30 kDa fragment, corresponding to the amino acid sequence starting at position 46 (see Fig. 2).

Fig. 2 shows nucleotide and deduced amino acid sequences of chicken CS. Regions confirmed by microchemical sequence analysis are also indicated.

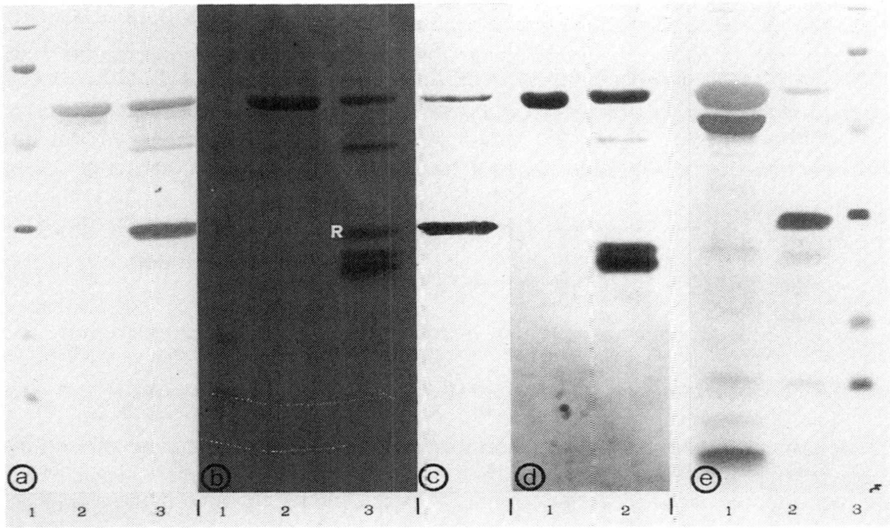

Fig. 1 Limited tryptic cleavage of chicken calsequestrin and differential peptide staining. *Panel a* Coomassie blue staining of: (1) molecular weight markers; (2) undigested calsequestrin; (3) calsequestrin fragments after tryptic cleavage. *Panel b* Stains-AII staining of: (1) molecular weight markers; (2) undigested calsequestrin; (3) calsequestrin after tryptic cleavage. R indicates the main red coloured tryptic fragment. *Panel c* Immunostaining of calsequestrin fragments after tryptic cleavage. *Panel d* 45Ca-overlay of: (1) untreated calsequestrin; (2) calsequestrin tryptic fragments. *Panel e* Coomassie blue staining of: (1) peptides obtained by tryptic digestion at low ionic strength (see below); (2) peptides obtained by tryptic digestion at high ionic strenght, as before (from panel a to d); (3) molecular weight markers. Molecular weight markers: Phosphorylase b (97 400 Da), Bovine serum albumin (66 200 Da), Ovalbumin (43 000 Da), Carbonic anhydrase (31 000 Da), Soybean trypsin inhibitor (21 500 Da) and Lysozyme (14 400 Da).

Tryptic digestion was performed at a trypsin/protein ratio of 1:20, at 25°C for 30 min, in a medium containing 20 mM MOPS, pH 7.0, and 150 mM NaCl (from panel a to d; panel e, lane 2) or 5 mM Tris-CL pH 7.5, without salts (panel e, lane 1). The reactions were stopped by adding soybean trypsin inhibitor in a ratio of 1:1 with trypsin. SDS-PAGE was carried out according to Laemmli (20), in a 12.5% polyacrylamide gel.

A multiple amino acid sequence alignment of known CSs is shown in Fig. 3. The homology of chicken CS is 73.1% and 82.2% for mature form of rabbit skeletal CS and dog cardiac CS, respectively, and even more (84.4% and 90.2%, respectively) taking into account conservative changes.

Fliegel *et al* (11) reported that rabbit CS binds the antipsychotic drug trifluoperazine (TFP), similarly to calmodulin and troponin C and indicated the binding site in the region 209–226, which displays a high degree of homology with TFP-binding regions in the other two Ca^{2+}-binding proteins. It is seen in Fig. 3 that the chicken CS region is very similar to that of rabbit CS (13/18 amino acid residues) and, more surprisingly, identical to that of canine cardiac CS. Fig. 4 shows that chicken CS is able to bind 3H-TFP to a similar extent, as compared to rabbit CS.

```
GAAGAAGGACTGAATTTCCCCACGTATGATGGGAAAGACCGAGTGATCGACCTGAACGAGAAGAACTACAAGCATGCCCTGAAGAAGTAT   90
 E  E  G  L  N  F  P  T  Y  D  G  K  D  R  V  I  D  L  N  E  K  N  Y  K  H  A  L  K  K  Y

GACATGCTTTGTCTGCTCTTCCATGAGCCTGTGAGCTCCGACAGGGTGTCCCAGAAGCAGTTCCAGATGACGGAGATGGTCCTGGAGCTG  180
 D  M  L  C  L  L  F  H  E  F  V  S  S  D  R  V  S  Q  K  Q  F  Q  M  T  E  M  V  L  E  L

GCACCTCAGGTTCTGGAGCCTAGGAGCATCGGATTTGGGATGGTGGATTCCAAGAAGGATGCCAAACTTGCCAAAAAGCTAGGCTTGGTT  270
 A  A  Q  V  L  E  P  R  S  I  G  F  G  M  V  D  S  K  K  D  A  K  L  A  K  K  L  G  L  V

GAAGAGGGAAGTCTCTATGTCTTTAAGGAGGAGCGGTTGATTGAATTTGACGGTGAACTGGCCACAGATGTCTTGGTGGAATTCCTCTTG  360
 E  E  G  S  L  Y  V  F  K  E  E  R  L  I  E  F  D  G  E  L  A  T  D  V  L  V  E  F  L  L

GATCTGCTAGAAGACCCTGTGGAGATCATAAACAGCAAGCTGGAGCTTCAAGCCTTTGACCAAATTGATGACGAAATCAAACTCATTGGC  450
 D  L  L  E  D  P  V  E  I  I  N  S  K  L  E  L  Q  A  F  D  Q  I  D  D  E  I  K  L  I  G

TACTTCAAAGGAGAAGACTCTGAACATTACAAGGCATTTGAAGAGGCTGCTGAACACTTCCAGCCATATGTCAAATTCTTTGCCACATTC  540
 Y  F  K  G  E  D  S  E  H  Y  K  A  F  E  E  A  A  E  H  F  Q  P  Y  V  K  F  F  A  T  F

GACAAAGGGGTTGCCAAGAAACTGGGCCTGAAGATGAACGAGGTGGACTTCTATGAACCATTCATGGATGAGCCTGTTCACATCCCCGAT  630
 D  K  G  V  A  K  K  L  G  L  K  M  N  E  V  D  F  Y  E  P  F  M  D  E  P  V  H  I  P  D

AAACCTTACACAGAAGAGGAGCTGGTTGAATTTGTGAAGGAACACAAAAGAGCCACCTTGCGGAAGCTGCGCCCAGAAGACATGTTTGAG  720
 K  P  Y  T  E  E  E  L  V  E  F  V  K  E  H  K  R  A  T  L  R  K  L  R  P  E  D  M  F  E

ACATGGGAGGATGACATGGAAGGAATCCATATCGTGGCCTTTGCTGAAGAAGATGACCCAGATGGCTTTGAGTTCCTGGAAATCCTGAAG  810
 T  W  E  D  D  M  E  G  I  H  I  V  A  F  A  E  E  D  D  P  D  G  F  E  F  L  E  I  L  K

CAGGTTGCCAGGGACAACACTGATAATCCTGACCTGAGCATTGTGTGGATTGACCCTGACGACTTTCCTCTGCTCATCACTTACTGGGAG  900
 Q  V  A  R  D  N  T  D  N  P  D  L  S  I  V  W  I  D  P  D  D  F  P  L  L  I  T  Y  W  E

AAGACCTTGAACATTGACCTGTTCAGGCCGCAGATTGGGATTGTGAATGTCACAGACGCTGACAGTGTCTGGATGGAGATCAGAGATGAT  990
 K  T  F  K  I  D  L  F  R  P  Q  I  G  I  V  N  V  T  D  A  D  S  V  W  M  E  I  R  D  D

GATGACCTGCCCACAGCCGAGGAGCTGGAGGACTGGATAGAGGATGTGCTTTCTGGGAAGATAAATACTGAAGACGATGACGATGATGAC 1080
 D  D  L  P  T  A  E  E  L  E  D  W  I  E  D  V  L  S  G  K  I  N  T  E  D  D  D  D  D  D

GATGATGATGACGACGACGATGATGATGACGATGATGACGATGATGACGACGATGATGACGACGATGACGATGATGATGACTAA
 D  D  D  D  D  D  D  D  D  D  D  D  D  D  D  D  D  D  D  D  D  D  D  D  D  D  D  *
```

Fig. 2 Nucleotide sequence and amino acid translation of mature chicken calsequestrin cDNA. Numbers refer to nucleotide position. The lines under amino acid sequences indicate regions determined by microchemical sequence analysis of purified chicken calsequestrin.

Discussion

Within the CS-family chicken skeletal CS has unusual biochemical properties (15), particularly a reduced Ca^{2+}-binding capacity (13, 15). Based on results of Fig. 1, chicken CS is composed of at least two domains of similar extent, the first of which is orthochromatically stained with Stains-AII and is the hydrophobic N-terminus. On the other hand, the highly acidic C-terminus is responsible for the metachromatic staining of CS and, as expected, Ca^{2+} binding. The reason why chicken skeletal CS and dog cardiac CS have a reduced Ca^{2+}-binding capacity, compared to rabbit CS, is not immediately evident from an examination of the primary structure of the proteins. On the contrary, the degree of homology of chicken CS is strong, and is more pronounced with canine cardiac CS rather than mammalian skeletal CS. The most striking difference is the presence in chicken CS of an uninterrupted poly-Asp tail. While the functional meaning of this homopolymeric structure has to be more carefully investigated, the possibility exists that the reduced Ca^{2+} binding is not directly related to the number of acidic residues, since Asp+Glu residues account for more than 30% of the total residues, in all CS so far investigated

```
           **** ** *** ***     ** *   ***  * *  **    *   * **   *  *** *****    **  **  *
RFCS       EEGLDFPEYDGVDRVINVNAKNYKNVFKKYEVLALLYHEPPEDDKASQRQFEMEELILELAAQVLEDKGVGFGLVDSEKD   80
NRCS       EEGLDFPEYDGVDRVINVNAKNYKNVFKKYEVLALLYHEPPEDDKASQRQFEMEELILELAAQVLEDKGVGFGLVDSEKD
CHCS       EEGLNFPTYDGKDRVIDLNEKNYKHALKKYDMLCLLFHEFVSSDRVSQKQFQMTEMVLELAAQVLEPRSIGFGMVDSKKD
DCCS       EEGLNFPTYDGKDRVVSLTEKNFKQVLKKYDVLCLYYHESVSSDKVAQKQFQLQEIVLELVAQVLEHKDIGFVMVDAKKE

           *  *****  ** * ** *    ** ***   * *******  ***** *     * ***  * * ***** **    ** *
RFCS       AAVAKKLGLTEEDSIYVFKEDEVIEYDGEFSADTLVEFLLDVLEDPVELIEGERELQAFENIEDEIKLIGYFKNKDSEHY   160
NRCS       AAVAKKLGLTEEDSIYVFKEDEVIEYDGEFSADTLVEFLLDVLEDPVELIEGERELQAFENIEDEIKLIGYFKNKDSEHY
CHCS       AKLAKKLGLVEEGSLYVFKEERLIEFDGELATDVLVEFLLDLLEDPVEIINSKLELQAFDQIDDEIKLIGYFKGEDSEHY
DCCS       AKLAKKLGFDEEGSLYVLKGDRTIEFDGEFAADVLVEFLLDLIEDPVEIINSKLEVQAFERIEDQIKLIGFFKSEESEYY

           *** **** * **  ******  ***** ** ** **** ** **   *****  *** * ** ** * *** * ** * *
RFCS       KAFKEAAEEFHPYIPFFATFDSKVAKKLTLKLNEIDFYEAFMEEPVTIPDKPNSEEEIVNFVEEHRRSTLRKLKPEEMYE   240
NRCS       KAFKEAAEEFHPYIPFFATFDSKVAKKLTLKLNEIDFYEAFMEEPVTIPDKPNSEEEIVNFVEEHRRSTLRKLKPEEMYE
CHCS       KAFEEAAEHFQPYVKFFATFDKGVAKKLGLKMNEVDFYEPFMDEPVHIPDKPYTEEELVEFVKEHKRATLRKLRPEDMFE
DCCS       KAFEEAAEHFQPYIKFFATFDKGVAKKLSLKMNEVDFYEPFMDEPIAIPDKPYTEEELVEFVKEHQRPTLRRLRPEDMFE

           *****  *********  **** ******* ** ********** **********  ****** ***  **** ******
RFCS       TWEDDMDGIHIVAFAEEADPDGYEFLEILKSVAQDWTDNPDLSIIWIDPDDFPLLVPYWEKTFDIDLSAPQIGVVNVTDA   320
NRCS       TWEDDMDGIHIVAFAEEADPDGYEFLEILKSVAQDWTDNPDLSIIWIDPDDFPLLVPYWEKTFDIDLSAPQIGVVNVTDA
CHCS       TWEDDMEGIHIVAFAEEDDPDGFEFLEILKQVARDWTDNPDLSIVWIDPDDFPLLITYWEKTFKIDLFRPQIGIVNVTDA
DCCS       TWEDDLNGIHIVAFAERSDPDGYEFLEILKQVARDWTDNPDLSIVWIDPDDFPLLVAYWEKTFKIDLFKPQIGVVNVTDA

           ******  *  *** ******* **** * *******
RFCS       DSVWMEMDDEEDLPSAEELEDWLEDVLEGEINTEDDDDE
NRCS       DSVWMEMDDEEDLPSAEELEDWLEDVLEGEINTEDDDDEDDDDDDDD
CHCS       DSVWMEIRDDDDLPTAEELEDWIEDVLSGKINTEDDDDDDDDDDDDDDDDDDDD    DDDD■DDDDDDDD
DCCS       DSVWMEIPDDDDLPTAEELEDWIEDVLSGKINTEDDDNEEGDDGDDDEDDDDDGNNSDEES■DDSDDDDE
```

Fig. 3 Multiple amino acid sequence alignment of mature forms of known calsequestrins. RFCS = rabbit fast-twitch skeletal muscle calsequestrin; NRCS = neonatal rabbit fast-twitch skeletal muscle calsequestrin; CHCS = chicken fast-twitch skeletal muscle calsequestrin; DCCS = canine cardiac muscle calsequestrin. Asterisk (*) indicates common identity.

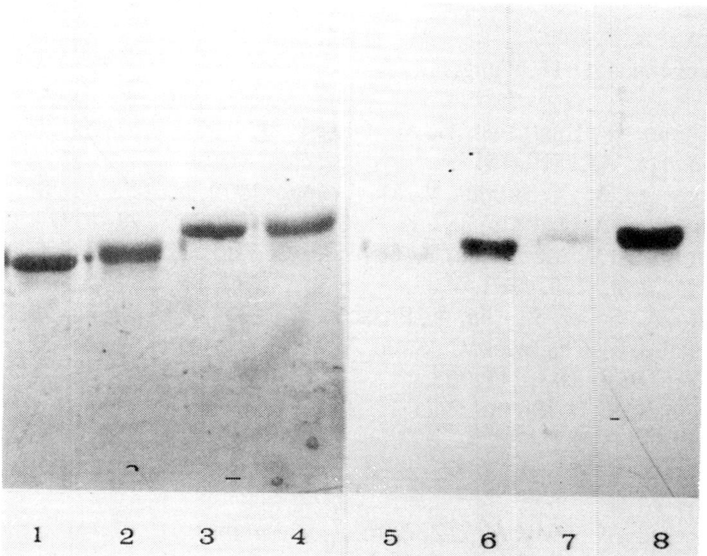

Fig. 4 Chicken and rabbit CSs photo-labelling with 3H-Trifluoprazine. 250 μg of purified CS was incubated with 3H-TFP and UV irradiated as described by Dabbeni-Sala *et al.* (5), in the absence (lanes 1, 3, 5 and 7) and in the presence (lanes 2, 4, 6 and 8) of 80 μM unlabelled TFP. Lanes: 1, 2, 5 and 6, purified chicken CS; 3, 4, 7 and 8, purified rabbit CS; from 1 to 4: Coomassie blue staining; from 5 to 8: autoradiography.

(2). An interesting correlation can be seen between the Glu/Asp ratio and the Ca^{2+}-binding capacity, with both values being higher in adult rabbit CS, lower in chicken CS and intermediate in dog cardiac CS.

The binding of Ca^{2+} induces a dramatic conformational change, possibly involving the hydrophobic region responsible for CS binding to phenyl-Sepharose. Since this region has been suggested to be likely involved in protein–protein interaction (8, 11), it could be of interest to test this hypothesis also in view of possible interactions of CS with itself. In fact, the hypothesis of dimer formation has been advanced by Mitchell et *al* (8) as a possible explanation for the apparently reduced Ca^{2+}-binding capacity of cardiac CS. Preliminary evidence obtained by non-denaturing PAGE, seems to indicate that chicken CS and rabbit CS, exist as a dimer in a low ionic strength solution (results not shown). The significance of possible CS oligomerization *in vivo* is under evaluation.

Acknowledgements

This research was supported by the National Foundation grants DCB 8613881 and INT 8515846, the California Metabolic Research Foundation, and Consiglio Nazionale delle Ricerche (CNR) grant 8800555.04. We thank Miss S. Furlan (CNR, Padova) for technical assistance.

References

1. MacLennan, D. H., Wong, P. T. S., 1971, *Proc. Natl. Acad. Sci. USA*, **68**,1231–1235.
2. Ikemoto, N., Bhatnagar, G. M., Gergely, J., 1971, *Biochem. Biophys. Res. Commun.*, **44**,1510–1517.
3. Jorgensen, A. O., Kalnins, V., MacLennan, 1979, *J. Cell Biol.*, **80**, 372–384.
4. Jorgensen, A. O., Campbell, K. P., 1984, *J. Cell Biol.*, **98**, 1597–1602.
5. Somlyo, A. V., McClellan, G., Gonzalez-Serratos, H., Somlyo, A. P., 1985, *J. Biol. Chem.*, **260**, 6801–6807.
6. Saito, A., Seiler, S., Chu, A., Fleischer, S., 1984, *J. Cell Biol.*, **99**, 875–885.
7. Costello, B., Chadwick, C., Saito, A., Chu, A., Maurer, A. Fleischer S., 1986, *J. Cell. Biol.*, **103**, 741–753.
8. Mitchell, R. D., Simmerman, H. B. K., Jones, L. R., 1988, *J. Biol. Chem.*, **263**,1376–1381.
9. Ikemoto, N., Ronjat, M., Meszaros, L. G., Koshita, M., 1989, *Biochemistry*, **28**, 6764–6771.
10. Damiani, E., Salvatori, S., Zorzato, F., Margreth, A., 1986, *J. Muscle Res. Cell Motil.*, **7**, 435–445.
11. Fliegel, L., Onishi, M., Carpenter, M. R., Khann, V., Reithmeier, R. A. F., MacLennan, D. H., 1987, *Proc. Natl. Acad. Sci. USA*, **84**, 1167–1171.
12. Scott, B. T., Simmerman, H. K. B., Collins, J. H., Nadal-Ginard, B., Jones, L. R., 1988, *J. Biol. Chem.*, **263**, 8958–8964.

13. Yap, J. L., MacLennan, D. H., 1976, *Can. J. Biochem.*, **54**, 670–673.
14. Cala, S. E., Jones, L. R., 1983, *J. Biol. Chem.*, **258**,11932–11936.
15. Damiani, E., Salvatori, S., Margreth, A., 1990, *J. Muscle Res. Cell Motil.*, **11**, 48–55.
16. Yazaki, P., Salvatori, S., Sabbadini, R., Dahms, S., 1990, *Biochem. Biophys. Res. Commun.*, **166**, 898–903.
17. Yazaki, P., Salvatori, S., Dahms, S., 1990, *Biochem. Biophys. Res. Commun.*, **170**, 1089–1095.
18. Clegg, D. O., Helder, J. C., Hann, D. E., Reichardt, L. F., 1988, *J. Cell Biol.*, **107**, 699–705.
19. Slupsky, J. R., Ohnishi, M., Carpenter, M. R., Reithmeier, R. A. F., 1987, *Biochemistry*, **26**, 6539–6544.
20. Laemmli, U. K., 1970, *Nature*, **227**, 680–685.
21. Matsudaira, P., 1987, *J. Biol. Chem.*, **262**, 10035–10038.
22. Dabbeni-Sala, F., Schiavo, G., Palatini, P., 1990, *Biochim. Biophys. Acta*, **1026**, 117–125.
23. Ohnishi, M., Reithmeir, R. A. F., 1987, *Biochemistry*, **26**, 7458–7465.
24. Cozens, B., Reithmeier, R. A. F., 1984, *J. Biol. Chem.*, **259**, 6248–6252.
25. Salvatori, S., Bravin, M., Margreth, A., 1988. In: Gerday, Gilles and Bolis, (eds), *Calcium and Calcium Binding Proteins*, pp. 228–235. Springer-Verlag, Berlin and Heidelberg.

5
Structure and Properties of High Molecular Weight Mitochondrial Creatine Kinase from Beef Heart

L. V. BELOUSOVA, S.N. FEDOSOV, V.J. STELMASCHUK* AND E.V. ORLOVA*

Department of Biochemistry, School of Biology, Moscow State University, Moscow 119899, USSR
**A. V. Shubnikov Institute of Crystallography, USSR Academy of Sciences, Moscow*

Mitochondrial creatine kinase (Mi-CK) is a typical oligomeric enzyme which exists in two interconvertible forms (1, 2, 4, 7–11, 14, 16). The form with low molecular weight (80–85 kDa) is a dimer consisting of two identical subunits. For a long time measurements of the molecular weight of the second form gave contradictory results. The M_r values of 200–250 kDa (2–5, 7, 8) as well as 300–350 kDa (5, 6, 9–11, 14–16) were reported. In the present paper we give the results of the investigation of two high molecular weight Mi-CK forms by gel-filtration, ultracentrifugation and electron microscopy methods.

Methods

High molecular weight Mi-CK forms were obtained by the method decribed earlier (2, 5). Dimer was obtained after incubation of these forms in 8 M urea during 1 hour (2, 5).

Molecular weights (M_r) were determined by gel-filtration on Sephacryl S-300 (0.1 M Tris-HC1 buffer 0.2 mM dithiothreitol pH 7.4). The column (K-16, LKB) was calibrated by markers proteins (MS-II set, Serva) (5).Sedimentation coefficients (S_{20}^{w}) were determined by the method of non-equilibrium sedimentation (Beckman E Spinco analytical ultracentrifuge) in the same buffer (2).

Muscle and Motility Vol. 2. Proceedings of XIXth European Conference in Brussels
© Intercept Ltd, PO Box 716, Andover, Hampshire, SP10 1YG, UK

For electron microscopy (EM) Mi-CK was dialysed against 0.01 M Tris-HC1 buffer pH 7.4 with 0.2 mM dithiotreitol over 8–12 h and then diluted with the same buffer to 10–50 μg/ml. EM specimens were negatively stained with 2% aqueous uranyl acetate pH 4.5 on carbon coated EM grids following the technique of Valentine *et al.* (18).

EM investigation was carried out in a Phillips EM-400 microscope operating at 80 kV. A program package for image processing with the use of Fourier filtering (up to 512 × 512 pixels), correlation analysis of individual particle images was developed at the Institute of Crystallography.

Results and Discussion

Characteristics of different Mi-CK oligomeric forms are given in Table 1. Straightforward comparison of M_r's of the various Mi-CK forms with the molecular weight of Mi-CK subunit equal to 43 kDa leads to the conclusion that Form I is most likely octameric; Form II, hexameric; Form III, dimeric. It is necessary to point out that sedimentation coefficients ($S^w{}_{20}$) for Form I with M_r = 320 kDa and Form II with M_r = 240 kDa are practically the same and equal to 12.3 S. This value is a characteristic of octamers (9–11, 17, 18). Therefore it is possible to suggest that Form II is not the hexamer, but actually a special conformation state of the octamer. This result poses a question whether Forms I and II have not only different M_rs but also some other different characteristics.

Table 1. M_r and $S_{20}{}^w$ (mech ± s.e.m.) of different oligomeric forms of Mi-CK from beef heart.

Characteristics	Form I (octamer)	Form II (hexamer?)	Form III (dimer)
M_r (kDA)	317±15 $n = 4$	244±15 $n = 4$	84±6 $n = 7$
$S^w{}_{20}$	12.3±0.3 $n = 4$	12.4±0.2 $n = 4$	5.3 $n = 2$

Indeed investigation of cysteine and arginine residues properties of these forms carried out in our previous studies revealed important differences between them (5, 6, 14, 15, 17). It was found out that only 50% of the essential arginine and cysteine residues in 320 kDa form were modified by specific reagents (5, 13) and only 50% of active sites could bind MgADP into E-MgADP-creatine, E-MgADP-NO^-_3-creatine complexes (6).

For electron microscopy we used two forms of Mi-CK with M_r = 240 kDa and M_r=320 kDa. Fig. 1, A, B presents fields of Mi-CK molecules with M_r=240 kDa and M_r=340 kDa correspondingly. In both cases one can see annular type images of approximately the same size 12–14 nm, but for M_r = 240 kDa one much more often observes images of other type (*x*-like or butterfly-like) which are probably fragments of the basic annular type particles. But this case has not as yet been analysed in detail.

Much more carefully we analysed the case of M_r = 320 kDa. As is seen from Fig. 1B, observed images in this case are mainly annular particles 12–14 nm in

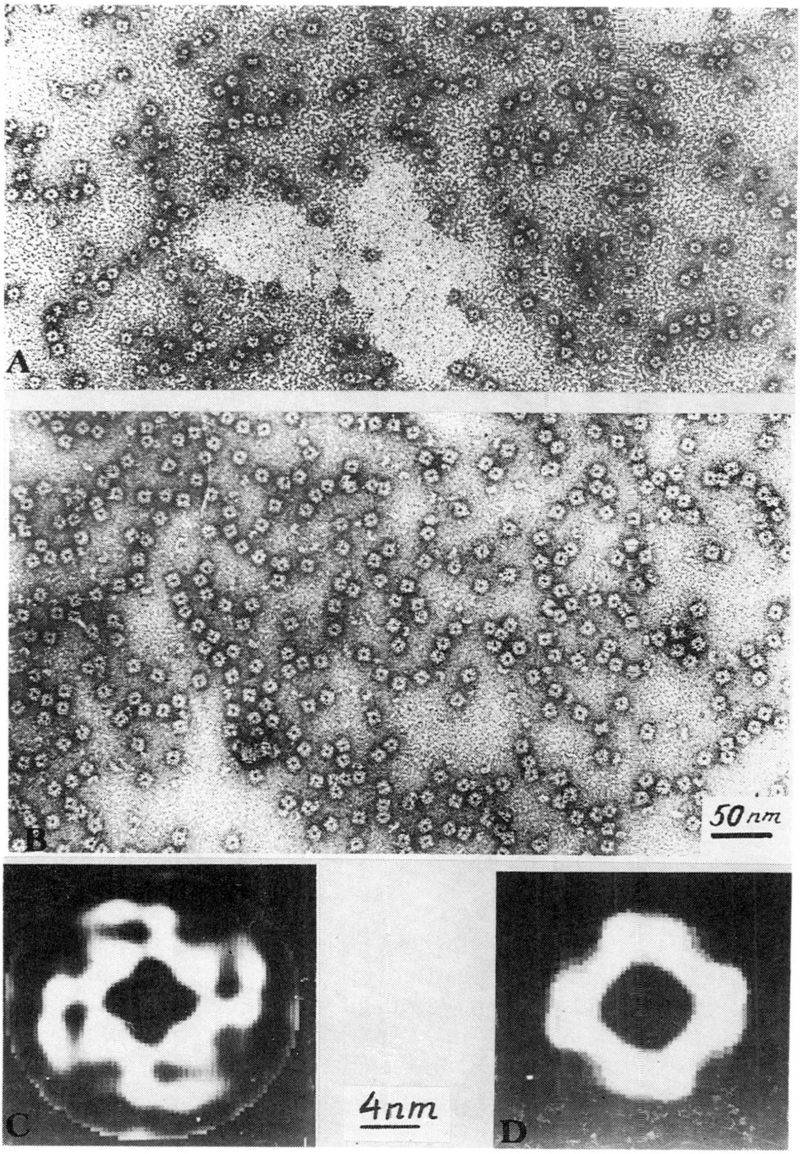

Fig. 1 (A) A field of Mi-CK molecules with molecular weight of 240 kDa; (B) a field of Mi-CK molecules with molecular weight of 340 kDa; (C, D) results of correlation averaging of individual creatine kinase molecules 320 kDa for various reference particles: 25 images were included in averaging of 75 scanned ones.

diameter with well detected subunit structure and with central cavity filled with stain (3–4 nm dia.). This is similar to the picture described for Mi-CK from

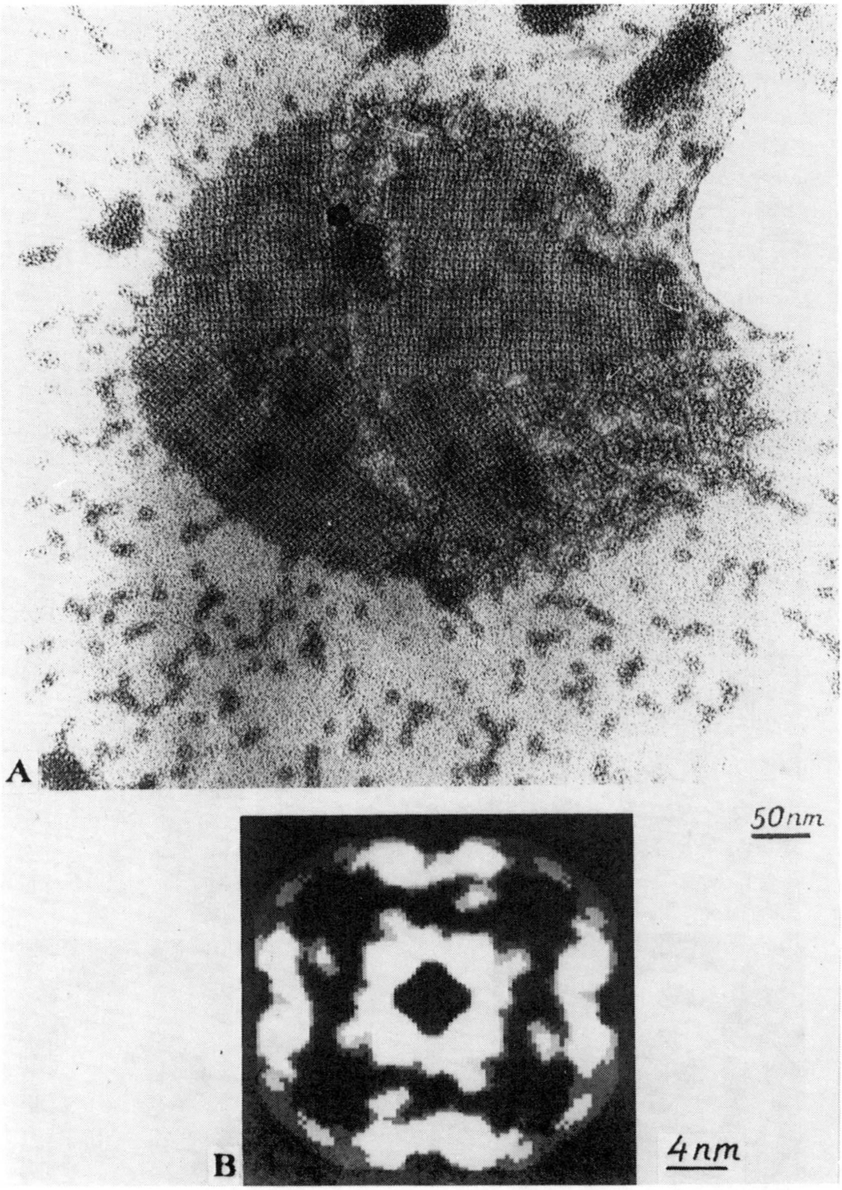

Fig. 2 (A) A micrograph of a two-dimensional imperfect crystal of Mi-CK with 340 kDa; (B) a correlation averaged particles from the crystal layer. A well ordered small part of the image (3 × 3 unit cells) was enhanced by band filtering. Four-fold symmetrized and used as a reference for the calculation of the correlation map. I, upper subunit layer; II, lower subunit layer.

chicken cardial muscle and brain (15, 16). One should emphasize however the absence of absolute similarity of the observed projections in Fig. 1B. Multiplicity of forms close to annular is the evidence of the great lability and structural mobility of subunits inside of CK molecules.

Averaging of the images has been made relative to the rotational axis of fourth order. The selected images were aligned to a reference one by rotational and translational shifts using a correlation function. The results of averaging and symmetrization are shown at Fig. 1C, D. One can see clear evidence of four subunits.

Much more distinct picture was obtained after analysis of the images taken from the region of imperfect crystal (Fig. 2A, B). The important point is using of filtering of high frequencies in image processing. One can now see evidence of two layer structure of each particle. Besides the clearly seen four subunits of upper layer, one can also see parts of the subunits relating to lower layer (Fig. 2B). The whole picture is in good accordance with octameric structure of Mi-CK molecule (320 kDa). It is important to note that each subunit has a nonglobular elongated bent dumb-bell form.

Such a structure of the octamer may explain the experimental fact of 50% decrease in the number of the essential for the enzyme activity cysteine and arginine residues in the process of the octamer formation from the dimers (5, 13). Such a decrease becomes possible if one assumes that the essential SH-group is localized, for example in the elongated part of the subunit. One can assume that the essential SH-group of each subunit is screened by the neighbouring subunit in the upper layer whereas in the lower layer this is not the case. The screening effect may take place if the functional dimer is formed by one subunit from the upper layer and underlying subunit from the lower layer.

References

1. Belousova, L. V., Lipskaya, T. Yu., Temple, V. J., Rostovtsev, A. P., 1980, *J. Mol. Cell Cardiol.*, **12** (suppl. I), 15.
2. Belousova, L. V., Lipskaya, T. Yu., Temple, V. J., Rostovtsev, A. P., 1983, *Adv. Microcardiol.* Baltimore: Univ. Parc Press, 585–595.
3. Cheneval, R. C., Carafoli, E., 1988, *Eur. J. Biochem.*, **171**, 1–9.
4. Farrel, E. C., Baba, N., Brierley, G. P., Prumer, H. D., 1972, *Lab. Invest.*, **27(2)**, 209–219.
5. Fedosov, S. N., Belousova, L. V., 1988, *Biokhimiya* (in Russian), **53**, 550–564.
6. Fedosov, S. N., Belousova, L. V., 1989, *Biokhimiya* (in Russian), **54**, 54–67.
7. Grace, A. M., Perryman, M. B., Roberts, R., 1983, J. Biol. Chem. 258, 15346–15354.
8. Hall, N., Addis, P., DeLuca, M., 1977, *Biochem. Biophys. Res. Commun.*, **76**, 950–956.
9. Jacobs, H. K., Graham, M., 1978, *Fed. Proc.*, **37**, 1574.
10. Lipskaya, T. Yu., Kedishvili, N. Yu., Kalenova, M. E., 1985, *Biokhimiya* (in Russian), **50**, 1571–1581.

11. Marcillat, O., Goldschmidt, D., Eichenberger, D., Vial, C., 1987, *Biochem. Biophys. Acta*, **890**, 233–241.
12. Moskvitina, E. L., Belousova, L. V., 1985, *Doklady Akademii Nauk*, USSR (in Russian), **281**, 209–213.
13. Moskvitina, E. L., 1987, PhD Thesis, Moscow University.
14. Quemeneur, E., Marcillat, O., Eichenberger, D., Vial, C., 1989, *Biochem. Intern.*, **18**, 365–371.
15. Schlegel, J., Zurbriggen, B., Wegmann, B., Wyss, M., Eppenberger, H. M., Walliman, T., 1988, *J. Biol. Chem.*, **263**, 16942–16953.
16. Schnyder, T., Engel, A., Lustig, A., Walliman, T., 1988, *J. Biol. Chem.*, **263**, 16942–16969.
17. Severin, S. E., Belousova, L. V., Moskvitina, E.L., 1983, *Biochem. Intern.*, **6**, 149–156.
18. Valentine, R. C., Shapiro, B. M., Stadtman, E. R., 1968, *Biochemistry*, **7**, 2143–2148.

SECTION TWO

Myogenesis and Regeneration

6
Peptide Growth Factors in Myogenesis and Regeneration

M. D. GROUNDS

University Department of Pathology, Queen Elizabeth II Medical Centre, Nedlands, Western Australia, 6009

Introduction

It is now known that many growth factors (GF) affect the behaviour of muscle precursor cells (mpc). Much of this information is derived from tissue culture studies using myogenic cell lines or mpc from embryonic or mature muscle and is reviewed in detail by Grounds (1). There is general correlation between the response to GF of mpc derived from embryonic or adult muscle in tissue culture (Table 1), although the response may be modified by interaction between these different factors.

Table 1 Factors affecting the behaviour of muscle precursor cells.

	Proliferation	Differentiation and Fusion
Fibroblast GFI	↑	↓
Platelet derived GF	↑	↓
Bischoff muscle GF	↑	–
Leukaemia Inhibitory Factor (LIF)	↑	–
Interleukin-6	↑	–
Adrenocorticotropin	↑	–
Insulin GF-1	↑	↑
Prostaglandin E1	–	↑
Transforming GF-b	↑	↓
Interferon	–	↓

Muscle and Motility Vol. 2. Proceedings of XIXth European Conference in Brussels
© Intercept Ltd, PO Box 716, Andover, Hampshire, SP10 1YG, UK

The precise relationship between mpc derived from embryonic and adult muscle is unclear; while they share many features their behaviour is not always completely equivalent (1). This paper is confined to myogenesis in mature muscle which is regenerating in response to injury, and summarizes the role of GF in this process. It must be emphasized that many other factors such as the availability of GF receptors, changes in composition of the extracellular matrix, and the complex role of other cells particularly macrophages, are also of central importance and are reviewed elsewhere (1).

Activation of Quiescent Satellite Cells

Muscle precursor cells of mature muscle are considered to be the satellite cells which lie between the sarcolemma and external lamina of muscle fibres. Whether other cells can also contribute to myogenesis of mature muscle *in vivo* is unclear. The satellite cells are normally quiescent and are considered to be in the Go stage of the cell cycle. It is not known what maintains this quiescent state: it may be the absence of available GF mitogens (such as fibroblast growth factor [FGF]) or of receptors for these mitogens, or it might be due to the presence of negative growth factors such as transforming growth factor-beta (TGF-β).

Experiments in our laboratory with probes to the recently described skeletal muscle specific regulatory genes MyoD1 and myogenin (2) show almost no *in situ* hybridization to tissue sections of uninjured tibialis anterior muscles of mature mice. However, when muscles are crush injured, mRNA for MyoD1 and myogenin is detected after 6 h and is conspicuous by 24 h after injury. Mononuclear cells co-expressing these genes are concentrated in a zone 1.6–2.6 mm distal to the site of injury (3). Although the expression of these genes enables the early identification of many skeletal mpc, it was not possible to confirm whether these were actually satellite cells. The rapid activation of mpc distant from the site of injury suggests a response to a change in the surface or internal environment of the muscle fibre as a result of the injury which is rapidly translated throughout the length of the fibre (e.g. altered electrical properties, changes in Ca++ levels or other ion fluxes, release of proteases or possibly muscle specific GF) rather than a sudden availability of mitogenic GF in the external environment of the satellite cells.

Whether these activated mpc are subsequently stimulated to move through G1 and synthesise DNA may well depend on changes in GF availability in their immediate environment. Our results suggest that some of the activated mpc migrate towards the injured site, where they will certainly encounter changes in GF levels. Large numbers of inflammatory cells accumulated near the site of injury will result in the enzymatic release of FGF bound to the extracellular matrix, plus the production of FGF and other GF by macrophages and other inflammatory cells, and the production of platelet-derived growth factor (PDGF) particularly by platelets. Both FGF and PDGF are potent mitogens for mpc and other mesenchymal cells. In addition, FGF is highly angiogenic and both factors are closely implicated in many aspects of general tissue repair. Leukaemia

Inhibitory Factor (LIF) and interleukin-6 which are also produced by macrophages, have recently been shown in tissue culture to be mitogenic for myoblasts, but not fibroblasts (4). LIF is a particularly potent myoblast mitogen and may well play a central role *in vivo.* The precise relationship of the muscle specific mitogen described by Bischoff (5) to FGF is unclear, and there is little information about the time course of this GF production by traumatized muscle fibres.

The association between the presence of inflammatory cells and the availability of mitogenic GF described above, is supported by our autoradiographic studies which show that the onset of DNA synthesis in mpc is delayed in transplanted as compared with injured muscles (6). These results indicate that the stimulus for replication is closely connected with diffusible substances produced by macrophages and revascularization. Similarly, other studies such as those by Bischoff (7) show that the onset of DNA synthesis in satellite cells is specifically dependent upon GF such as FGF.

Cossu and his colleagues (8) have reported that adrenocorticotrophic hormone (ACTH) is a specific mitogen for mpc. Although circulating levels of ACTH increase very rapidly and transiently in response to stress, it does not seem physiologically sensible that fluctuations which would occur in response to any general stress in mature animals would activate mpc. It seems more appropriate that local changes in ACTH or related neuropeptides might be important as appears to be the case during embryonic myogenesis. Similarly it is unlikely that changes in circulating glucocorticoid levels after stress are important for myogenesis of mature muscle *in vivo*, and the normal levels *in vivo* are probably not limiting. Tissue culture studies show that glucocorticoids such as dexamethasone potentiate the effects of GF mitogens by upregulating cellular receptors for these GF.

Replication of mpc

Once mpc are activated other growth factors are required for the cells to move through G1 to the DNA synthesis phase; these include the insulin growth factors (IGFs) and epidermal growth factor (EGF). IGF-1 and IGF-2 are produced by mpc and IGF-1 stimulates mpc replication in an autocrine and/or paracrine manner. IGF-2 appears to act later during myogenesis and may be involved in synaptogenesis. Although IGFs mediate the effects of growth hormone, studies with growth hormone deficient (*lit/lit*) mice (Grounds and McGeachie, unpublished) and rats indicate that the circulating levels of growth hormones and IGF are unimportant during muscle regeneration; it appears that only the local production of IGF is critical.

EGF is produced by endothelial cells (and other cell types) and is associated with revascularization. Like thrombin it acts in conjunction with FGF to increase the proliferation and inhibit the differentiation of mpc in tissue culture.

Proliferation will also be enhanced by high levels of GF which inhibit mpc differentiation and thus prevent mpc leaving the cell cycle. In tissue culture mpc differentiation is inhibited by high levels of FGF and PDGF (which complements their mitogenic role) and by TGF-β and possibly interferon. The potent inhibition of mpc differentiation by TGF-β is not readily overcome by combinations of FGF and IGF-1 (9, 10). TGF-β also inhibits proliferation but this can be overcome by FGF (9). TGF-β has an important role in enhancing the formation of the extracellular matrix which is important in new muscle formation: it is produced in a biologically inactive form by many cell types including platelets.

In summary, it seems likely that in mature muscle *in vivo*, mpc replication is enhanced by high levels of FGF, PDFG, Bischoff GF, TGF-β, IGF-1, EGF and thrombin.

Differentiation and Fusion of Myogenic Cells

At some point during the early part of the G1 phase of the cell cycle mpc start to differentiate and become committed to fusion. Differentiation results in the expression of a range of muscle specific genes including those for the contractile proteins (actins, myosins, tropomyosins): some of these genes are expressed before fusion and others after fusion.

Detailed ultrastructural studies of regenerating murine muscle carried out in our laboratories indicate that the process of fusion *in vivo* between immature myogenic cells with closely apposed plasmalemmae is slightly different to that where one or both myogenic cells are well differentiated and surrounded by some form of glycocalyx (10). Fusion of mpc to form myotubes *in vivo* is normally not seen before 2.5 to 3 days after injury. Fusion is a highly complex process and it is difficult to assess the role that growth factors play *in vivo*.

In tissue culture, fusion between mpc is partially dependent upon their density: this affects the balance of GF and leads to a relative increase in those produced by the mpc themselves, in particular IGF-1 and prostaglandin E1 (PGE1). The prostaglandins are a family of compounds derived from essential fatty acids such as linoleic acid and are closely associated with changes in many factors which are important for cell fusion (e.g activation of protein kinases, turnover of phospholipids and intracellular Ca++ levels). It seems likely that the autocrine and/or paracrine effect of prostaglandins may be central to the extensive membrane reorganization associated with myogenic fusion.

The receptors for IGF (unlike other GF) are not down-regulated during differentiation, and IGF-I may play a central role in stimulating mpc differentiation. The potent inhibition by TGF-β of mpc differentiation (and subsequent fusion) suggests that levels of this GF should decrease for differentiation and fusion to occur: however, the combination of high levels of IGF-1 and PGE1 might be able to overcome this inhibition in tissue culture. In tissue culture FGF, PDGF and interferon also inhibit mpc differentiation. There may be a decrease *in vivo* in the availability of FGF from extracellular matrix or a decreased production of these GF by infiltrating cells which facilitates differentiation. Regardless of changes in these GF levels, it seems likely that the local increases in IGF-1 and PGE1 produced by myogenic cells, are of critical importance for differentiation and fusion respectively.

In summary, the balance of GF which probably favour differentiation and fusion of myogenic cells *in vivo* are increased IGF-1 and PGE1, and decreased TGF-β, FGF, and PDGF.

Acknowledgements

The contribution of my various colleagues to these ideas is gratefully acknowledged. Research was supported by grants from the National Health and Medical Research Council of Australia.

References

1. Grounds, M. D., 1991, *Pathol. Res. Pract.*, in press.
2. Wright, W. E., Sassoon, D. A., Lin, V. K., 1989, **Cell**, 56, 607–617.
3. Grounds, M. D., Garrett, K. L., Lai, M. C., Wright, W. E., Beilharz, M. W. Submitted for publication.
4. Austin, L., Burgess, A. W., 1991, *J. neurol. Sci.*, in press.
5. Bischoff, R., 1986a, *Dev. Biol.*, **115**, 140–147.
6. Grounds, M.D., McGeachie, J. K., 1990. *Muscle & Nerve*, **123**, 305–313.
7. Bischoff, R., 1986b, *Dev. Biol.*, **115**, 129–139.
8. Cossu, G., De-Angelis, L., Vivarelli, R., Vella, S., Bouche, M., Boitani, C., Molinari M., 1989, *Dev. Biol.*, **131**, 331–336.
9. Allen, R. E., Boxhorn, L. A., 1989, *J. Cell Physiol.*, **138**, 311–315.
10. Bischoff, R., 1990, In: Eastwood, A.B., Karpati, G., Griggs, R. (eds), *Myoblast Transfer Therapy*. Plenum Press, New York, in press.
11. Robertson, T. A., Papadimitriou, J. M., Mitchell, C. A., Grounds, M. D., (1991), *Pathol. Res. Pract.*, in press.

7
Mechanical Signals Involved in the Expression of Slow Type Genes and the Determination of Muscle-fibre Phenotype in Mammalian Muscles

TH. JAENICKE, J. MARTINDALE, A. SCUTT, G.-F. GERLACH, P. LOUGHNA, P. WILLIAMS* AND G. GOLDSPINK

Unit of Molecular and Cellular Biology, The Royal Veterinary College, London University, London NW1 0TU, UK
**University of Hull, Kingston upon Hull, UK*

Introduction

Myosin is a double molecule that consists of two heavy chains (myosin hcs) each of about 200 000 molecular weight. The actin attachment site and the ATPase site are located in the S_1 (head) region of the myosin crossbridge of each heavy chain. Associated with the S_1 fragment are smaller polypeptides or light chains which are believed to modulate the crossbridge ATPase activity (1). The intrinsic velocity of contraction (V_{max}) of muscle fibres has been related to the specific activity of their myosin ATPase (2). Fast muscle fibre type subsets have been identified histochemically and these exist because of different combinations of myosin heavy chains and myosin light chains. We have focused our attention on the myosin hc genes because of the strong correlation between the V_{max} of the muscle fibres and their fast and slow myosin hc content (3). The isoforms of myosin have been shown to be the products of a multigene family and their expression is tightly regulated in a stage specific and tissue specific manner (4, 5). Phenotypic expression of muscle genes is known to be influenced by thyroid hormone (6) and altered patterns of innervation (7). However, the influence of physical activity at the gene level was unclear. We

Muscle and Motility Vol. 2. Proceedings of XIXth European Conference in Brussels
© Intercept Ltd, PO Box 716, Andover, Hampshire, SP10 1YG, UK

have studied changes in gene expression of the fast and slow genes in response to stretch and force generation.

Methods

Muscle stretch was altered by immobilization of the limb with the muscle in its lengthened or shortened position using a plaster cast as described elsewhere (8). To investigate the role of early activity on slow muscle differentiation, the limbs of one-week-old mice and one-week-old rabbits were immobilized for three weeks with the foot in its extended position (soleus in the shortened - position and inactive) using adhesive tape. To investigate the role of force generation, the fast tibialis anterior (TA) in the adult Netherland dwarf rabbit was stimulated using teflon covered stainless steel electrode wires implanted into the popliteal fossa (8). The electrodes wires were externalized at the back of the neck and attached to a miniature stimulation circuit which was held in position by a small saddle. Several designs of stimulator circuit were used including some which gave low-frequency continuous trains and some which gave higher-frequency intermittent trains. 30 Hz and 60 Hz intermittent stimulation circuits were designed to give the same number of pulses per minute as 2 Hz and 10 Hz continuous stimulation circuits and in this way the hypothesis that it is the number of pulses that is important, could be tested. In all cases the pulses were biphasic and the pulse voltage was adjustable up to a maximum of 3 V.

To study the switches in gene expression Northern blots were carried out using total RNA (10 μg per lane). Hybridization was carried out under different stringency conditions using cDNA fragments for the slow β-type myosin hc, originally cloned and characterized in the laboratory of Dr Zak (8, 9) and a fragment specific for the fast type IIb myosin hc generously made available by Dr Whittinghofer (10). A myosin light chain 1 and 3 cDNA was also obtained from Dr Whittinghofer.

Results

Stretch combined with electrical stimulation was found to induce very rapid hypertrophy of TA in the adult animal. Increases of up to 30% wet weight were recorded in a period as short as four days. Both force generation and stretch are major factors in activating protein synthesis and the combination of these stimuli apparently have an synergistic effect. Associated with this very significant increase in muscle size there was a marked increase (up to 250%) in RNA content of the muscles which was found to peak after two days of the commencement of stretch. Northern analyses using cDNAs specific for slow myosin hc and fast myosin hc transcripts showed that there was a marked switch in the levels of mRNA of the myosin genes (Fig. 1). In responding to both stretch and electrical stimulation, the fast TA becomes reprogrammed to transcribe slow myosin hc and to repress the expression of the fast myosin hc gene within only four days. Brownson *et al.* (11) also reported repression of the fast myosin hc gene with chronic stimulation but on a longer time-scale. Interestingly, our hybridisation studies using the 3' non-translated region of the slow β-myosin gene indicated that this cardiac slow gene is not the same as the skeletal slow myosin gene as previously supposed (12) although the β-myosin hc is strongly expressed in the soleus muscle (Fig. 1).

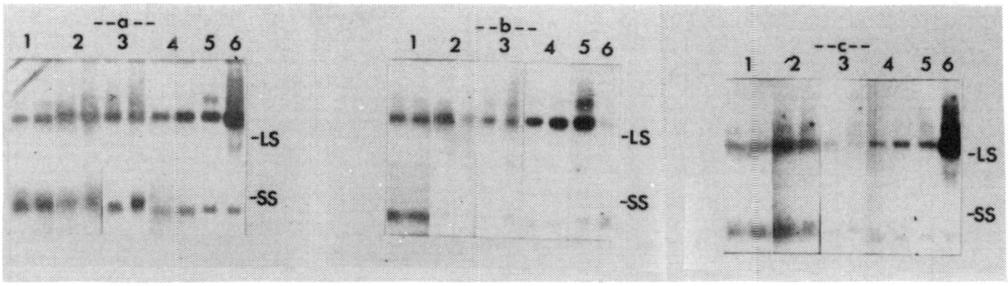

Fig.1 Northern blots for RNA extracted from the rabbit tibialis anterior (TA) muscle using (a) the slow β-cardiac cDNA probe under low stringency conditions, (b) the fast IIa cDNA probe under moderate stringency conditions, (c) the slow beta cDNA under high stringency conditions. Each membrane was simultaneously blotted with an actin probe (lower band) to check that the same amount of message was loaded in each lane. The large ribosome (IS) and the small ribosome (SS) RNA is shown as size markers. The lanes were (1) stretched TA; (2) stimulated only TA; (3) stimulated and stretched TA; (4) control TA; (5) normal EDL; (6) normal soleus.

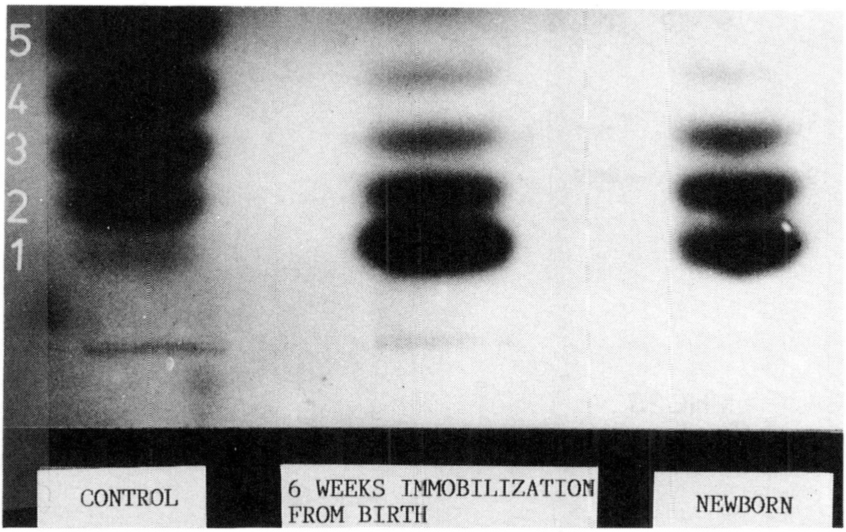

Fig. 2 Shows the lactic dehydrogenase isoform profiles for mouse soleus muscles that have been immobilized in the shortened position and allowed to recover from this immobilization.

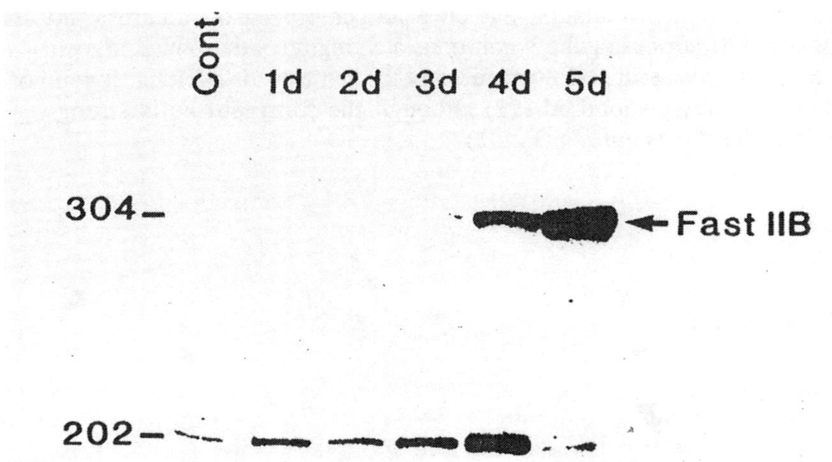

Fig. 3a S1 nuclease analysis using the rate fast IIb myosin hc probe of RNA from rat soleus muscles that have been immobilized in the shortened position. Strong expression of the fast myosin hc gene, which is not appreciably expressed in this slow postural muscle, is seen within a few days of immobilization. From Loughna *et al.* (13).

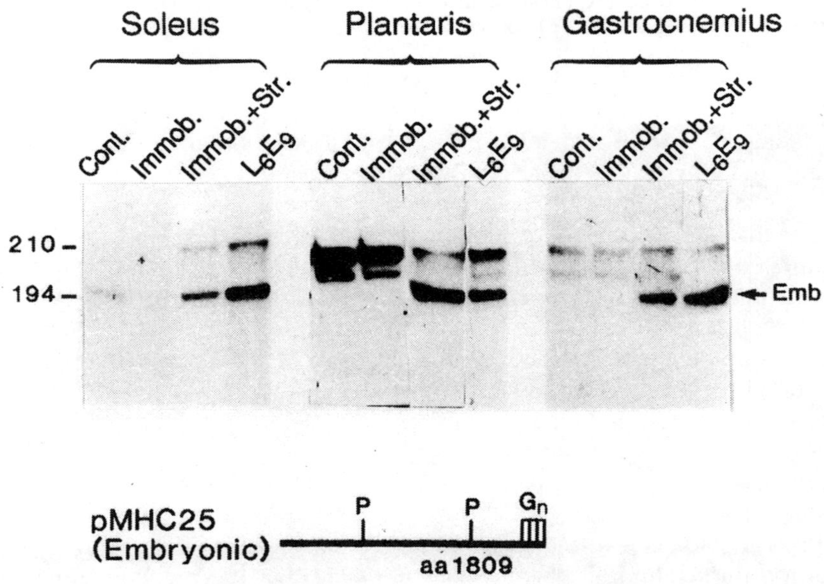

Fig. 3b Shows S1 analysis of RNA from the immobilized soleus, plantaris and gastrocnemius muscle using the rat embryonic myosin hc probe. Immobilization was carried out in both the stretched and shortened positions. L_6E_9 represents RNA extracted from cells in culture. From Loughna *et al.* (13)

In the experiments in which the young soleus muscle was immobilized it was found that the young muscle could be prevented from expressing slow myosin and developing the typical lactic dehydrogenase isoform pattern. When the cast was removed and the muscle subjected to activity and stretch during walking the slow postural features of the soleus developed within three weeks (Fig. 2). Even a mature, fully differentiated soleus muscle which does not normally express fast type IIb and embryonic myosin hc, will strongly express these genes (Fig. 3) if it is subjected to immobilization in the shortened position (13).

Discussion

The interconvertibility of fibre types was demonstrated by cross-innervation (14) and by chronic stimulation (15). Using 10 Hz stimulation, the fast tibialis muscle of the rabbit becomes slow contracting and it was generally accepted that it was motor impulse pattern, that is to say low frequency repetitive stimulation, that was responsible for slow muscle cell-phenotypic expression. However, more recently, it was shown (16) that intermittent 60 Hz chronic stimulation for five weeks was very effective in transforming fast muscles into slow muscles as judged by the appearance of slow myosin light chains and histochemical staining. The conclusions drawn from our data are in accord with these latter findings and indicate that it is those stimulation regimes that cause maximal force generation particularly when they are combined with stretch, that switches on the slow genes and represses the fast genes.

Our work differs from the use of chronic stimulation in that protein synthesis is activated by stretch combined with stimulation and the muscle is rapidly producing new actin and myosin filaments in series and in parallel (17). A change in the type of new protein that is being synthesized in abundant quantities can be therefore readily detected in days rather than in weeks (18). That is to say that the rates are not limited, as they presumably are in chronically stimulated muscle, to the protein turnover rates that are in operation in mature muscle tissue. Indeed, chronic stimulation experiments in which the muscle is not stretched result in considerable atrophy. The work described here strongly suggests that the isometric mechanical overload of a muscle that is developed by stretch or force generation not only induces rapid hypertrophy but also elicits slow muscle cell type-phenotypic expression. This makes physiological sense as it can be argued that the muscle cells by responding to isometric overload are adapting to a new pattern of activity that is akin to an increased postural role.

Protein studies have indicated that there is sequential myosin hc gene expression during muscle development (19) and that the predominant protein isoforms change from embryonic, to neonatal, to adult fast. It seems that all muscle fibres stay phenotypically fast unless they are subjected to stretch and isometric force development. As shown by the soleus muscle that is immobilized in the shortened position, the muscle reverts expressing the fast myosin genes unless it is receiving these mechanical stimuli under which circumstances the fast myosin genes are repressed. when the muscle is not subjected to stretch or force generation, the fast myosin genes are expressed by default which is essentially the same conclusion drawn by Swynghedauw (20). He reviewed the conditions under which the fast myosin hc gene is expressed except that he refers to the fast gene as the endogenous gene and we prefer the term the 'default gene'. Certainly, the main regulation in the expression of the slow phenotype seems to depend as much on the repression of the fast type as much as the activation of the slow type genes. The switch in expression from the fast gene to the slow gene can be looked on as an adaptation for postural activity as the slow myosin hc has a lower specific ATPase activity and is more economical in maintaining isometric force (21) .

Acknowledgements

This work was funded by a grant from the Wellcome Trust. Dr Gerlach was supported by a grant from NERC.

References

1. Greaser, M. L., Moss R. L., Reiser, P. J. 1988, *J. Physiol.*, **406**, 85–913.
2. Barany, M., 1967, *J. Gen. Physiol.*, **50** 197–218.
3. Reiser P. J., Moss, R. L., Giulian ,G. C., Greaser, M. L,. 1985, *J. Biol. Chem,* **206,** 14403–14405.
4. Whalen, R. G., 1985, *J. Exp. Biol.*, **115**, 43–53.
5. Mahdavi, V., Strehler, E., Periasamy, M., Wieczyorek, D. F., Izwno, S., Nadal-Ginard, B., 1986, *Med . Sci . Sports Exerc.*, **18**, 229–308.
6. Izumo, S., Nadal-Ginard, B., Mohdavi., V. J., 1986, *Science*, **231**, 597–600.
7. Pette, D., Vrbova, G., 1986, *Muscle and Nerve*, **8**, 676–589.
8. Umeda, P. K., Sinha, A. M., Jakovcic, S., Merten, S., Hsu, H.-J., Subramanian, K. N., Zak, R., 1981, *Proc. Natl. Acad. Sci. USA,* **78**, 2843–28847.
8. Williams, P. E., Watt, P., Bicik, V., Goldspink, G., 1986, *Exp. Neurol*, **93**, 500–509.
9. Sinha, A. M., Umeda, P. K., Kavinsky, C. J., Rajamanickam, C., Hsu, H.-J., Jakovcic, S,. Rabinowitz,, M., 1982, *Proc. Natl. Acad. Sci. USA*, **79**, 5847–5851.
10. Maeda, K., Sckzakiel, G., Whittinghofer, G., 1987, *Eur. J. Biochem.*, **167**, 97–1025.
11. Brownson, C., Isenberg, H., Brown, W., Stanley, S., Edwards, Y., 1989, *Muscle and Nerve*, **11**, 1183–1189.
12. Lompré, A. M., Nadal-Ginard, B., Mohdavi, V. J., 1984, *J. Biol Chem.,* **259**, 6437–6446.
13. Loughna, P. T., Izumo, S., Goldspink, G., Nadal-Ginard, B., 1990, *Development*, **109**, 217–223.
14. Buller, A. J., Eccles, J. C., Eccles, R. M., 1962, *J. Physiol.*, **150**, 399–416.
15. Salmons, S , Vrbová, G., 1969, *J. Physiol*, **201**, 535–549.

16. Streter, F. A., Pinter, K., Jolesz, F., Mabuchi, K., 1982, *Exp. Neurol.*, **75**, 95–102.
17. Goldspink, G, 1984, *Handbook of Physiology — Skeletal muscle*, Peachy, Adrian, Geiger (eds), pp. 593–554.
18. Goldspink, D. F., Goldspink, G., 1986, *Electrical Stimulation and Neuromuscular Disorders*, Nix, W. A., Vrobva, G. (eds), pp. 92–100.
19. Butler-Browne, G. S., Whalen, R. G.,. 1984 , *Dev. Biol* , **102**, 324–334.
20. Swynghedauw, D., 1986, *Physiological Reviews*, **65**, 710–771.
21. Goldspink, G ,1977, *Mechanics and Energetics of Animal Locomotion* (Alexander, R. McN., Goldspink, G. eds), pp. 57–81. Chapman and Hall, London. Hnik, P., Vejsada, R., Goldspink, D. F., Kasiciki, S., Krekule, I., 1985, *Exp. Neurol.*, **88**, 515–528

8
Development of Muscle fibres in a 'Pure' Fast-twitch Muscle

F. MASCARELLO*, A. ROWLERSON** AND E. BALLOCCHI*

**Istituto di Anatomia degli Animali Domestici con Istologia ed Embriologia, Via Celoria, 10-20133 Milano, Italy*
***Department of Physiology, UMDS, St. Thomas's Hospital, Lambeth Palace Road, SE1 7EH London, UK*

Introduction

The initial differentiation of skeletal muscle fibres into fast and slow types occurs at a very early stage. Myoblastic cells can be distinguished on the basis of their chronological appearance and the type of muscle fibres they give rise to (1, 2, 3). In the first stage of muscle fibre production, early or embryonic myoblasts fuse independently of innervation to form immature fibres (primary myotubes) which express initially embryonic and slow myosins. These myotubes then differentiate into slow or fast fibres depending on their position in the muscle. In the second stage of fibre production, after a delay of two or more days, late or foetal myoblasts form secondary myotubes in a nerve-dependent process. These myotubes contain initially embryonic and then neonatal myosins, but no slow myosin. Subsequently, depending upon the muscle, most secondary myotubes become fast fibres, and some (for example in the soleus muscle) become slow fibres. Some myogenic cells remain in the muscle as satellite cells, which are responsible for regeneration and contribute some nuclei to existing fibres during normal post-natal growth.

Muscle and Motility Vol. 2. Proceedings of XIXth European Conference in Brussels
© Intercept Ltd, PO Box 716, Andover, Hampshire, SP10 1YG, UK

Recent work has described the development of fibre types in mixed fast and slow muscles, but so far as we know, pure fast muscles have not been examined. We therefore studied the development of a pure fast-twitch muscle in rats (E17 foetuses to seven days post-partum) by immunohistochemistry using antibodies specific for slow and neonatal/embryonic myosins. The retractor bulbi muscle was chosen because, in the adult, this is a pure, fast muscle, containing no slow fibres at all. Its development was compared to the adjacent oculorotatory muscles which contain several types of both fast and slow fibres (4).

Methods

Foetal (gestational ages E17–E21) and neonatal (day of birth and seven days post-natal) rats were killed by decapitation. 10 μm cryostat sections were cut and were stained immuno-histochemically by indirect immunoperoxidase method with polyclonal antibodies specific for slow (anti-S) and neonatal/embryonic (anti-NE) myosins.

Results

SEVENTEEN DAYS' GESTATION

At this age the retractor bulbi and oculorotatory muscles are composed only of very immature primary myotubes. However they clearly differ already in their myosin composition. In the retractor bulbi primary myotubes give a stronger reaction with anti-NE and a weaker reaction with anti-S than the primary myotubes of all other muscles (Fig. 1a, b).

EIGHTEEN DAYS' GESTATION

In oculorotatory muscles primary and secondary myotubes are now visible arranged in clusters. The primary myotubes give a strong reaction with anti-S serum and are negative with anti-NE serum: on the other hand secondary myotubes are positive with anti-NE and negative with anti-S serum. The differences in myosin composition between primary myotubes of retractor bulbi compared to other muscles are increased, as primary myotubes of retractor bulbi are now positive only with anti-NE serum.

NINETEEN TO TWENTY-ONE DAYS' GESTATION

The myosin composition of fibres in retractor bulbi remained the same in this period as at eighteen days' gestation, but morphological changes occur gradually with increasing maturity. Very rarely, it is possible to find one or two small fibres with a moderate or weak reactivity with anti-S serum in retractor bulbi of this age, probably derived from secondary myotubes. The ring-shape profile, characteristic of myotubes in transverse section, is no longer evident in retrac-

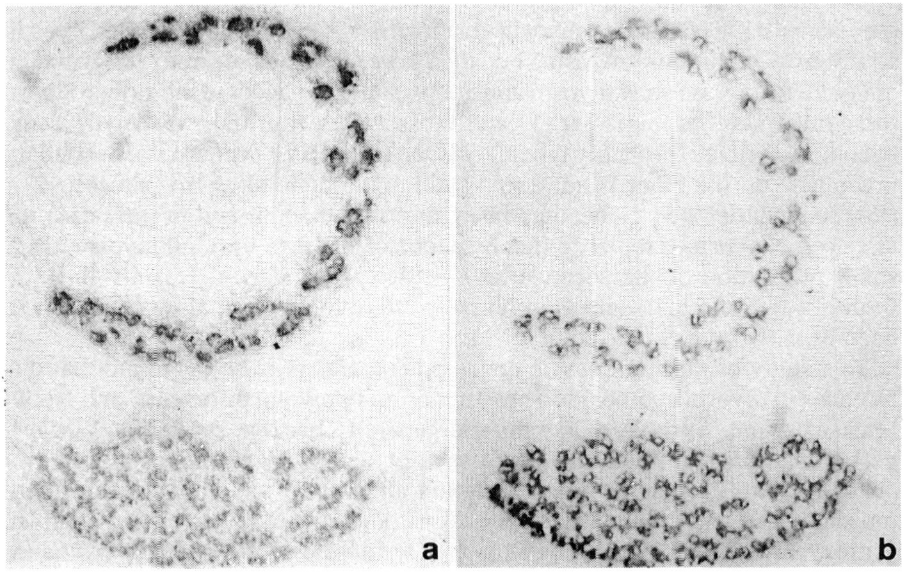

Fig. 1 Seventeen days' gestation. Serial sections of retractor bulbi (crescent shaped) and oculorotatory muscle. Immunoperoxidase with (a) anti-NE and (b) anti-S. All these muscles are composed only of very immature primary myotubes. However, they clearly differ already in their myosin composition: retractor bulbi primary myotubes give a stronger reaction with anti-NE, and a weaker reaction with anti-S, than the primary myotubes of oculorotatory muscle.

tor bulbi fibres. By contrast, in other muscles, classical clusters consisting of an anti-S positive primary myotubes with a ring profile surrounded by anti-S negative and anti-NE positive secondary myotubes are still present.

BIRTH TO SEVEN DAYS

During this period all myotubes acquire the normal adult morphology and the fibre diameter increases. The fibre type distribution remains unchanged with retractor bulbi consisting only of fast fibres, still reacting with anti-NE. Within the oculorotatory muscles the distribution of fast and slow fibres is unchanged (very similar to that seen in the adult) and type differentiation within the slow fibre population has begun.

Conclusions

In the retractor bulbi muscle of the rat, fibre type specialization is established very early with all primary myotubes being detectably different from those of adjacent mixed muscles before the appearance of secondary myotubes (visible at light microscope level).

In mammalian muscle, all primary myotubes pass through a stage in which they contain the isoform of myosin characteristic of adult slow-twitch fibres in addition to embryonic myosin, but then soon differentiate into those which continue to express slow myosin and are presumed to become the slow fibres in the adult ('slow primaries') (5) and those which lose the slow myosin and become fast fibres (probably type IIA) in the adult ('fast primaries'). Secondary myotubes, on the other hand, do not initially contain slow myosin. Most — 'fast secondaries' (5) — become fast fibres in the adult, and in their case the embryonic myosin is replaced first by neonatal and then by adult myosin (6). A small proportion of the secondaries — 'slow secondaries' (mainly those in future slow muscles such as soleus) — start to express slow myosin and become slow fibres in the adult (5, 7).

In a study of rat limb muscle development, Harris *et al.* (8) found that differences in myosin composition between primary myotubes of 'fast' and 'slow' lineages within a single mixed muscle generally became evident at 18 days' gestation, coincident with the appearance of secondary myotubes at light microscope level, but at 17 days' gestation all primary myotubes within these muscles were still identical (anti-S positive and anti-N positive). By contrast, primary myotubes of soleus (a future slow muscle) could already be distinguished by their different myosin profile (anti-S positive but anti-N negative) at 16 days gestation (i.e. 2 days before the appearance of secondary myotubes). In our case a future *pure fast* muscle, the primary myotubes were also different from other primary myotubes before the appearance of secondary myotubes, but had the opposite myosin profile (anti-S weak but anti-NE positive), consistent with their 'fast' fate.

It is thought that muscle innervation is not important in this early specialization of primary myotube lineages (3, 9), but it remains to be seen whether this differentiation is determined by specifically myogenic or other factors.

References

1. Miller, J. B., Stockdale, F. E., 1986, *J. Cell Biol.*, **103**, 2197–2208.
2. Hoh, J .F. Y., Hughes S., Chow, C., Hale, P. T., Fitzsimons, R. B., 1988, *J. Musc. Res. Cell Mot.*, **9**, 48–58.
3. Harris, A. J., Duxson, M. J., Fitzsimons, R. B., Rieger, F., 1989, *Development*, **107**, 771–784.
4. Sartore, S., Mascarello, F., Rowlerson, A., Gorza, L., Ausoni, S., Vianello, M., Schiaffino, S., 1987, *J. Musc. Res. Cell Mot.*, **8**, 161–172.
5. Hoh J.F.Y., Hughes S., Hale P.T., Fitzsimons R.B., *J. Musc. Res. Cell Mot.*, **9**, 30–47.
6. Whalen, R. G., Sell, S., Butler-Browne, G., Bouveret, P., Pinset-Harstrom, I., 1981, *Nature*, **292**, 803–809.
7. Narusawa, M., Fitzsimons, R .B., Izumo, S., Nadal-Ginard, B., Rubinstein, N. A., Kelly, A. M., 1987, *J. Cell Biol.*, **104**, 447–459.
8. Harris, A. J., Fitzsimons, R. B., McEwan, J. C., 1989, *Development*, **107**, 751–769.

9. Condon, K., Silberstein, L., Blau, H. M., Thompson ,W. J.,1990, *Dev. Biol.*, **138**, 256–274.

9

Changes in Muscle Fibre Types of Rat Skeletal Muscles during Growth as Examined by Anti-neurofilament Antibody and Myosin ATPase Reactions

M. TAKAGI*, T. NAKAMURA**, N. TSUZUKI, AND K. KOHAMA†

**Department of Orthopedic Surgery, Saitama Medical Centre, 1981 Tsujido-cho, Kamoda, Kawagoe City, 350 Japan*
***Department of Orthopedic Surgery, University of Occupational and Environmental Health, 1-1 Iogaoka, Yawatanishi-ku, Kitakyusyu City, 807 Japan*
†Department of Pharmacology, Gunma University, School of Medicine, 3-39-22 Showa-machi, Maebashi City, 371 Japan

Introduction

We have reported that monoclonal anti-neurofilament (NF) antibody (200 kDa) reacted with not only nervous tissue but also type 1 (slow muscle) and type 2C (foetal muscle) fibres in skeletal muscles (1) and that the substance reacted with the antibody in the muscles is an isoform of troponin-T in type 1 and type 2C fibres (2,3).

In this study we examined muscle differentiation by using this antibody as a probe.

Materials and Methods

Twenty-eight Sprague-Dawley rats were produced by one breeding pair over three deliveries at our laboratory. Their ages ranged from 1, day, 2 days, 3 days, 5 days, 7 days, 8 days, 10 days, 14 days, 17 days and 21 days to 4 weeks, 5 weeks, 7 weeks and 9 weeks.

Muscle and Motility Vol. 2. Proceedings of XIXth European Conference in Brussels
© Intercept Ltd, PO Box 716, Andover, Hampshire, SP10 1YG, UK

Four speciments of soleus muscle and anterior tibial muscle were obtained from two animals and cryosectioned serially at 10μm thickness at -20°C.

The sections were stained either by myosin ATPase reaction at pH 9.4 following pre-incubation at pH 10.4 or pH 4.6 (4) or with the anti-NF antibody using anti-mouse IgG equipped with the avidin-biotin-peroxidase complex as a second antibody (5).

The identification of fibre types was followed by the differences of staining pattern, as shown in Table 1. Fibre type compositions were determined by counting more than 100 fibres.

Table 1 Identification of fibre types using with myosin ATPase reaction and anti-NF antibody (4).

	1	2C	2B	2A
Myosin ATPase (pH 10.4)	–	+	+	+
Myosin ATPase (pH 4.6)	+	+	±	–
Anti-NF (200 kd) antibody	+	+	–	–

Results

At neonatal stages we identified two new fibre types and named them type $2C_1$ and type $2C_{AB}$ fibres, according to their staining patterns (Table 2).

Table 2 Identification of fibre types using with myosin ATPase reaction and anti-NF antibody (II).

	$2C_1$	$2C_{AB}$
Myosin ATPase (pH 10.4)	±	+
Myosin ATPase (pH 4.6)	+	+
Anti-NF (200 kd) antibody	+	+

All fibres of soleus muscles were of type 2C fibres at birth. Type $2C_1$ fibres represented 10% of all fibres at 5 days after birth. Type 1 fibres were 50% of all fibres at 17 days continued to represent 40% of all fibres up to 9 weeks. On the other hand, type $2C_{AB}$ fibres were 40% of all fibres at 2 weeks. Type 2B and type 2A fibres stood at30% in weeks 4 and 5 respectively. This proportion was observed up to 9 weeks. The chronological changes of fibre type composition of soleus muscles are shown in Fig. 1.

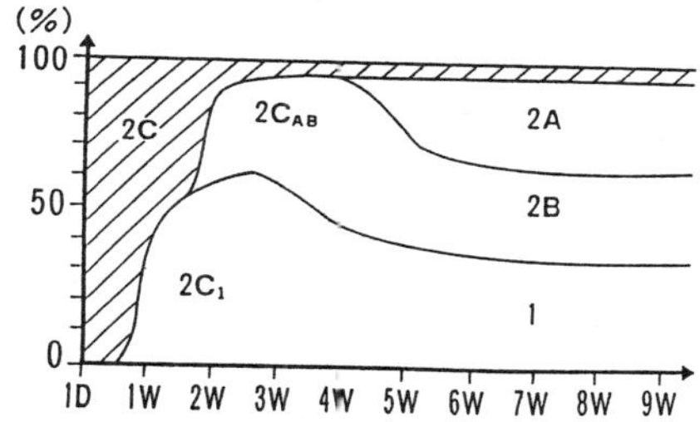

Fig. 1 Chronological changes of fiber type composition of soleus muscle

Anterior tibial muscles included type 2C and type $2C_{AB}$ fibres at birth. Type 2C1 fibres represented 5% of all fibres at 3 days. Type 1 fibres were 5% of all fibres at 8 days. This continued up to 9 weeks. On the other hand, type 2B and type 2A fibres stood at 40% by 2 to 3 weeks. This proportion was observed up to 9 weeks. The chronological changes of fibre type composition of anterior tibial muscle are shown in Fig. 2.

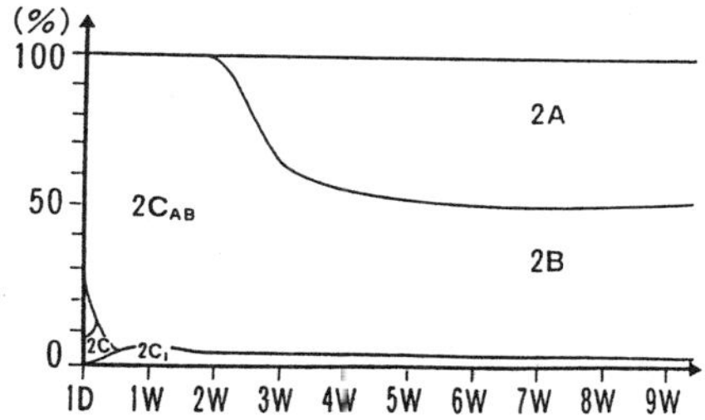

Fig. 2 Chronological changes of fiber type composition of anterior tibial muscle

Discussion

Based on the difference of fibre type composition between soleus muscle and anterior tibial muscle during growth, the next two points are clear:

1. Anterior tibial muscle develops earlier than soleus muscle.
2. Fibre type differs as follows: type 2C fibres develop to type $2C_1$ and type$2C_{AB}$ fibres, whereas type $2C_1$ fibres develop to type 1 fibres. On the other hand, type $2C_{AB}$ fibres develop either to type 2B or type 2A fibres. (Fig. 3).

Two kinds of fibres which we evaluated as type $2C_{AB}$ fibres were included. One comprised the fibres that do not react with anti-NF antibody, the other the fibres that react a little with anti-NF antibody and then the latter shows a faint staining as shown in Table 2. Thus we speculate that there exists a transitional form

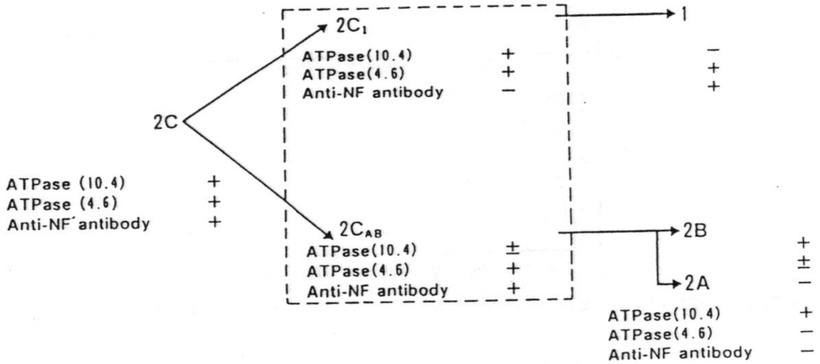

Fig. 3 Transitional forms of muscle fiber types during growth

between type $2C_{AB}$ fibres and type B/A fibres. To ascertain this, however, further studies are needed.

References

1. Nakamura, T., Kawahara, H., Myashita, H., Watarai, K., Takagi, M., Tachibana, S., 1987, *Histochem.*, **87**, 39–45.
2. Takagi, M., Takano-Ohmuro, H., Nakamura, T., Kohama, K., 1988. In: Carraro, U. (ed.), *Sarcomeric and Non-sarcomeric Muscles: Basic and Applied Research Prospects for the 90s*, pp. 131–135. Unipress Padova, Italy.
3. Takagi, M., Takano-Ohmuro, H., Nakamura, T., Kawahara, H., Shimizu, T., Obinata, T., Kohama, K., 1989, *Muscle Nerve*, **12**, 827–832.
4. Dubowitz, V., Brooke, M. H., 1973, *Muscle Biopsy: a Modern Approach*, pp. 98–102. W. B. Saunders, London.
5. Hsu, S. M., Rame, L., Panger, A., 1981, *Am. J. Clin. Pathol.*, **75**, 734–738.

10
The Masseter, a Very Specialized Muscle

N. SOUSSI-YANICOSTAS*, E. M. BREUER**, D. C. DANG* AND G.S. BUTLER-BROWNE*

**INSERM U262, Clinique Universitaire Baudelocque, 123 Bd Port-Royal, 75674 Paris, Cedex 14, France.*
***Institute for Veterinary Pathology, Freie Universität Berlin, Strasse 518, No 15, 1000 Berlin 37, Germany.*

Introduction

The physiological and dynamic properties of skeletal muscle are largely determined by the enzymatic composition of the muscle (1, 2, 3). Thus by using enzyme-histochemical techniques skeletal muscle fibres can be classified into different types: type I fibres which are slow contracting , have a high oxidative capacity and a high resistance to fatigue and type II which are fast contracting. These fast fibres are subdivided into type IIA which have both high oxidative and high glycolytic activity and a high resistance to fatigue and type IIB which have a high glycolytic capacity , a low oxidative capacity and are sensitive to fatigue. These three fibre types are functionally specialized by virtue of the differential expression of genes coding for myosin and other muscle proteins. It has been shown that there are three adult myosin heavy chains (MHC) corresponding to the three types of muscle fibre: slow, fast IIA and fast IIB, each of which is encoded by a different gene (for review see 4). Two additional MHC's, embryonic and foetal/neonatal and one MLC, MLC1emb, have been shown to be expressed during muscle development (5–11).

Muscle and Motility Vol. 2. Proceedings of XIXth European Conference in Brussels
© Intercept Ltd, PO Box 716, Andover, Hampshire, SP10 1YG, UK

The majority of these studies have concentrated their attention upon a few well defined limb and trunk muscles in a restricted number of species. The muscles which have been studied are those which are involved mainly in the maintenance of posture or in locomotion. This has resulted in a relative neglect of the more specialized body muscles; for example sphincter, eye, ear, and masticatory muscles. In our laboratory we have decided to look in more detail at the masticatory muscles.

The jaw apparatus has been divided by Turnbull (13) on different taxonomical criteria into four adaptive types: generalized chewing (primates and other mammalian omnivors), ungulate grinding, rodent gnawing and nibbling and carnivorous biting and shearing. Interestingly enough this is very closely correlated to their histochemical properties (13) since it has been shown that ungulates which masticate slowly and regularly have only slow type I fibres, small rodents which nibble fast have a mixture of predominantly IIA and IIB; carnivores have a special superfast myosin present in the type IIM fibres which is associated with the aggressive bite and omnivores have a mixture of type I and type II fibres.

Early histochemical studies carried out by Ringqvist *et al.*, (14, 15) demonstrated in humans that the masseter muscle differs from the limb muscles in several respects. First, there is a marked difference between the diameter of the type I and type II fibres, with type II fibres being considerably smaller than the type I fibres. Secondly, when sections were processed by the standard method for myofibrillar ATPase at pH 9.4 and following pre-incubation at 4.3, two additional fibre types, IIC and IIM, were shown to be present in the adult human masseter muscle. Biochemical and immunohistochemical investigations (16) have confirmed a predominance of slow myosin, which is localized mainly in the type I fibres and the presence of adult fast myosin, which is localized in the type IIA and IIB fibres,. In addition we demonstrated that the adult human masseter expresses variable amounts of a MHC isozyme indistinguishable from the foetal/neonatal MHC found during development. This MHC was generally found in the type IIC and IIM fibres. Another protein characteristic of muscle development, MLC1emb, was also expressed to the same extent as the other MLCs, MLC1f, MLC1s and MLC2s, in these adult muscles.

Recently we have shown (17) that the adult phenotype develops much later in the masseter than in the limb muscles. This observation is similar to the finding of d'Albis *et al.*, (18) in the rat. In addition, there is a persistence of these developmental isoforms throughout development and even an increase in expression of MLC1emb after birth (17) .

Table 1 Summary of myosin light chains (MLC) and heavy chains (MHC) expression in adult masseter muscles in pig, monkey and human.

	MLC1e	MLC1s	MLC1f	MLC2s	MLC2f	MLC3f	MHCe	MHCf	MHCs	MHCf
Pig	–	+	+	+	+	+	–	–	+	+
Monkey	–	–	+	–	+	+	–	–	+	+
Human	+	+	+	+	–	–	+	+	+	+

Due to the atypical pattern of expression of myosin that we have found in the human masseter, we decided to extend our study to compare this pattern of expression with that of the pig, the macaque monkey and the pygmy lemur to see if the expression of developmental isoforms was typical to humans or whether it could be found in other species.

Materials and Methods

SPECIMENS

Muscles biopsies were obtained from the superficial masseter muscles and the vastus lateralis muscles of pig, monkey and human.

TWO-DIMENSIONAL GEL ELECTROPHORESIS

Muscle samples, weighing 20 mg, were scissor minced in two volumes (*w/v*) of sample buffer: 9.95 M urea, 4% NP40, 2% ampholines LKB pH 5–7, 100 mM DTT, 1 mg of solid urea per mg of tissue was then added. The muscles were extracted for 15–20 min at room temperature and were vortexed several times during the extraction. Samples were then stored at -80°C. Prior to electrophoresis the samples were centrifuged for 5 min in a microfuge to remove any tissue fragments.

Electrophoresis was carried out according to O'Farrell (19) as modified by Garrels (20).

NATIVE PYROPHOSPHATE GEL ELECTROPHORESIS

Muscles were extracted in three volumes of sample buffer: 300 mM sodium chloride, 100 mM dihydrogen phosphate, 50 mM disodium phosphate, 1 mM magnesium chloride, 10 mM sodium pyrophosphate, 10 mM EDTA, 0.1% 2-mercaptoethanol, pH 6.5. The tissue was scissor minced and extracted for 30 min on ice; the extract was clarified by centrifuging for 15 mins in a microfuge. The supernatants were diluted ten-fold into diluting buffer; 50% glycerol, 40 mM sodium pyrophosphate, 0.1% 2-mercaptoethanol pH 6.5 and were stored at -20°C.

Electrophoresis was carried according to d'Albis *et al.*, (21) for 20–22 h at 2°C.

Results and Discussion

EXPRESSION OF MLC TYPES

The MLC types present in the masseter and in the quadriceps muscles of each of these species were compared by two-dimensional gel electrophoresis. In the pig muscle extracts a similar pattern of MLCs was found in both the masseter and the quadriceps (Fig. 1A, B), i.e., $MLCs_1$, $MLCf_1$, $MLCs_2$, $MLCf_2$ and $MLCf_3$.

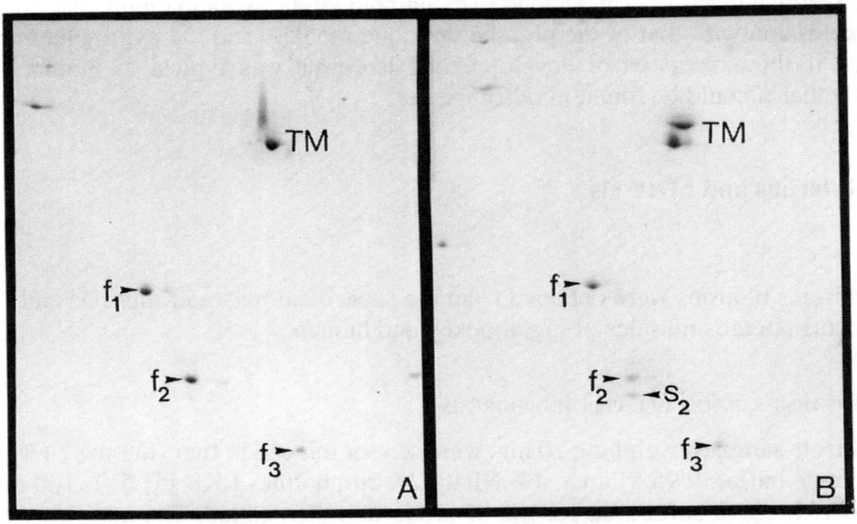

Fig. 1 Two-dimensional gel electrophoresis of adult pig muscles. Total extracts from adult pig muscle: masseter (A) ; quadriceps (B) . Isoelectrofocusing in first dimension (IEF), was carried out between pH 4 and p H6. Proteins were separated in the second dimension on 12,5% SDS-polyacrylamide slab gels which were stained with coomassie blue. The various myofibrillar proteins are indicated as follow: tropomyosin (TM); embryonic light chain (e) slow myosin light chains (s_1 , s_2) ; fast myosin light chains:(f_1, f_2 and f_3).

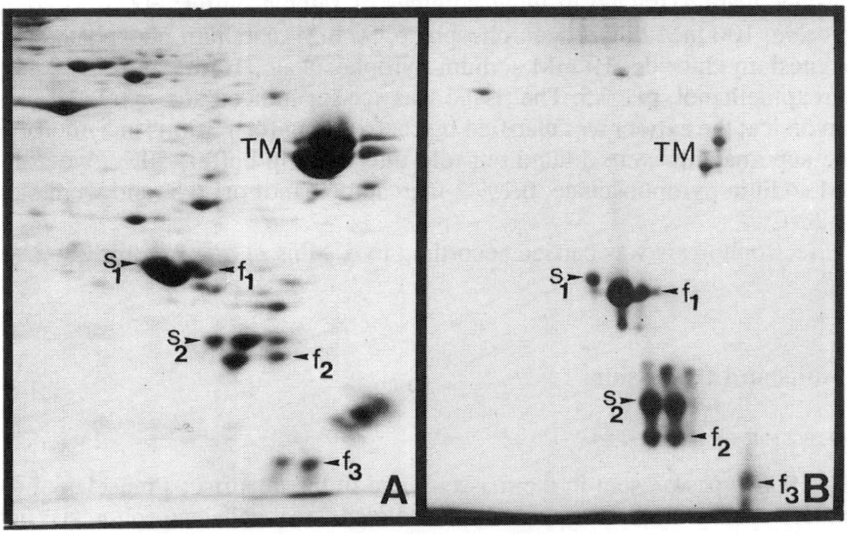

Fig. 2 Two-dimensional gel electrophoresis of adult monkey muscles. Total extracts from adult monkey muscles: masseter (A) ; quadriceps(B) .

In the monkey both muscle extracts showed a predominance of fast MLCs ($MLCf_1$, $MLCf_2$ and $MLCf_3$) together with a small amount of $MLCs_2$ (Figs 2A and 1B). In contrast, Fig. 3 shows that the pattern of MLCs observed in the adult human quadriceps muscle was different to that observed in the masseter muscle. The adult masseter muscle expresses the embryonic MLC which normally disappears at about 30 weeks of gestation in trunk and limb muscles together with $MLCs_1$, $MLCf_1$ and $MLCs_2$ (Fig. 3A), these three types of MLCs having been described by Billeter *et al.*, (22) in single human slow type I fibres.

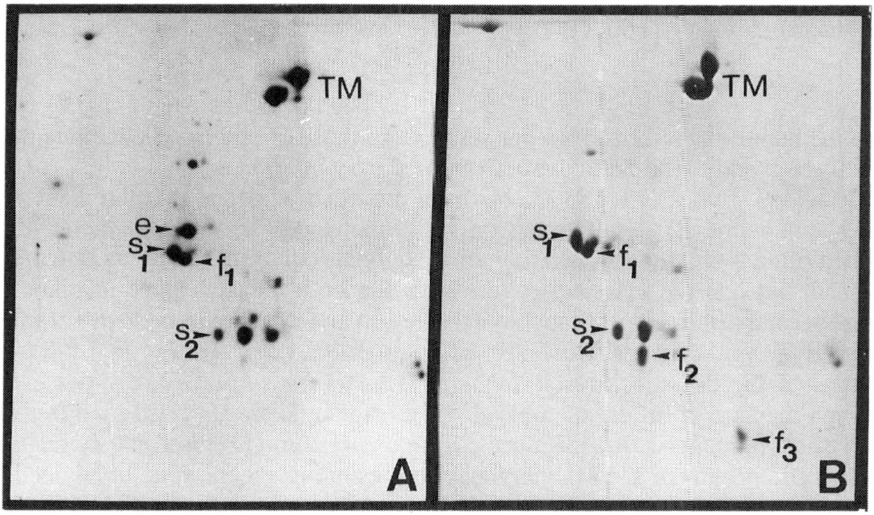

Fig. 3 Two-dimensional gel electrophoresis of adult human muscles. Total extracts from adult monkey muscles: masseter (A) ; quadriceps(B) .

EXPRESSION OF MHC ISOFORMS

The different MHC isoforms were analysed by native pyrophosphate gel electrophoresis. In pig, a similar pattern of migration was found in both masseter and quadriceps muscles with one band corresponding to the slow MHC and three bands corresponding to fast MHC (Fig. 4e, f.) In monkey, a similar mixture of slow and fast myosin was observed in both muscles although in the masseter there was a predominance of fast myosin (Fig. 4c, d). In humans, the masseter muscle displays a predominance of slow MHC when compared to the quadriceps muscle and smaller amounts of fast and foetal MHC (Fig. 4a,b). The masseter of the pygmy lemur was found to contain exclusively fast myosin (Fig. 4g).

Immunocytochemical studies have also been carried out in these species (data not shown) and foetal MHC was found in neither the macaque nor the pig

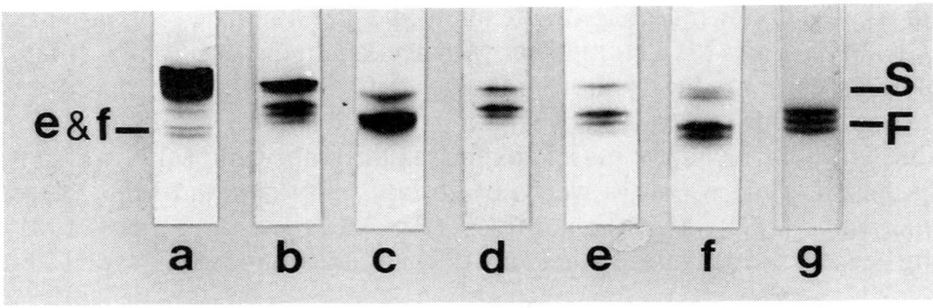

Fig. 4 Pyrophosphate gel analysis. Electrophoresis of total adult muscle extracts from masseter (a) and quadriceps (b) of human; masseter (c) and quadriceps (d) of monkey; masseter (e) and quadriceps (f) of pig; and a masseter muscle of pygmy lemur (g). The bands corresponding to slow, fast , embryonic and fetal myosin are indicated as: fast myosin (F), slow myosin (S), embryonic and fetal myosin (e & f) .

nor the lemur masseters. From our studies and those of others we conclude that MLC1emb and foetal MHC were expressed only in human.

Previous studies (13, 18, 23, 24) have provided evidence that the masseter muscle in mammals are different from the limb muscles and have characteristics which often seem to be related to their functional requirements. It has also been pointed out that the masseter differs from the limb muscles since it is not of mesodermal origin but of branchiomeric origin and it is innervated with cranial instead of spinal nerves. However, Eriksson *et al.*, (25) showed that the two bellies of the digastric muscle in humans had very similar fibre types even though they are of different embryological origin and are supplied by different nerves. This suggests that the muscle fibre composition is not primarily related to specific origins or specific nervous supplies but to special functional needs.

Our results are in agreement with previous histochemical studies (13) and confirm that the expression of different myosin isoforms is related to the different functional requirements of this muscle. The pig is an omnivore and has a mixture of both fast and slow type fibres which adequatly meet its functional chewing and biting requirements. The macaque masseter on the other hand has been previously shown by Rowlerson *et al.*, (13) to contain predominantly the special superfast myosin typical of animals with an aggresive bite. This was confirmed by the predominantly fast biochemical phenotype which we found in our samples. The lemur, which is a nibbler, like the rat and mouse expressed only fast myosin. None of these species expressed either MLC1emb or foetal MHC.

In humans, the masseter, unlike the limb muscles, has minimal postural activity. Support of the jaw against gravity is achieved largely by passive tension and with some activity of the temporal motor units. At rest no activity can be detected in the masseter (26). Electrophysiological studies on the masseter (26, 27) have revealed conflicting data when compared to hand and limb muscles since there is a poor correlation between twitch tension and contraction speeds resulting in an unimodal distribution of relatively fast contraction times. In addition Nordström and Miles (26) were able to identify relatively few fatigue resistant motor units which is unusual since the masseter contains predominantly type I fibres.

The masseter is involved in a wide variety of activities such as mastication, speech, swallowing and facial expression, tasks which require rapid intermittent activity of motor units. Maybe all those unique features which have been revealed in the masseter are the adaptation of this muscle to its complex functional requirements.

Acknowledgements.

This work is supported by the Institut National de la Santé et de la Recherche Médicale and the Association Française contre les Myopathies.

References

1. Barany, M., Close, I., 1971, *J. Physiol. Lond.*, **213**, 455–474.
2. Burke, R. E.,Levine, D. N., Tsaris, P., Zajac, F. E., 1973, *J. Physiol. Lond.*, **234**, 723–748.
3. Edström, L., Kugelberg, E., 1968, *J. Neurol. Neurolsurg. Psychiat.*, **31**, 424–433.
4. Swynghedauw, B., 1986, *Physiol. Rev.*, **66**, 710–771.
5. Butler-Browne, G. S., Whalen, R. G., 1984, *Dev. Biol.*, **102**, 324–334.
6. Fitzsimons, R. B. , Hoh, J. F. Y., 1981, *J. Neurol. Sci.*, **52**, 367–384.
7. Lyons, G. E., Haselgrove, J., Kelly, A. M., Rubenstein, N., 1983, *Differentiation*, **25**, 168–175.
8. Periasamy, M., Wieczorek, D. F., Nadal-Ginard, 1984, J. *Biol. Chem.* **259**,13573–13578.
9. Periasamy, M., Wydro, R. M., Strehler-Page, A. M., Strehler, E. E., Nadal-Ginard, B., 1985, *J. Biol. Chem.* **260**, 15856–15862.
10. Whalen, R. G., Sell, S., Butler-Browne, G. S., Schwartz, K., Bouveret, P., Pinset-Härström, I., 1981, *Nature*, **292**, 805–809.
11. Whalen, R. G. , Butler-Browne, G. S., Gros, F., 1978, *J. Mol. Biol.*, **126**, 415–431
12. Turnbull, W. D., 1970, *Fieldiana, Geology* **18**, 149–356.
13. Rowlerson, A., Mascarello, F., Vegeti, A., Carpene, E., 1983, *J. Muscle Res. Cell Mot.*, **4**, 443–472.
14. Ringqvist, M., 1971, *Scand. J. dent. Res.* **79**, 336–368.
15. Ringqvist, M., Ringqvist, I., Eriksson, P-O., Thornell, L-E., 1982, *J. Neurol. Sci.*, **53**, 273–282.
16. Butler-Browne, G. S., Eriksson, P. O., Laurent, C., Thornell, L. E., 1988, *Muscle & Nerve.*, **11**, 610–620.
17. Soussi-Yanicostas, N., Barbet, J. P., Laurent-Winter, C., Barton, P. and Butler-Browne, G. S., 1990, *Development.*, **108**, 239–249.
18. d'Albis,A., Janmot, C., Bechet, J.-J., 1986, *Eur. J. Biochem.*, **156**, 291–296.
19. O'Farrel, P. H., 1975, *J. Biol. Chem.*, **250**, 4007–4021.

20. Garrell J. I., 1979, *J. Biol. Chem.*, **254**, 7961–7977.
21. d'Albis, A,. Pantaloni, C., Bechet, J. J., 1979, *Eur. J. Biochem.*, **99**, 261–272.
22. Billeter, R., Heizmann, C. W., Howald, H., Jenny, E., 1981, *J. Biochem.*, **116**, 389–395.
23. Bredman, J. J., Weijs, W.A., Moorman, A. F. M., Brugman, P., 1990, *J. Anat.*, **168**, 31–47.
24. Shelton, G. D., Bandman, E., Cardinet, G. H., 1985, *Am. J. Vet. Res.*, **46**, 493–498.
25. Eriksson, P.-O., Eriksson, A., Ringqvist, M., Thornell, L.-E., 1982, *Archs. Oral Biol.*, **27**, 207–215.
26. Nordström, M. A., Miles, T. S., 1990, *J. of Applied Physiol.*, **68**, 26–34.
27. Yemm, R., 1977, *J. Physiol.*, **265**, 163–174.

11
Expression of a 'Cardiac-specific' Myosin Heavy Chain in Intrafusal Fibres of the Developing Human Muscle Spindle

A. WESSELS, C. A. S. SOFFERS, J. L. M. VERMEULEN, T. A. MIJNDERS, J. J. BREDMAN AND A. F. M. MOORMAN

Department of Anatomy and Embryology, University of Amsterdam, Meibergdreef-15, 1105 AZ, Amsterdam, The Netherlands

Introduction

The intrafusal fibres of mammalian muscle spindles differ in several respects: in their morphology, ATPase staining and innervation, from normal, extrafusal, skeletal and cardiac muscle fibres. The intrafusal fibres can be divided into three types: bag_1, bag_2 and the nuclear chain fibres. Immunohistochemical studies have revealed that these intrafusal fibres often contain a great variety of myosin heavy chain (MHC) isoforms. Furthermore it has been demonstrated that, between different species, differences in the MHC expression patterns with respect to the fibre types do exist (1) and that along one single intrafusal fibre a nonuniform expression of MHC isoforms can be observed (2, 3). Therefore, these fibres are highly interesting from the point of view of MHC expression regulation. In addition, the aberrant MHC expression of the intrafusal fibres, compared with 'normal' extrafusal fibres, provides a valuable tool in the study of the muscle spindle and its development. For example, the expression of slow tonic MHC allows the detection of primary myotubes that become the

Muscle and Motility Vol. 2. Proceedings of XIXth European Conference in Brussels
© Intercept Ltd, PO Box 716, Andover, Hampshire, SP10 1YG, UK

primordial cells of the muscle spindle fibres in human muscle as early as at 10–12 weeks of development (4). In this contribution evidence is presented that shows that, early in human muscle development, a subpopulation of the developing intrafusal fibres is also characterized by the expression of the 'cardiac-specific' α-MHC.

Methods

TISSUE SOURCE AND PREPARATION

Human foetal, neonatal and adult muscles were obtained at autopsy within 24 hours of postmortem. Tissue specimens were fixed in a mixture of methanol:acetone:acetic acid:water (35:35:5:25), dehydrated in ethanol, embedded in Paraplast Plus (Monoject, Ireland) and cut into 5μm thick serial sections. Sections were mounted on microscope slides coated with either gelatin-chromium sulphate or poly-L-lysine (5).

MONOCLONAL ANTIBODY PRODUCTION

Anti-α-MHC (249-5A4) was raised against adult human atrial myosin, anti-β-MHC (169-ID5) against chicken heart myosin. The production of these antibodies has been described elsewhere (6, 7). Monoclonal antibody 330-5B4 was raised against myosin extracted from the leg of a 15-week-old foetus and dubbed anti-embryonic MHC (*vide infra*). The specificity of the antibodies was examined by Western blot analysis and with immunohistochemical studies on a variety of tissues.

IMMUNOHISTOCHEMISTRY

To detect the binding of the antibodies on sections, the indirect unconjugated immunoperoxidase technique (PAP technique) was applied (5). To optimize the binding of the antibodies with the antigens in the paraplast sections, sections were occasionally pretreated with pronase (0.1 mg/ml, 30 min) (8).

Results

SPECIFICITY OF THE ANTIBODIES

The immunohistochemical characterization of anti-α-MHC and anti-β-MHC has been described previously (6, 7). However, given the fact that in the muscle spindle a great variety of MHCs is expressed, the possibility of crossreaction of anti-α-MHC with other MHC isoforms such as slow-tonic MHC, super-fast MHC and embryonic/foetal-MHC was examined in detail during the study at hand. No reaction of anti-α-MHC with any of these MHC isoforms could be detected (data not shown). The antibody 330-5B4 reacted strongly with the developing muscles of human embryos from four weeks of development onward (Fig. 1). In the adult extrafusal fibres reaction was never observed (see Fig. 3). Therefore this antibody was dubbed anti-embryonic-MHC. However, the possibility of crossreaction of this antibody with other developmental MHC isoforms (e.g. foetal/neonatal MHC) can formally not be excluded.

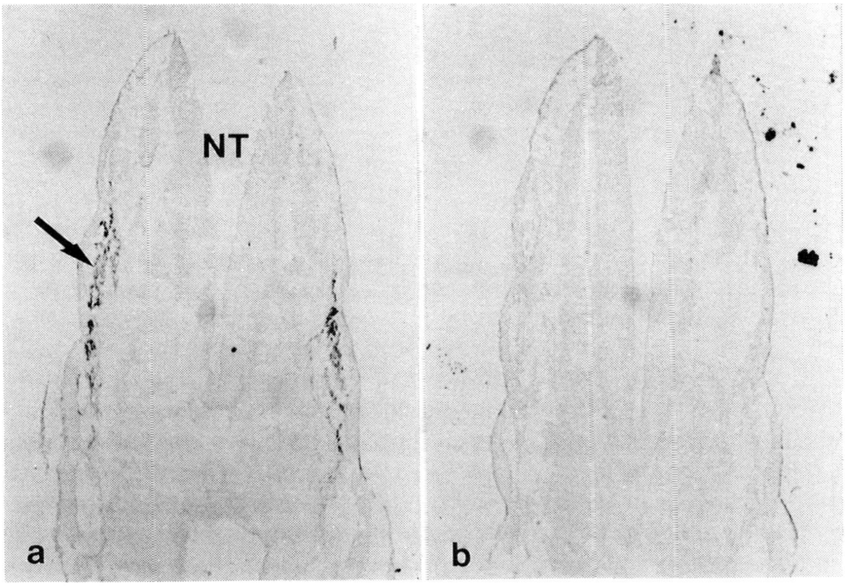

Fig. 1 Serial sections of a human embryo of approximately four weeks of development incubated with 330-5B4 (a) and anti-α-MHC (b). In the dorsal region of this embryo strong reaction of 330-5B4 is observed in the myotomes (arrow). No reaction is observed with anti-α-MHC. NT = neural tube.

MHC EXPRESSION IN THE DEVELOPING HUMAN MUSCLE SPINDLE

In foetal muscle (15–20 weeks of development), anti-embryonic-MHC reacts with all extra and intrafusal fibres (Fig. 2a). In these muscles, anti-β-MHC reacts with the relatively large extrafusal primary myotubes, while in the developing muscle spindle reaction with anti-β-MHC is observed with at least one, but occasionally also with more intrafusal fibres (Fig. 2b). With anti-α-MHC no reaction is observed in any of the extrafusal fibres. However, in the muscle spindle strong reaction of anti-α-MHC is detected, in particular in those intrafusal fibres that also react with anti-β-MHC (Fig. 2c).

In neonatal skeletal muscle anti-embryonic MHC does not react with extrafusal muscle fibres under standard conditions. However, if the sections are pretreated with pronase, a protease that makes the antigens better accessible for the antibodies, reaction of this antibody is observed in a subpopulation of the extrafusal fibres (Fig. 2d). In the muscle spindle expression of embryonic-MHC is observed in relatively large intrafusal fibres, probably nuclear bag fibres. Anti-β-MHC reacts with the slow (type I) extrafusal fibres and with some of the intrafusal fibres that have a relatively large diameter (Fig. 2e). As

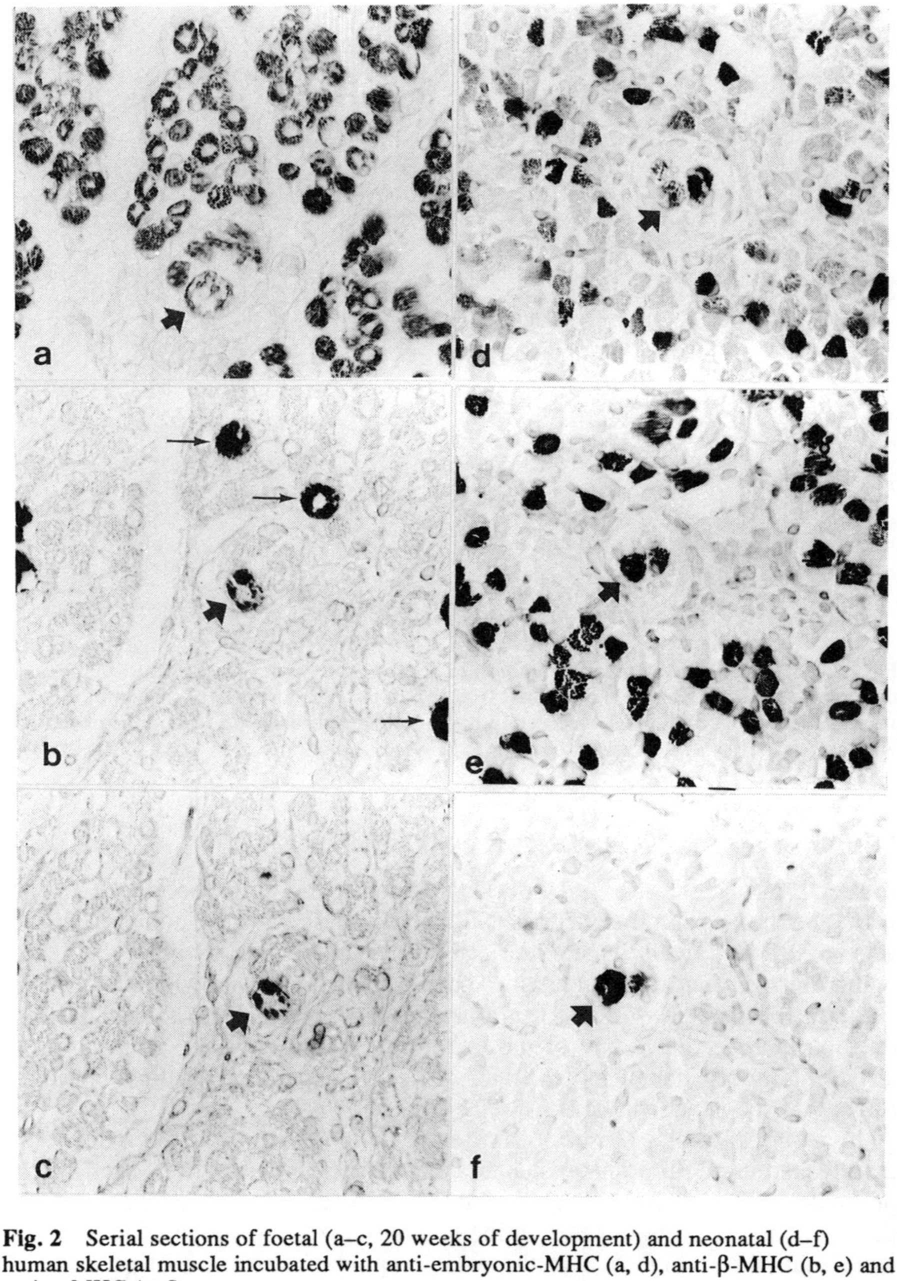

Fig. 2 Serial sections of foetal (a–c, 20 weeks of development) and neonatal (d–f) human skeletal muscle incubated with anti-embryonic-MHC (a, d), anti-β-MHC (b, e) and anti-α-MHC (c, f).

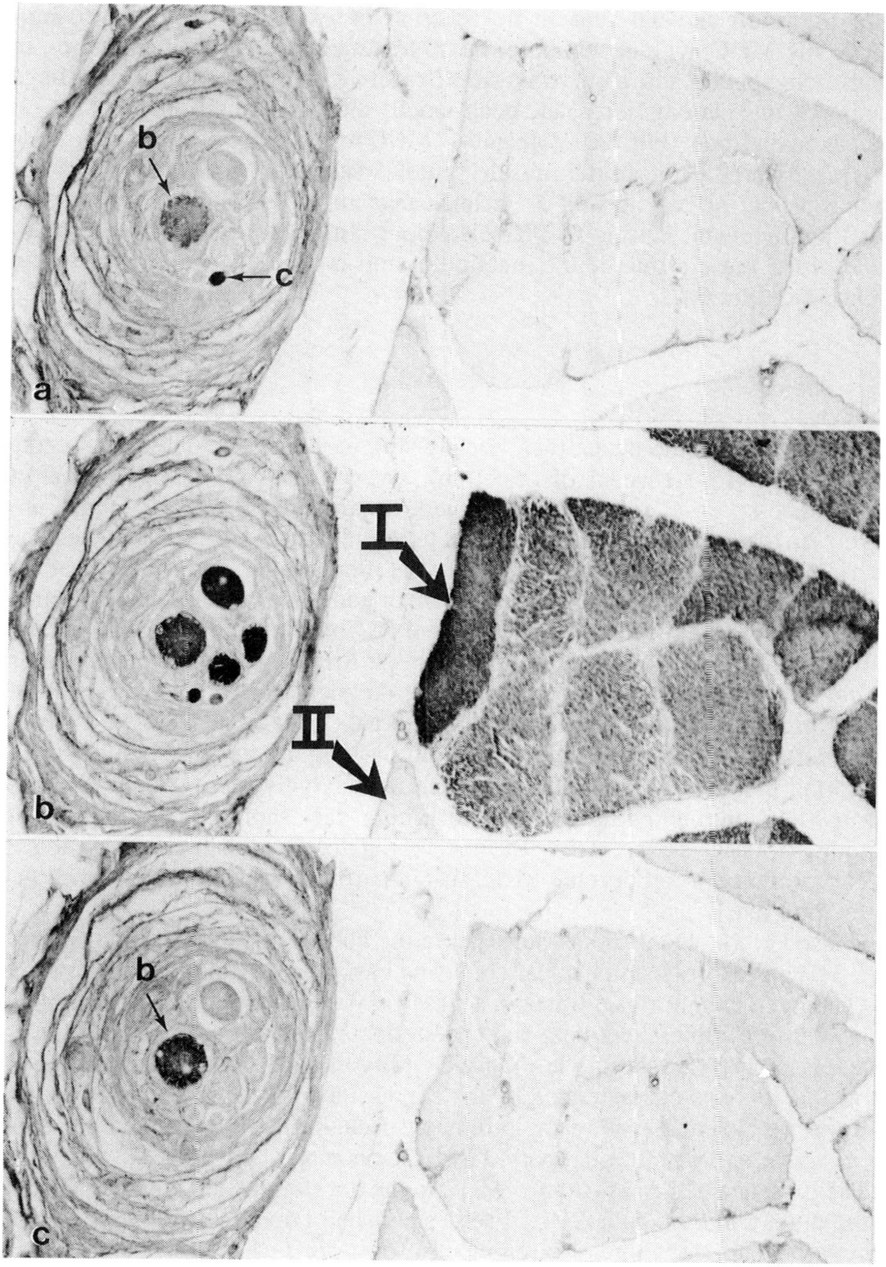

Fig. 3 Serial sections of adult human soleus muscle incubated with anti-embryonic-MHC (a), anti-β-MHC (b) and anti-α-MHC (c). b = nuclear bag fibre, c = nuclear chain fibre, I = type I fibre, II = type II fibre.

in foetal muscle, no reaction of anti-α-MHC is observed in the extrafusal fibres of the neonatal muscle. In the muscle spindle the reaction of anti-α-MHC is again restricted to those fibres that also express β-MHC (Fig. 2f).

In the adult skeletal muscle no reaction of extrafusal fibres with anti-embryonic-MHC, with or without pronase treatment, is observed any more. In the muscle spindle anti-embryonic-MHC reacts with fibres of relatively large diameters, nuclear bag fibres, and occasionally also with small-diameter fibres, nuclear chain fibres (Fig. 3a). With anti-β-MHC reaction is observed in the slow extrafusal fibres while in the muscle spindle reaction of this antibody is observed in nuclear bag as well as in nuclear chain fibres (Fig. 3b). As in the foetal and neonatal muscle anti-α-MHC does not react with extrafusal fibres while in the muscle spindle the reaction of anti-α-MHC is again restricted to nuclear bag fibre(s).

Discussion

In this paper we demonstrate that, already very early in human muscle spindle development (15–20 weeks of gestation), a very characteristic expression of the 'cardiac specific' α-MHC is observed. Expression of α-MHC is also observed in the neonatal and adult muscle spindle. In all these muscles the reaction seems to be restricted to nuclear bag fibres. Interestingly, Te Kronnie *et al.* (9, 10) reported that in skeletal muscle of adult mice and rat, in addition to the type I extrafusal fibres, also the nuclear bag fibres reacted with a polyclonal antibody that was raised against chicken-heart MHC. After adsorption of this serum with acetone powder prepared from the slow twitch soleus muscle of guinea pig they observed that only a very weak staining of the type I fibres remained whereas the strong staining of the bag fibres was not effected, indicating that the anti-β-MHC antibodies were largely removed and that the anti-α-MHC antibodies have remained. These data support our finding that 'cardiac' α-MHC is expressed in the developing mammalian nuclear bag fibres.

The presence of embryonic-MHC in human intrafusal fibres has been reported previously.

Erikson *et al.* (11) demonstrated that in the human masseter muscle embryonic-MHC is expressed in nuclear chain and bag_2 fibres. Although we have not characterized the intrafusal fibres in a classical way (ATPase activity cannot be measured in paraplast sections) and cannot distinguish between bag_1 and bag_2 fibres, it is obvious that in adult muscle embryonic-MHC is expressed in bag fibres that are also characterized by the expression of β-MHC and α-MHC. In addition we also observed reaction in nuclear chain fibres. Therefore our data are in agreement with those reported by Erikson *et al.* (11).

The presence of slow-twitch(β) MHC in the developing and adult bag fibres of the human muscle spindle has been established previously (4). Our results with respect to the expression of this isoform are in agreement with these reports.

Although it should be stressed that the results presented in this paper are by no means complete, only a few muscle specimens have been examined, and no systematic study on the spatial distribution of the MHC isoforms along the muscle spindles have been performed, yet our findings do contribute to the growing body of evidence that certain muscles are capable of expressing an extreme large scala of MHCs. Apart from the 'ordinary' fast- and slow-twitch MHCs found in normal extrafusal skeletal muscle fibres, embryonic-, 'cardiac'-α-MHC and slow-tonic MHC, are not only detected in intrafusal fibres but also, in various combinations, in extraocular, tongue and masticatory muscles. Interestingly, all these muscles are characterized by multiple innervation and complex contractile properties. This might indicate that for the precise regulation of the contractile apparatus of these muscles, the expression of nearly the whole complement of MHC isoforms is required.

Acknowledgements

The authors gratefully acknowledge Mr C. Gravemeijer for excellent photography. This work was partly supported by grant 76-086h of the Dutch Heart Foundation. The study was approved by the local medical-ethical committees.

References

1. Maier, A., Gambke, B., Pette, D., 1988, *Histochem.*, **88**, 267–271.
2. Kucera, J., Walro, J. M., 1989, *Histochem.*, **92**, 291–299.
3. Pedrosa, F., Butler-Browne, G. S., Dhoot, G. K., Fischman, D. A., Thornell, L.-E., 1989, *Histochem.*, **92**, 185–194.
4. Thornell, L. E., Grove, D. A., Pedrosa, F., Butler-Browne, G. S., Dhoot, G. K., Fischman, D. A., 1989. In: *Cellular and Molecular Biology of Muscle Development*, pp. 471–480. Alan R. Liss, Inc.
5. Wessels, A., Vermeulen, J. L. M., Virágh, Sz., Kálmán, F., Morris, G. E., Nguen thi Man, Lamers, W. H., Moorman, A. F. M., 1990, *Anat. Rec.*, **228**, in press.
6. Wessels, A., Vermeulen, J. L. M., Virágh Sz., Moorman, A. F. M., 1990. In: *Embryonic Origins of Defective Heart Development*, vol. 588 of the Annals of the New York Academy of Sciences.
7. Wessels, A., Vermeulen, J. L. M., Virágh, Sz., Kálmán, F., Lamers, W. H., Moorman, A. F. M., 1990, *Anat. Rec.*, in press.
8. Christensen, L., Strange, L., 1987, *J. Histotech.*, **10**, 11–15.
9. Kronnie, G. te, Donselaar, Y., Soukup, T., Van Raamsdonk, W., 1981, *Histochem.*, **73**, 65–74.
10. Kronnie, G. te, Donselaar, Y., Soukup, T., Zelena, J., 1982, *Histochem.*, **74**, 355–366.
11. Eriksson, P.-O., Butler-Browne, G., Fischman, D., Grove, B. K., Schiaffino, S., Virtanen, I., Thornell, L. E., 1988. In: Hnik, P., Soukup, T., Vejsada, R., Zelena, J. (eds), *Mechanoreceptors*, pp. 273–274. Plenum, New York.
12. Bredman, J. J., Weijs, W. A., Moorman, A. F. M., 1990, *Proc. XIX Eur. Conf. Muscle & Mot.*, Brussels.

12
Creatine Kinase Isozyme Expression in Pre-natal Rat Heart: a 3D Analysis of its Distribution and a Morphometric Analysis of the Apical Ventricular Trabeculae

H. D. J. HASSELBAINK*, A. C. G. WENINK**, A. F. M. MOORMAN* AND W. H. LAMERS*

*Department of Anatomy and Embryology, University of Amsterdam, The Netherlands
**Department of Anatomy and Embryology, University of Leiden, The Netherlands

Introduction

The developmental changes in the distribution pattern of the creatine kinase (EC 2.7.3.2) isozymes in pre-natal rat heart have been described in a previous publication (1). Creatine kinase M (CK-M) is heterogeneously expressed during embryonic day (ED) 12–18. A pronounced staining of CK-M is initially observed in the outflow tract, the trabeculae of the right ventricle, the venous valves, the interatrial septum and the sinoatrial node. From ED18 onwards, a homogeneous expression of CK-M is observed due to an increase in isozyme concentration in the remaining part of the myocardium. By contrast, creatine kinase B (CK-B) is homogeneously expressed throughout the heart between ED 11–14. Thereafter, a decrease of CK-B content is observed in the inflow tract, the atria, the interventricular septum and in the papillary muscle of the

Muscle and Motility Vol. 2. Proceedings of XIXth European Conference in Brussels
© Intercept Ltd, PO Box 716, Andover, Hampshire, SP10 1YG, UK

left ventricle. At birth, a considerable amount of CK-B remains present in the atrial and ventricular walls.

On ED16, when the decrease in CK-B content is first observed, two types of areas can be distinguished: areas lacking CK-B and areas with a low content of CK-B. In this contribution, a more precise description of these areas is made with the help of a computeraided 3D-reconstruction programme (2). Furthermore, the initial difference in CK-M staining intensity between the left and right ventricular trabeculae raised the question when in development the difference in the apical trabecular pattern that is typical for the 'morphologically' left and the 'morphologically' right ventricle, can first be observed (4).

Material and Methods

3D-RECONSTRUCTION

3D-reconstruction of the CK-B distribution pattern on ED16 was performed on serial sections of a transversely and a sagitally sectioned rat heart. The contours of the hearts were traced on acetate sheets using a camera lucida set up. By use of a digitizer tablet the input of the contours was realized into a 3D-reconstruction database. With help of these computer-aided 3D-reconstructions, schematic drawings of the hearts were made.

MEASUREMENT OF THE SIZE OF THE APICAL VENTRICULAR TRABECULAE

The apical ventricular trabeculae of the different developmental stages were measured by use of a counting grid by established morphometric methods (3). Briefly, by use of a drawing tube, fixed to the microscope, the counting grid was projected on a section. In such a section the number of intersections through the trabeculae and the number of points within the trabeculae were counted. The size of the apical ventricular trabeculae was calculated with standard formulae (3).

Results and Discussion

CREATINE KINASE DISTRIBUTION PATTERN

CK-B negative areas

On ED16, two separate areas lacking CK-B are temporarily observed in the ventricles. Anteriorly, in the free wall of the left ventricle an isolated area without CK-B, comprising less than 5% of the left ventricular surface, is found. Posteriorly, on the adjacent walls of the left and right ventricle at both sides of the interventricular sulcus, a large 'butterfly'-shaped area, comprising approximately 15–20% of the ventricular surface, is observed (Fig. 1a). Possibly, this distribution is related to the developmental appearance of the collagen skeleton of the rat heart that was previously described by one of us (5). This area continues anteriorly in the upper part of the interventricular septum where it ends under the orifice of the aorta (Fig. 1b).

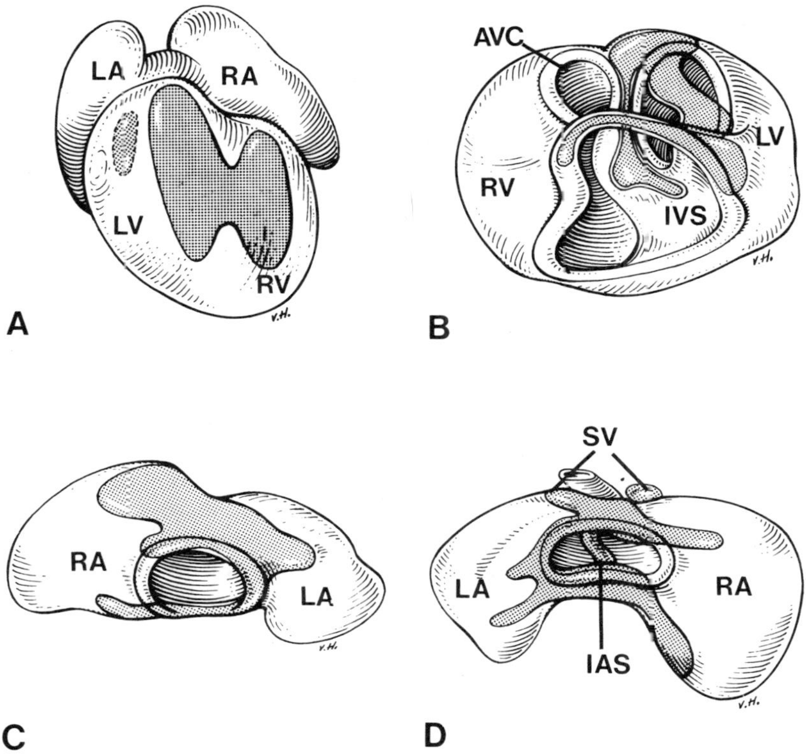

Fig. 1 Schematic drawings of the ED16 rat hearts. (A) Posterior view of the heart. On the posterior adjacent walls of the left and right ventricle at both sides of the interventricular sulcus, a 'butterfly'-shaped area lacking CK-B is observed. Another area lacking CK-B is seen (encircled with ? a dotted line) in the anterior wall of the left ventricle. (B) Anterior view of the ventricles. The atria are removed at the level of the atrioventricular (AV) junction and the outflow tract at the level of its junction with the ventricles. The area lacking CK-B on the posterior wall of the ventricles continues anteriorly via the interventricular septum as far as the orifice of the aorta. Areas with a low content of CK-B are observed in the myocardium of the lesser curvature of the heart and in the left AV funnel. (C) Anterior view of atria. The ventricles are removed at the level of the AV junction. An area with a low CK-B content is seen in the anterior, left and posterior portion of the AV junctional myocardium. (D) Posterior view of the atria. The ventricles are removed at the level of the AV junction. The area with a low content of CK-B in the anterior, left and posterior portion of the AV junctional myocardium continues, posteriorly, in upward direction with the CK-B low areas in the sinus venosus and with the interatrial septum. ao, aorta; AVC, atrioventricular canal; IAS, interatrial septum; IVS, interventricular septum; LA, left atrium; LV, left ventricle; RA, right atrium; RV, right ventricle; SV, sinus venosus.

Weakly CK-B positive areas

An area with a low CKB content is found in the myocardium of the lesser curvature of the heart, i.e in the posterior myocardial wall of the aorta. A low CK-B content is also found in the myocardium of the left atrioventricular funnel, i.e in the valve primordium. Anteriorly, this area is continuous with the area in the lesser curvature that is already mentioned and posteriorly with the CK-B-negative area described above (Fig. 1b).

A low CK-B content is also observed in the anterior (continuous with that in the lesser curvature), left and posterior portion of the atrioventricular junctional myocardium. Anteriorly, the area extends over the middle part of the atrial free wall; posteriorly, the area extends upwards over the posterior wall of the atria and is continuous with the CK-B low areas in the sinus venosus on the one hand and in the interatrial septum on the other (Fig. 1c, d). The right atrioventricular junction seems not to differ in CK-B content from the surrounding myocardium.

DEVELOPMENTAL CHANGES IN THE SIZE OF THE APICAL VENTRICULAR TRABECULAE

Fig. 2 illustrates the average diameter of the left and right apical ventricular trabeculae as a function of the developmental stage. In the embryonic period (ED12–16) there seems only little difference in size between the apical trabeculae of the ventricles. Possibly, the right apical ventricular trabeculae are somewhat smaller than the left apical ventricular trabeculae. At the beginning of the foetal period, on ED18, the diameter of the apical trabeculae increases and a difference in size between the left and right ventricle emerges. On ED22, the day of birth, the right apical ventricular trabeculae are bigger in size than those of the

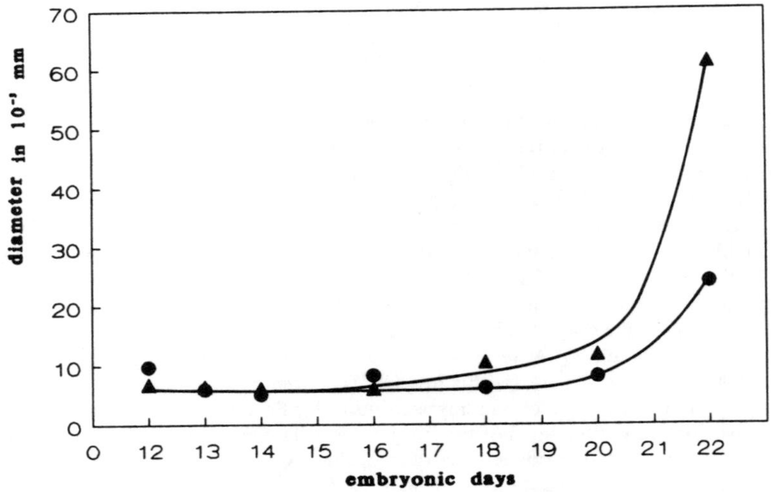

Fig. 2 Developmental changes in the diameter of the apical ventricular trabeculae. Triangles, right ventricle; circles, left ventricle.

left ventricle. Another interesting feature in the development of the trabeculae is the volume percentage of the trabeculae within the lumen of the ventricle, which increases with ageing (not shown). This increase is more pronounced in the right ventricle than in the left ventricle. This could reflect the incorporation of the intertrabecular spaces into the coronary vascular bed (6).

The difference between the right and left apical ventricular trabeculae in adult human heart is used by morphologists to distinguish the 'morphologically' right ventricle (with coarse apical trabeculae) from that of the 'morphologically' left ventricle (with fine apical trabeculae) (4). This study indicates that this phenomenon is not noticeable in rats until the late foetal period.

Acknowledgements

This work was supported by the Foundation for Biological research (BION grant no. 431161). The authors wish to thank Mr C. Gravemeyer, Mr C. Hershbach and Mr A. A van Horssen for their excellent art work.

References

1. Hasselbaink, H. D. J, Labruyère, W. Th, Moorman, A. F. M., Lamers, W. H., 1990, *Anat. Embryol.*, **182**, 195–203.
2. Verbeek, F. J., Schoutsen, N. J. C., Huysmans, D. P., 1990, *Cytometry, suppl.* **4**, 108.
3. Gundersen, H. J. G., 1988, *APMIS*, 379–394.
4. Anderson, R. H., Shinebourne, E. A., 1978, *Paediatric Cardiology*, 1977, Churchill Livingstone, Edinburgh.
5. Wenink, A. C. G., Spliet, W. G. M., Mansoer, J. R., 1988, *Wiener klinische Wochenschrift*. Springer-Verlag, Wenen, New York.
6. Virágh, S., Challice, C. E., 1981, *Anat. Rec.*, **201**, 157–168.

13
Distribution of the Cardiac Gap Junction Protein, Connexin 43, in the Neonatal and Adult Human Heart

P. W. OOSTHOEK, M. J. A. VAN KEMPEN, A. WESSELS, W. H. LAMERS, AND A. F. M. MOORMAN

Department of Anatomy and Embryology, University of Amsterdam, Meibergdreef 15, 1105 AZ Amsterdam, The Netherlands

Introduction

Gap junctions are thought to be responsible for the electrical coupling of the cardiomyocytes and are involved in the spread of the impulse over the myocardium (1). A gap junction consists of an array of cell to cell channels, each composed of six subunits, dubbed connexins. Biochemical and molecular studies have revealed several tissue specific variants. They comprise a family of closely related proteins. In the heart gap junctions are composed of connexin 43 (CX43) that can be visualized immunohistochemically (2). Recently we have described that in developing rat hearts (up to 20 days of gestation) the distribution of CX43 parallels the pattern of conduction velocities, as derived from the chicken condition, i.e. the virtual absence of CX43 in areas of low conduction velocity (atrioventricular junction and outflow tract) and the presence of CX43 in areas of relatively high conduction velocities (atrial and ventricular working myocardium) (3, 4). Therefore, we hypothesized that the selective formation of gap junctions might explain the gradual appearance of a more rapid conduction in the developing heart. Paradoxically in the adult rat heart the atrioventricular (AV) bundle and the top of the bundle branches did not express

Muscle and Motility Vol. 2. Proceedings of XIXth European Conference in Brussels
© Intercept Ltd, PO Box 716, Andover, Hampshire, SP10 1YG, UK

CX43 (4), indicating low conduction velocities, similar to the AV node. This raises the question of whether this phenomenon is found in human hearts as well.

In this paper we describe that in the neonatal human heart the AV bundle and bundle branches contained almost no CX43 in contrast to the working myocardium. In the adult human heart CX43 was present in the AV bundle and top of the bundle branch, indistinguishable from the working myocardium. Evidence will be discussed that the occurrence of gap junctions and, hence, the conduction velocity in the bundle depends on the size of the heart, and are in line with the notion of Meijler (5) that the length of the AVdelay is similar in hearts with very different size.

Materials and Methods

Human hearts were obtained from the department of Pathology, University of Amsterdam. Serial sections of the neonatal heart and of parts of the adult heart were incubated (see 4) with an antibody against CX43 (2), antibodies against atrial myosin heavy chain (α-MHC) and ventricular myosin heavy chain (β-MHC) (6), and an antibody directed against the intercalated discs (antiID). Binding was visualized by the unconjugated peroxidaseantiperoxidase method. Only one adult and one neonatal heart could be used for the study at hand, as they revealed a clear and reproducible staining both with antiCX43 and with the control antibodies anti-α-MHC, anti-β-MHC and anti-intercalated disc. Difficulties in obtaining high quality autopsy material that allows unambiguous detection of CX43 staining have so far prevented study of additional hearts.

Results

THE NEONATAL HEART

CX43 was found as small dots at the cell membrane of the myocytes of both the atria and the ventricles. In the ventricles a clear staining pattern of CX43 was observed (Fig. 1c). Interestingly, staining was prevalent at the elongated lateral sides of the myocytes, whereas only in some parts of the ventricle the distribution had an intercalated disc-like pattern. In the sinoatrial node (SAN) CX43 could not be detected, whereas in the AV bundle and bundle branches only a faint staining of the intercalated discs was visible (Fig. 1b, arrows). As the ventricular conduction system expresses α-MHC it could be easily distinguished from the ventricular working myocardium (Fig. 1a).

THE ADULT HEART

CX43 was predominantly expressed in the intercalated discs of both the atria and the ventricles (Fig. 1d and f). This staining pattern coincides with the distribution pattern of the intercalated discs in the working myocardium as determined by the staining with antiID (Fig. 1e). In the SAN no staining was

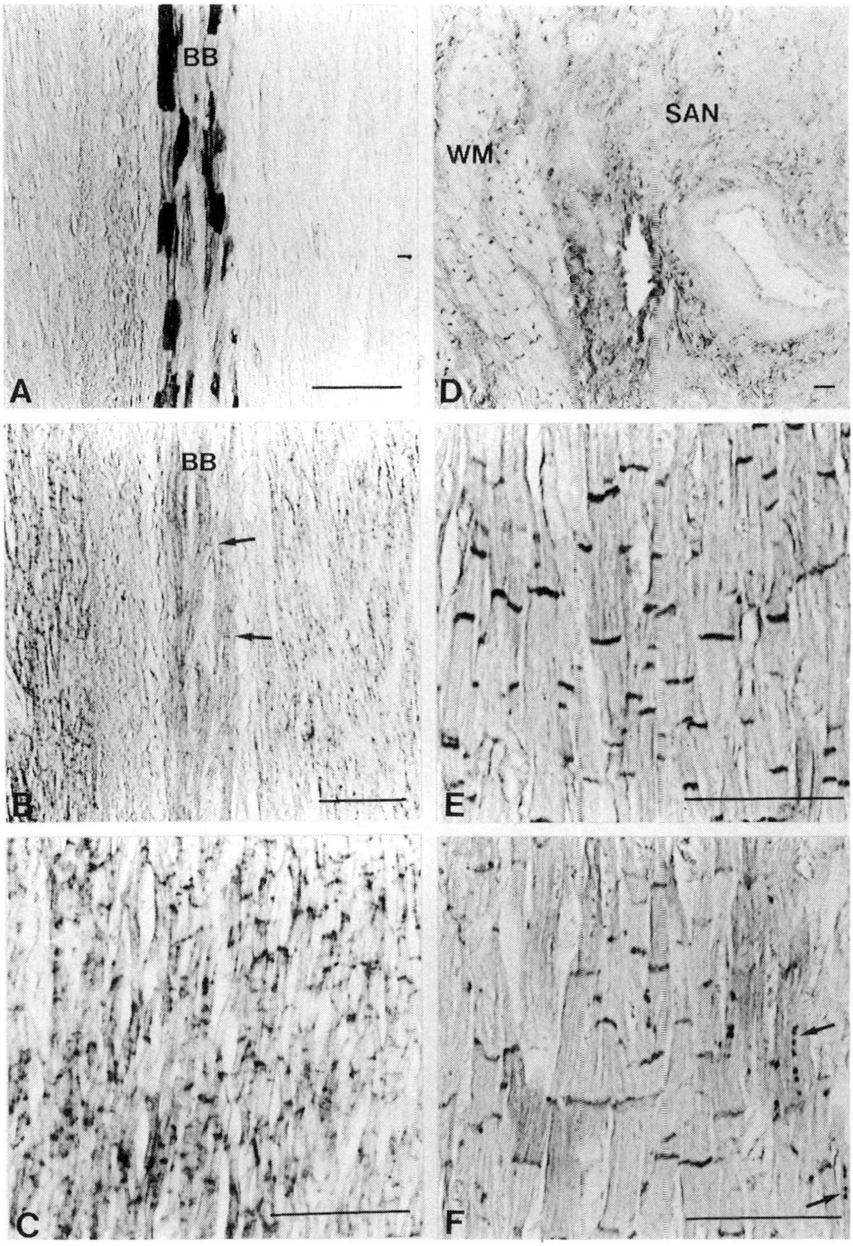

Fig. 1 Neonatal (A,B,C) and adult (D,E,F) human heart. (A,B) Expression of α-MHC and CX43, respectively, in the bundle branch of the human neonatal heart. (C) Part of the ventricular wall, incubated with anti-CX43. D: SAN and surrounding atrial wall of the adult heart, incubated with anti-CX43. (E,F) Part of the interventricular septum, incubated with antiID and antiCX43 respectively. BB-bundle branch, SAN-sinoatrial node, WM-working myocardium. Bar = 0.05 mm.

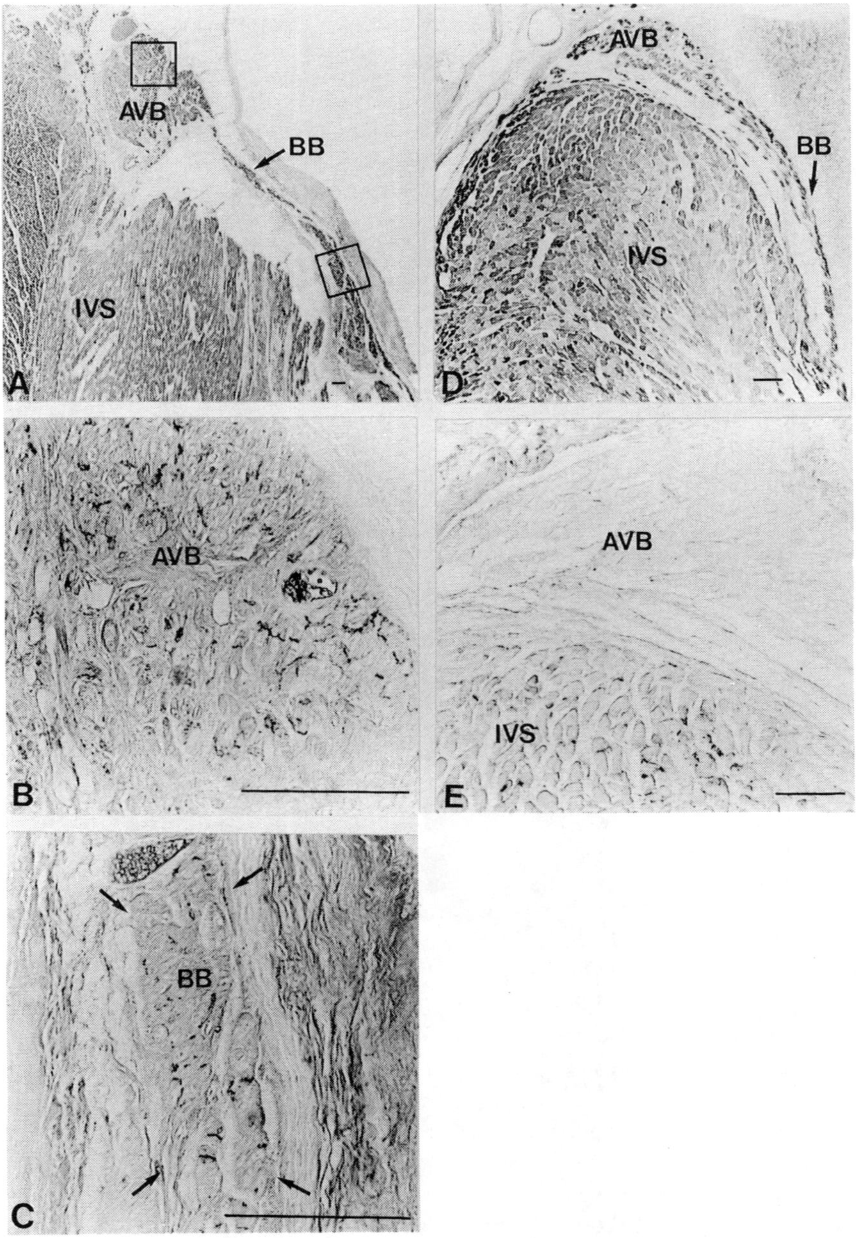

Fig. 2 Adult human (A,B,C) and rat (D,E) heart. (A) Expression of β-MHC, marking the myocytes of the AVB, BB and IVS in the human. Squares showing analogous regions, shown in B and C. (B) Expression of CX43 in the AVB. (C) Expression of CX43 in the BB. (D) Expression of MHC, marking the myocytes of the AVB, BB and IVS in the rat. (E) Expression of CX43 in the AVB and IVS. AVB, atrioventricular bundle; BB, bundle branch; IVS, interventricular septum. Bar = 0.1 mm.

observed (Fig. 1d), similar to the neonatal condition, whereas the AVbundle (Fig. 2b) and bundle branches (Fig. 2c) were now CX43positive. Remarkably, in the ventricular working myocardium staining at the lateral sides of some myocytes was visible (Fig. 1f, arrows).

Discussion

During development of the human heart there is a change in the distribution of CX43 in the conduction system and a change in distribution of CX43 across the cell membrane.

In the embryonic human heart, in which the intercalated discs are not formed, CX43 was randomly distributed across the membrane of the myocytes. (data not shown). In the neonatal heart CX43 is still randomly distributed. These results correspond with the findings of Chalice and Virgh (7), who described that, during development, gap junctions are mainly located at the lateral contacts between cells as isolated junctions. In the working myocardium of the neonatal heart intercalated discs may not be fully developed as they were not distinguishable not even after incubation with antiID. Nevertheless, in some parts of the ventricle CX43 already had an intercalated disclike pattern indicating the formation of intercalated discs, corresponding the findings of McNutt (8) who described that the formation of gap junctions precedes the appearance of intercalated discs. In the human adult heart CX43 was mainly restricted to the intercalated discs. The question of whether lateral gap junctions exist in the myocytes of the adult heart is controversial (9, 10). The antiCX43 staining definitely shows the presence of CX43 at the lateral sides of the myocytes indicating the presence of lateral gap junctions in the adult heart.

As mentioned above we hypothesized that gap junctions play a role in the high conduction velocity in the heart. In the neonatal human heart almost no CX43 could be detected in the AV bundle and bundle branches, while the working myocardium was clearly positive. In the adult human heart no differences in distribution of CX43 between the ventricular conduction system and the working myocardium could be found. These observations differ from those in the adult rat heart in which CX43 could not be detected in the AV bundle and top of the bundle branches (Fig. 2e) (4).

Two possibilities can be envisaged to explain this apparent discrepancy. First, the AV bundle and top of the bundle branches have a fast conduction velocity, due to the existence of gap junctions that are, in rat, composed of connexins other than CX43. Although formally not excluded, this idea is not very likely as no connexin 32 and as so far no other connexins have been found in the heart (4). Second, the conduction velocity in the AV bundle and bundle branches is.different in rat and human. This hypothesis is in line with the findings of Meijler (5), that hearts of different sizes have a similar AVdelay. This notion would imply that in smaller hearts the propagation of the impulse is slower and, hence high densities of gap junctions are not required. This explains also the relatively low amount of CX43 in the AV bundle and bundle branches of the 'small' neonatal human heart, compared to the 'large' adult human heart. To substantiate this notion a combined electrophysiological and immunohistochemical study of hearts of different size is indicated.

Acknowledgements

We like to thank Mr C. Gravemeijer for the excellent photography. Ms J. L. M. Vermeulen and Mr T.A. Mijnders for their technical assistance, and Mr D. Gros for kindly providing antiCX43. This work was supported by grants nos 86.076, 88.092 and 088.093 from the Netherlands Heart Foundation. The study was approved by the local medical-ethical committee.

References

1. Barr, L., Dewey, M. M., Berger, W., 1965, *J. Gene Physiol.*, **48**, 797–823.
2. Aoumari, A. E., Fromaget, C., Dupont E., Reggio, H., Durbec, P., Briand, J. P., Boller, K., Kreitman, B., Gros, D., 1990, *J. Membrane Biol.*, in press.
3. De Jong, F., Op't Hof, T., Wilde, A .A. M., Janse, M. J., Charles, R., Lamers, W. H., Moorman, A. F. M., 1990, *Circ. Res.*, in press.
4. Van Kempen, M. J. A., Aoumari, A. E., Gros, D., Moorman, A. F. M., Lamers, W. H., 1990, submitted.
5. Meijler, F. L., Janse, M. J., 1988, *Physiol. Rev.*, **68**, 608–647.
6. Wessels A., Vermeulen, J. L. M., Virágh, Sz., Kálmán, F., Lamers, W. H., Moorman, A. F. M., 1990, *Anat. Rec.*, in press.
7. Challice, C. E., Virágh, Sz., 1973. In: *Ultrastructure in Biological Systems*, vol. 6, pp. 91–126, Academic Press, New York and London.
8. McNutt, N. S., 1970, *Am. J. Cardiol.*, **25**, 169–183.
9. Canale, E. D., Campbell, G. R., Smolich, J. J., Campbell, J. H., 1986, *Cardiac Muscle*, pp. 20–26. Spinger-Verlag, Berlin.
10. Hoyt, R. H., Cohen, M. L., Saffitz, J. E., 1989, *Circ. Res.*, **64**, 563574.

14
Neural Tissue Antigen Identifies the Evolutionary Origin of the Left and Right Ventricle

WOUTER H. LAMERS AND ANTOON F.M. MOORMAN

Department of Anatomy and Embryology, University of Amsterdam, Meibergdreef 15, 1105 AZ, Amsterdam, The Netherlands

Introduction

The differentiation of the working myocardium from the myocardium of the primary heart tube, which accompanies the onset of cardiac function, leads to the appearance of cardiac segments and junctional zones. Morphologically, this basic architecture is revealed by an alternation of bulging segments and constricted junctional zones. The identification of (additional) functional and biochemical criteria can provide this classification with a more objective basis. Such criteria are necessary if one wants to trace the fate of the embryonic segments during development. Here we focus on the development of the junction between the left and right ventricle.

Cardiac Segments and Conduction Velocities

The measurement of conduction velocities in avian embryos (1, 2, 3; see Table 1) has shown that conduction is slow and nearly uniform in the early tubular heart, resulting in a peristaltoid contraction pattern. Whereas conduction velocity markedly increases in the developing ventricles and atria, it remains low in the

Muscle and Motility Vol. 2. Proceedings of XIXth European Conference in Brussels
© Intercept Ltd, PO Box 716, Andover, Hampshire, SP10 1YG, UK

Table 1 Impulse propagation (mm/s) in the cardiac compartments and junctional regions of embryonic chickens (at 48 h and 96 h of incubation) and of adult amphibians.

	Chicken[1, 2]			Urodela[5]	Anura[6, 7]
	40 h	48 h	120 h		
SAN junction	—	—	<15	—	—
Atrium	—	18	158	115	—
AV junction	—	7	15	4	<5
Ventricle	8	17	113	37	250
Outflow tr./ VB junction	—	8	13	<1	<5
bulbus	—	—	—	9	150

transitional zones between the cardiac compartments, *viz.* in the sinoatrial and atrioventricular junctions, and in the embryonic outflow tract. Due to the length of the embryonic outflow tract, slow conduction results in a peristaltoid contraction wave that exceeds the cycle during which the activation was initiated. As a result the outflow tract substitutes for the function of the future semi-lunar valves (2).

The alternation of rapid conduction in the working myocardium and slow conduction in the myocardium of the junctional zones and the outflow tract that is observed in avian development prior to the emergence of the structures of the definitive conduction system (chick, second week of incubation; 4) appears to be an phylogenetically old pattern as it is also observed in adult amphibians, in particular in urodeles (5, 6, 7; see Table 1). Common functional properties of the cardiomyocytes in the junctional zones of embryonic avian and of adult amphibian hearts are also suggested by the similarity in configuration, amplitude and duration of the action potentials.

The Development of a Left and Right Ventricle

These electrophysiological measurements support the morphological distinction of four cardiac components (sinus venosus, atrium, ventricle and bulbus) in lower vertebrates and early avian embryos. However, due to a lack of adequate cell markers, the developmental events leading to the formation of a right and left ventricular segment within the ventricular component of the heart of reptiles, birds and mammals have proven to be more difficult to trace. Therefore, two of the major questions concerning this remodelling of the heart are the mechanism involved in connexion of the right atrium to the right ventricle, and of the left ventricle to the aorta. Since it is well accepted that, during early stages of development, the atrial segment of the heart is connected only to the proximal (= inlet = left) segment of the ventricular component of the primary heart tube, while the outflow segment takes origin exclusively from the distal (= outlet = right) segment of the ventricular component, these mechanisms of transfer must be completed before the ventricular mass can properly be septated. Furthermore, disagreements about these processes have coloured many of the arguments concerning evolutionary adaptations and the classification of congenital malformations.

In Mammals a 'Primary' Ring of Myocytes Demarcates the Left and Right Ventricle

We now believe that we can demonstrate the developmental origin of the muscular ventricular septum between the left and right ventricle. This is because a monoclonal antibody to an extract of the *Ganglion Nodosum* (GlN) of the chick has been shown to demarcate the junction between the proximal and distal segments of the ventricular component of the heart tube (8). This 'primary' ring of cardiomyocytes is situated between both *embryonic* ventricles and surrounds the 'primary' interventricular foramen. By tracing the fate of this tissue during the period of septation of the human heart, we have been able to conclude that the remodelling of the myocardium that is responsible for development of the connection of the right atrium to the right ventricle, and of the left ventricle to the aorta, occurs mainly in the lesser curvature of the cardiac loop, and that the cardiac loop, and the associated wedge shape of the cardiac segments, is an essential feature of shape for cardiac septation. The entire muscular part of the ventricular septum, the right component of the atrioventricular junction and the junction between the left ventricle and its subaortic outlet are derived from the primary ring (Fig. 1). The primary ring eventually surrounds the inlet to the definitive right ventricle and the outlet from the definitive left ventricle. Although the entire junction between both embryonic ventricles originates within the ventricular myocardium (as defined by the expression of the ventricular isoform of the myosin heavy chain), the part that surrounds the right atrioventricular connexion topographically becomes an atrial structure upon the formation of the fibrous junction between atria and ventricles. The fate of the GlN-positive junctional cells as the precursors of the ventricular conduction system explains why the ventricular conductive tissue itself is a useful tissue marker for the description of ventricular malformations, that are due to septational defects.

The 'Primary' Ring in Birds

The developmental fate of the 'primary' (GlN) ring is conserved in evolution. Whereas in most, and perhaps all, mammals large parts of the 'primary' ring disappear in development, virtually the entire ring persists and is identifiable as bundles of so called Purkinje cells in the chicken (4, 9–11). It should be stressed that the topographical position of the ring of Purkinje cells in the adult chicken is completely homologous to that which we described in the human embryo (Fig. 2; cf. Fig. 1), indicating that not only the development of the conduction system, but also cardiac septation are completely comparable in both taxonomic classes. The demonstration of a phylogenetically conserved origin of the ventricular conductive system from a single ring of cardiomyocytes between two cardiac segments (the embryonic left and right ventricles) is

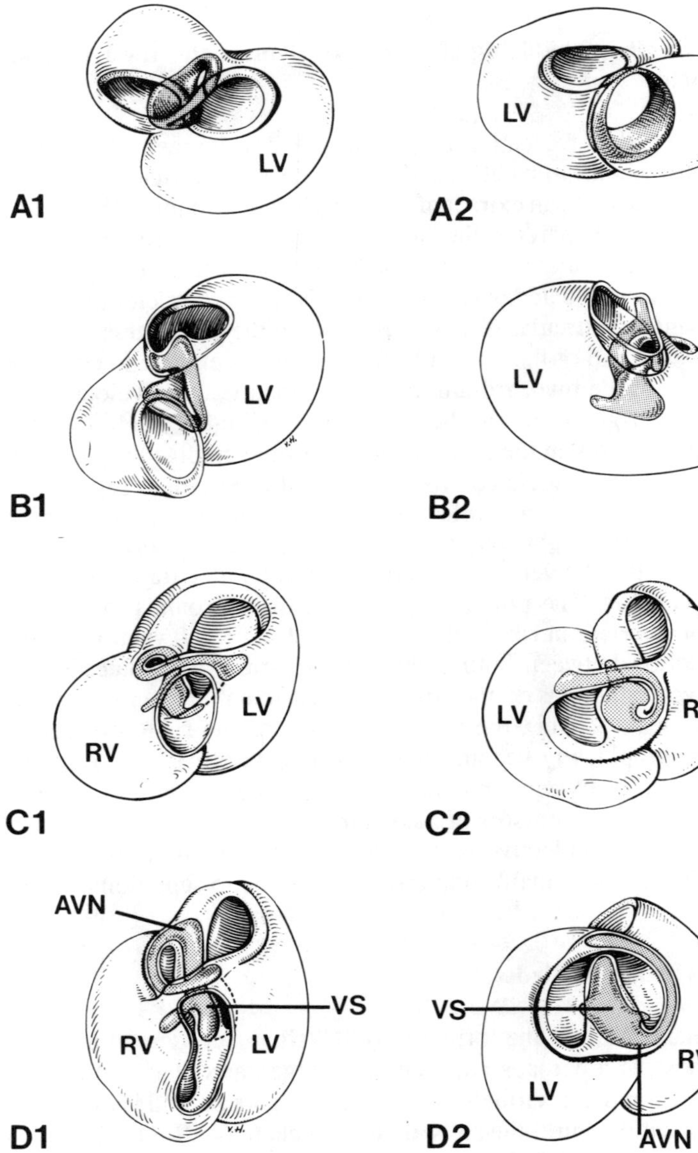

Fig. 1 The spatial distribution of GlN in human embryos at Carnegie stage XIV (A; approx. 32 days), stage XV (B, approx. 35 days), stage XVII (C, approx. 42 days) and stage XIX (D, approx. 49 days). In panel 1 of each set, an anterosuperior view and in panel 2 a posterosuperior view is shown. In these ventricular reconstructions, the atrial and outflow segments were removed and the spatial distribution of GlN is indicated. AVN, atrioventricular node; LV, left ventricle; RV, right ventricle; VS ventricular septum, ▒ 'primary' ring.

certainly in line with classical comparative studies on this subject (12) and suggests that the position of this ring in bird hearts can be used to analyse the possible origin of the ventricular septum in reptiles.

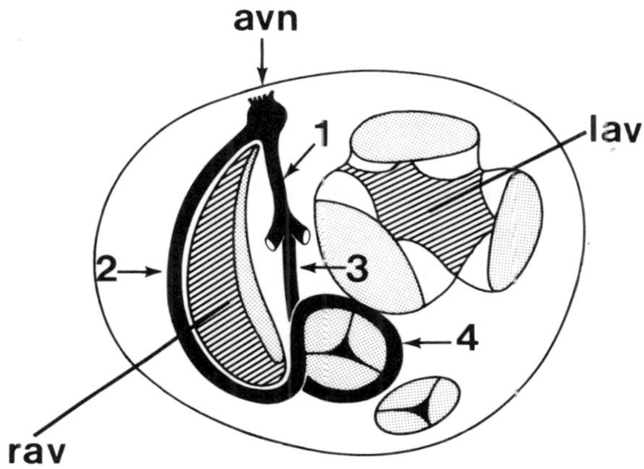

Fig. 2 The ventricular conductive system of birds as seen from above. The atria are removed. Note that the septal branch passes below the junction of the retroaortic root branch and the atrioventricular ring bundle. 1, atrioventricular bundle; 2, atrioventricular ring bundle; 3, septal branch; 4, retroaortic root branch; avn, atrioventricular node; lav, left atrioventricular communication; rav, right atrioventricular communication

Is there a 'Primary' Ring in Reptiles?

In reptiles, a (partial) septum, the 'vertical' septum, arises as a condensation of trabeculae in the posterior wall of the ventricle (13). The developmental formation of this septum is not unlike that described in the chick embryo (14). Furthermore, the presence of an atrioventricular bundle, consisting of cells with the characteristics of Purkinje cells, has been described in at least one species (15), leaving little doubt that this ventricular septum in reptiles is homologous with the posterior part of the ventricular septum of birds and mammals. Opinions diverge on the origin of its anterior part, however, in particular concerning the issue whether or not the so called 'horizontal' septum, which divides the non-crocodilian heart in a dorsal and a ventral rather than a right and a left compartment, contributes to its formation. Shaner's (16) accurate reconstructions of developing crocodilian hearts show that the muscular ventricular septum in these species develops as a single structure, at a position that is exactly predicted by the position of the primary ring, suggesting that cardiac development in this group of reptiles closely resembles that in birds, indeed. In this view an incomplete (vertical) septum in less evolved reptile hearts would merely represent a lesser degree of development of this structure, while the horizontal septum would represent a specialization of the non-crocodilian reptilian hearts. This explanation does not require any of the major shifts in position of cardiac structures that were hypothesized by Holmes (17).

Conclusion

The identification of a cell marker for the muscular ventricular septum in higher vertebrates (the 'primary' ring) has allowed to trace the origin of this septum to the reptilian class of vertebrates. Since the same cell marker also reveals the precursor cells of the ventricular conductive tissue, it is likely that a ventricular conduction system first becomes organized in reptiles (see also 18). The slow conduction velocity that is characteristic of other junctional zones, may not be a characteristic of the cells of the primary ring.

Acknowledgements

Human embryos were collected after legal abortions at the University of Amsterdam and at the Postgraduate Medical Of Budapest. The studies were approved by the local medical-ethical committees.

References

1. Arguello, C., Alanis, J., Pantoja, O., Valenzuela, B., 1986, *J. Mol. Cell. Cardiol.*, **18**, 499–510.
2. De Jong, F., Op't Hof, T., Wilde, A. A. M., Janse, M. J., Charles, R., Lamers, W. H., Moorman, A. F. M., 1990, submitted.
3. Hirota, H., Sakai, T. M., Fujii, S., Kamino, K., 1983, *Dev. Biol.*, **99**, 517–523.
4. Vassall-Adams, P. R., 1982, *Anat.* **134**, 164–183.
5. Alanis, J., Benitez, D., Lopez, E., Martinex-Palomo, A., 1973, *Jap. J. Physiol.*, **23**, 149–164.
6. Irisawa, H., Hama, K., Irisawa, A., 1965, *Circ. Res.*, **17**, 1–10.
7. Kanno, T., 1963, *Jap. J. Physiol.* **13**, 97–111.
8. Wessels, A., Vermeulen, J. L. M., Virágh, Sz., Kálmán, F., Lamers, W. H., Moorman, A. F. M., 1990, *Anat. Rec.*, in press.
9. Davies, F., 1930, *J. Anat.*, **64**, 129–146.
10. De Groot, I. J. M., Hardy, G. P. M. A., Sanders, E., Los, J. A., Moorman, A. F. M., 1985, *Anat. Embryol.*, **172**, 239–245.
11. Szabo, E., Viragh, Sz., Challice, C. E., 1986, *Anat. Rec.*, **215**, 1–9.
12. Benninghoff, A., 1933. In: L. Bolk, E. Göppert, E. Kallius & W. Lubosch, (eds), *Handbuch der vergleichende Anatomie der Wirbeltiere*, pp. 467–555, Urban & Schwarzenberg.
13. Van Mierop, L. H. S., Kutsche, L. M., 1984. In: Nora, J. J., Takao, A. (eds), *Congenital Heart Disease: Causes and Processes*, pp. 459–479, Futura, Mount Kisco, New York.
14. Harh, J. Y., Paul, M. H., 1975, *J. Embryol. Exp. Morph.*, **33**, 13–28.

15. Prakash, R., 1960, *Anat. Rec.*, **136**, 469–475.
16. Shaner, R. F., 1962, *Anat. Rec.*, **142**, 519–529.
17. Holmes, E. B., 1975, *J. Morphol*, **147**, 209–228.
18. Robb, J. S., 1965, *Comparative Basic Cardiology*. Grune & Stratton, New York.

15
Muscle Regeneration after Chronic Lidocaine Application in Innervated and Denervated Rat Soleus Muscle

D. DANIELI-BETTO, A. MEGIGHIAN, C. CATANI*, M. MIDRIO AND U. CARRARO*

Institute of Human Physiology, Via Marzolo 3, Padova, Italy
**Institute of General Pathology, Via Trieste 75, University of Padova, I-35100 Padova, Italy*

Introduction

Several studies have been devoted to the analysis of muscle regeneration after acute local application of an anaesthetic as Bupivacaine or Lidocaine (1–5). This experimental model, after the initial fibre degeneration, enables the observation of the development of new fibres, their morphological, histochemical and biochemical changes and the influence of many factors, mainly reinnervation of regenerated myofibres. In our study the muscle regeneration was analysed in innervated and denervated rat soleus muscle exposed for a long time to the effect of Lidocaine studying if any difference with the analogous acute experiment exists.

Materials and Methods

Experiments were carried out on 30 adult male Wistar rates. Under either anaesthesia in both forelimbs of the rates, Lidocaine mixed (20% w/w) with unpolymerized sylastic (MDX4-4210 Down-Corning) was applied on the posterior surface of soleus muscle through a little cutaneous incision. Thereafter,

Muscle and Motility Vol. 2. Proceedings of XIXth European Conference in Brussels
© Intercept Ltd, PO Box 716, Andover, Hampshire, SP10 1YG, UK

one forelimb was immediately denervated by cutting the sciatic nerve at the level of the trochanter.

At 1-, 2-, 4-, 7-, 14-, 21,- and 28-days after drug application, rats were killed and the two solei were rapidly removed and frozen for histochemical and biochemical analyses. Histochemical analysis was carried out on serial cross-sections stained for myofibrillar ATPase after acid (pH 4.35 and 4.6) or alkaline pre-incubation, using the Padikula and Hermann method as modified by Guth and Samaha (6, 7).

Electrophoretic study of heavy chains of myosin from longitudinal cryostat sections, cut from the degenerating–regenerating layer of the muscle, was carried out on 6% SDS polyactylamide gel as described in detail in (8) and (9).

Results and Discussion

Histochemical analysis

At one day after drug application in both innervated and denervated muscles, a layer of necrotic cells appeared. Throughout the experiment the necrotic aspects as well as the following regenerative events are restricted both in innervated and denervated muscles to the superficial layer of the muscle underlying the sylastic patch (Fig. 1A, B). The necrosis of the myofibers is well documented by the negative ATPase staining (results not shown). Up to four days after drug application both in innervated and denervated muscle the removal of necrotic cells by inflammatory reaction prevails.

At seven days after drug application in innervated and denervated muscles, myotubes or small round new fibres with a positive uniform staining for alkaline ATPase began to appear within the necrotic region (Fig. 1C, D). Two weeks after drug application in innervated muscle the new regenerating fibres are larger than after seven days, angulated and different in size, shape and ATPase staining. This evidence of re-innervation appears clearly when the regenerating cells in innervated limbs are compared with those of the denervated muscle. These latter, in fact, are smaller and very uniform in size, shape and staining for ATPase (Fig. 1E, F; note that these panels show a higher magnification). Twenty-one days after drug application in innervated muscle, the regenerating cells are quite similar in size and shape to the fibres of the undamaged region of the muscle. Moreover their differentiation in ATPase staining appears improved (Fig. 1G). On the other hand, the regenerating fibres remain smaller than the regenerating innervated ones and also than the already atrophic fibres in the undamaged part of the denervated soleus. Histochemical analysis of the growing denervated myofibres showed an uniform fast-like staining pattern (Fig. 1H). Twenty-eight days after drug application in innervated muscle the regeneration was fully developed and the region of new myofibres could be only detected by the presence of superficial reactive connective tissue formation due to the sylastic patch. In denervated muscle the regenerating fibres are even smaller than the very atrophic fibres of the undamaged region of the denervated soleus. They maintain a fast-like ATPase staining pattern (Fig. 1I).

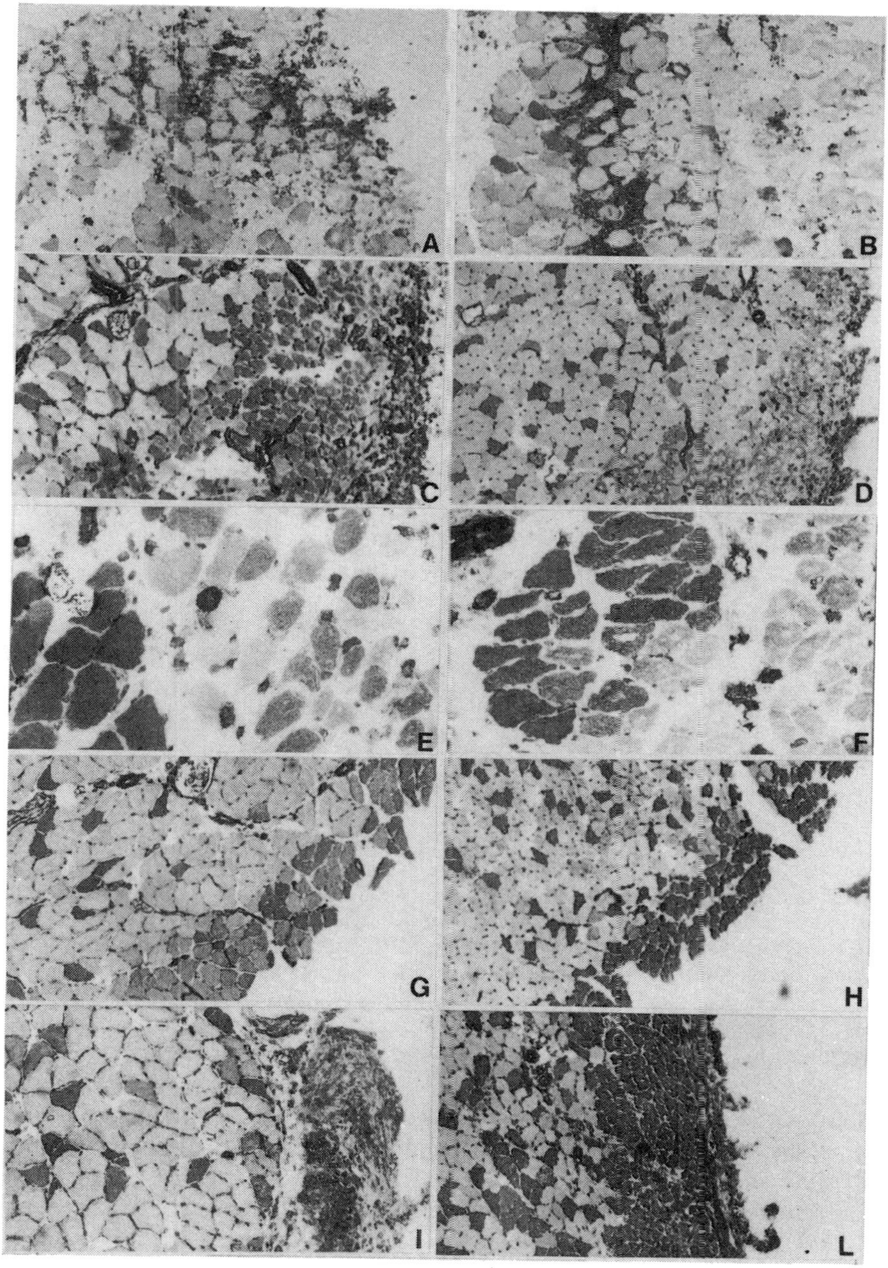

Fig. 1 Cross-sections of long-term Lidocaine treated soleus muscle. (A, C, E, G, I) Contralateral innervated muscles 1, 4, 7, 14, 21 days after Lidocaine injury. (B, D, F, H, L) Denervated muscles 1, 4, 7, 14, 21 days after denervation and Lidocaine injury. (A, B, C, D, G, H, I, L) Alkaline ATPase staining, (E, F) Acid ATPase staining.

BIOCHEMICAL ANALYSIS

Electrophoresis of myosin heavy chains (MHC) with SDS 6% polyacrylamide gel in innervated and deverated regenerating layers, revealed up to day four the presence of an almost normal soleus pattern which presents a large prevalence of MHC 1 (Fig. 2, lanes A, B, C and D). Embryonic myosin does not appear as a discrete band since its content is under the detection level of the Coomassie blue staining. The MHC pattern confirms that the peculiar ATPase staining of myofibres underneath the sylastic patch is due to their necrosis. At day 14 the MHC pattern of the regenerating layer is dramatically changed: MHC1 is decreased to about half and is substituted by a band running at the MHC2A position and by a new band in the region of MHC2B. On naked-eye inspection the MHC2B band is really a double band, one of which is embryonic/neonatal MHC. At day 22 of treatment, while the innervated sample shows a pattern very similar to that of day 14, in the denervated sample MHC1 almost disappears substituted by MHC2A and, as a minor component by a MHC 2B band (really a double band). Since MHC2B, MHC emb and MHC neo almost comigrate, the isomyosin pattern of regenerating soleus at different times needs further study.

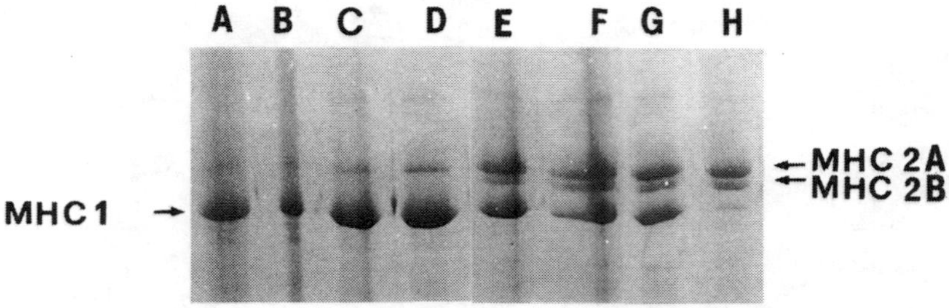

Fig. 2 SDS 6% PAGE of myosin heavy chains of long-term Lidocaine treated soleus muscle. (A, C, E, G) Contralateral muscles 1, 4, 14, 22 days after Lidocaine injury. (B, D, F, H) Denervated muscles 1, 4, 14, 22 days after denervation and Lidocaine injury. MHC1, myosin heavy chain of slow type fibres; MHC2A, myosin heavy chain of fast-oxidative fibres; MHC2B, myosin heavy chain of fast-glycolytic fibres.

Conclusion

The morphological aspects of muscle regeneration after long-term Lidocaine application appear quite similar to the analogous acute experiments with Lidocaine or Bupivacaine, even if the time of exposition to the drug was considerably longer. However, minor differences in respect to acute experiments may be detected during the process of regeneration in reinnervated fibres which seem to show a delay in the sequential expression of adult MHC genes. This could be the consequence of the more invasive surgical approach but, since Lidocaine-mixed sylastic remains soft for the first 10 days, it seems likely that a purely mechanical interference in regneration is reduced to the minimum.

Thus, Lidocaine in itself seems to interfere in some way with the differentiation of regenerating innervated fibres even if further experiments are necessary to confirm this point.

For the unambiguous interpretation of these experiments, it remains to be verified if the local concentration of Lidocaine is high enough to block nervous activity between experimental days 10 and 14, i.e. when differentiation of innervated regenerating fibres begins to appear.

References

1. Benoit, P., W., Duane Belt, W., 1970, *J. Anat.*, **107**, 547–556.
2. Hall-Craggs, E. C. B., 1974, *Exp. Neurol.*, **43**, 349–358.
3. Hall-Craggs, E. C. B., Singh Seyan, H., 1975, *Exp. Neurol.*, **46**, 345–354.
4. Mussini, I., Favaro, G., Carraro, U, 1987, *J. Neuropath. Exp. Neurol.*, **46**, 315–331.
5. Smith, E. R., Feagans, W. M., Belt, W. D., Blair M. R., 1964, *I.A.D.R. 47th General Meeting*, Houston.
6. Guth, L., Samaha, F. S., 1969, *Exp. Neurol.*, **76**, 331–346.
7. Padikula, H. A., Hermann, E., 1955, *J. Histochem. Cytochem.*, **3**, 161–165.
8. Carraro, U., Catani, C., Degani, A., Rizzi, C., 1990. In: D. Pette (ed.) *Dynamic State of Muscle Fibers*, pp. 247–262. W. de Gruyter, Berlin.
9. Danieli-Betto, D., Zerbato, E., Betto, R., 1986, *Biochem. Biophys. Res. Commun.*, **138**, 981–987.

16
Rat Intrafusal Fibres Regenerated after Autotransplantation of the EDL Muscle do not Express their Typical Myosin Heavy Chain Isoenzymes in the Absence of Sensory Innervation

T. SOUKUP*, F. PEDROSA** AND L.-E. THORNELL**

Institute of Physiology, Czechoslovak Academy of Sciences, 142 20 Prague, Czechoslovakia
***Department of Anatomy, University of Umea, 5-901 87 Umea, Sweden*

Introduction

Intrafusal fibres of muscle spindles have characteristic physiological, morphological and histochemical properties (for review see 1, 2) and express specific types of myosin heavy chains (MHC) (3–11). We have studied the MHC composition of regenerated intrafusal fibres in autotransplanted rat extensor digitorum longus (EDL) muscle. Free grafting (12, 13, 14) disrupts the neurovascular supply to the muscle, which results in degeneration and subsequent regeneration of both extra- and intrafusal muscle fibres. Fibre regeneration starts from satellite cells located under the basal lamina of original muscle fibre. In muscle spindles, however, the satellite cells are derived from three different types of intrafusal fibres each with distinct immunocytochemical properties. The question arises whether the muscle spindle satellite cells are preprogrammed to differentiate into intrafusal fibres with distinct MHC com-

Muscle and Motility Vol. 2. Proceedings of XIXth European Conference in Brussels
© Intercept Ltd, PO Box 716, Andover, Hampshire, SP10 1YG, UK

position, or whether the presence of the sensory innervation is necessary for the differentiation of the specific MHC pattern.

Methods

SURGICAL PROCEDURE

Under Nembutal (sodium pentobarbital 40 mg/kg i.p.) anaesthesia, the right EDL muscle of four 2-month-old female rats (Wistar I pcv SPF) was excised, its nerve and blood supply were cut and the muscle was sutured back to its original place (15).

TISSUE PROCESSING

One to nine months after surgery, both control and operated EDL muscles were excised, stretched to their resting length and frozen in propane chilled in liquid nitrogen. Serial cryostat sections, 10 μm thick, were either stained with haematoxylin and eosin, processed for myofibrillar ATPase activity (16), or processed for immunocytochemistry.

ANTIBODIES AND LABELLING

We used mouse IgGs monoclonal antibodies (mAb) against slow-tonic MHC (mAb ALD 19 and mAb ALD 58), slow-twitch MHC (mAb 9812), fast-twitch MHC (mAb F2 and F39), neonatal and fast MHC (mAb MF30), human alpha- and beta-cardiac MHC (Sera Lab., England) and a rabbit polyclonal antibody specific for neonatal MHC (NN5) (for detailed description of mAbs see 5, 8, 11). MAbs against synaptophysin (Progen, Heidelberg, FRG), neurofilament 68 kD subunit Dako-NF (Dakopatts, Copenhagen, Denmark) and laminin (E-4 Labs Inc., San Mateo, CA, USA) were used to detect nerve endings, neurones and extracellular basement membrane components, respectively. A standard indirect peroxidase-anti-peroxidase (PAP, Dakopatts) technique was used for the immunolabelling. Sections were examined and photographed using an Olympus Vanox microscope.

Results

IMMUNOHISTOCHEMICAL CHARACTERISTICS

Twenty-eight regenerated spindles of four grafted muscles were studied. The regenerated intrafusal fibres did not exhibit the typical ATPase reactions seen in control bag_1, bag_2 and chain fibres (Fig. 3a,b). In 23 regenerated spindles, all intrafusal fibres were of the same histochemical and immunocytochemical type exhibiting high alkali- and low acid-stable ATPase activity (Fig. 1a,b), they were stained with anti-'fast' (Fig. 1c),but unstained with anti-slow tonic (Fig. 1d), anti-α (Fig. 1e) and anti-β-cardiac (Fig. 1f) and with anti-slow twitch MHC

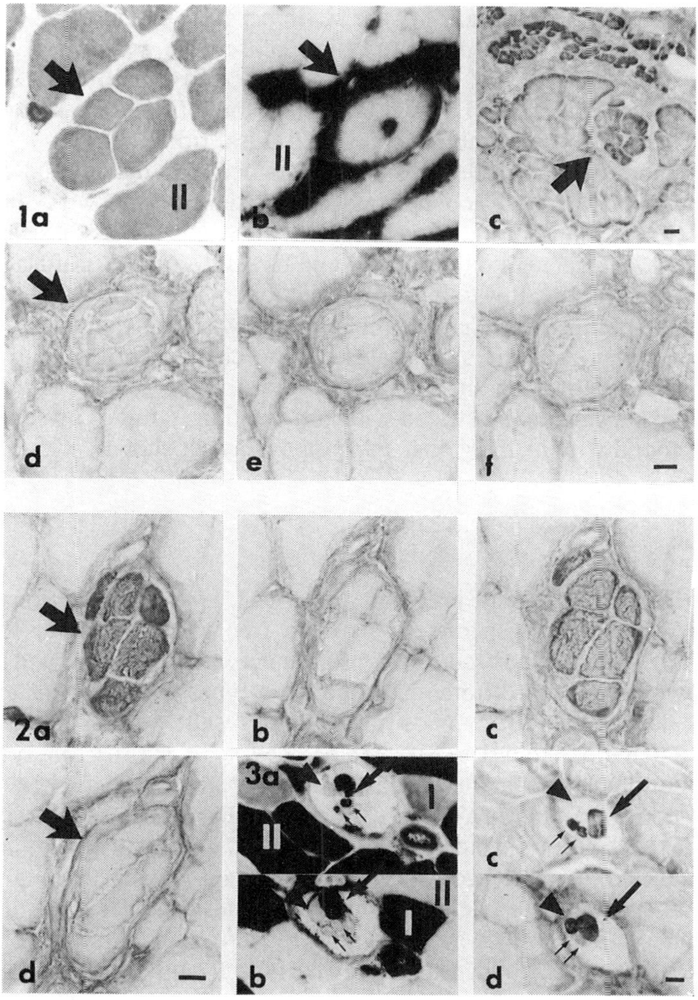

Fig. 1a–f Serial sections of a regenerated muscle spindle (thick arrow) in the EDL muscle of a 10-month-old rat, 8 months after autotransplantation. Sections were stained for mATPase at pH 10.3 (a) and 4.3 (b) at the opposite polar ends and with mAbs MF 30 (C)7 ALD 19 (d), anti-α cardiac MHC (e) and anti-β cardiac MHC (f) in the central part of the spindle. Note the variable number of intrafusal fibres, all exhibiting similar ATPase activity, reversible after acid and alkaline preincubations, corresponding to type II fibres and immunostaining only with mAb MF 30. a–b, d–f ×400; c ×200; Bars indicate 10 um.
Fig. 2a–d Serial sections of a regenerated spindle from the same rat as in Fig. 1. Sections were stained with mAbs 9812 (a), anti-α cardiac MHC (b), anti-β cardiac MHC (c) and ALD 58 (d). Note that all five intrafusal fibres exhibit ATPase activity and immunostaining corresponding to type I fibres. a–d ×320; Bar indicates 10 μm.
Fig. 3a–d Serial sections of a normal spindle from the control leg of the same rat as in Fig. 1. Sections were stained for mATPase at pH 10.3 (a), 4.3 (b) and with mAbs MF 30 (c) and ALD 19 (d). Note the fibre type differences between nuclear bag$_2$ (large arrow), nuclear bag$_1$ (arrow head) and nuclear chain (small arrow) fibres. a–d ×270; Bar indicates 10 μm.

I and II in figs1–3 indicate extrafusal type I and type II fibres.

mAbs, thus resembling extrafusal type II fibres. In five regenerated spindles, one to all intrafusal fibres displayed high acid-stable and low alkali-stable ATPase activity and staining with mAbs against slow-twitch (Fig. 2a) and β-cardiac MHCs (Fig. 2c),but not with anti-slow tonic (Fig. 2d) and anti-α-cardiac MHC mAbs, thus resembling type I extrafusal fibres.

The regenerated intrafusal fibres were unstained with the antibodies against slow tonic, alpha-cardiac and neonatal MHC, in contrast with the control intrafusal fibres (Fig. 3d), which were specifically stained with these antibodies (cf. 11, 12, 17). Regenerated intrafusal and newly formed extrafusal fibres in muscle grafts exhibited low to moderate reaction to MF 30 (Fig. lc), whereas in control adult spindles, extrafusal and bag_1 fibres were not stained (Fig. 3c).

Anti-neurofilament staining revealed the presence of nerve axons in regenerated spindles. However, in the spindle central zone (characterized by the multilayered capsule demonstrated with anti-laminin staining) no sensory endings were found with mAb against synaptophysin (not shown).

Discussion

Our experiments corroborate previous studies (15, 17–19) and demonstrate that intrafusal fibres can regenerate in standard free grafts, contrary to minced muscles (20), even in the absence of sensory innervation. Regenerated intrafusal fibres, however, do not differentiate into fibre types and they fail to express their typical original MHC composition. Similar results were obtained in nerve-intact grafts, in those spindles that remained devoid of sensory innervation (21, 22). On the other hand, regenerated intrafusal fibres innervated by afferents exhibited non-reversible mATPase (22) and the bag fibres re-expressed slow tonic MHC (21). In our standard grafts we found no sensory nerve terminals, either in the electron microscope (19) or after synaptophysin staining. The lack of periaxial space and of nuclear accumulations also indicates the absence of sensory innervation in regenerated spindles. The finding that most regenerated intrafusal fibres corresponded to extrafusal type II and some to type I fibres suggests that they are innervated by collaterals of extrafusal motor axons. Furthermore, the existence of regenerated intrafusal fibres of the same type (I or II) within individual spindles suggests clustering of fibre types typical for reinnervation, resulting from sprouting of terminal motor axons innervating groups of adjacent fibres (23). In regenerated spindles devoid of afferent innervation, the skeletomotor axons may play a decisive role determining the mATPase activity of regenerated intrafusal fibres, as it is the case for extrafusal fibres. We conclude, in agreement with previous reports (17, 19, 21, 22), that intrafusal fibres regenerated in the absence of sensory innervation, but innervated by motor terminals, do not develop their typical morphological, histochemical and immunocytochemical characteristics and they resemble extrafusal fibres. Thus, it seems that the intrafusal satellite cells are not genetically pre-programmed to differentiate into specific intrafusal fibre types independently of sensory innervation. However, it is still not clear to what extent the presence of extrafusal motor terminals could influence the MHC differentiation of intrafusal satellite cells.

Acknowledgements

We are grateful to Drs G. S. Butler-Browne, INSERM, Paris, France, G. K. Dhoot, University of London, England, and D. A. Fischman, Cornell University, New York, USA, for providing the antibodies, and to Mrs M. Krupkova and Ms A.-K. Nordlund for excellent technical assistance.

This work was supported by grants from the Swedish Medical Research Council (12X-3934), K. O. Hanssons Fund and the Medical Faculty of the University of Umeå, Umeå, Sweden.

References

1. Barker, D., Banks, R. W., 1986. In: Engel, A. G., Banker, B. Q. (eds) *Myology*, pp. 309–341. McGraw-Hill, New York.
2. Boyd, I. A., 1980, *Trends Neurosci.*, 3, 258-265.
3. Kucera, J., Walro, J. M., 1988, *Histochemistry*, **90**, 151–160.
4. Maier, A., Gambke, B., Pette, D., 1988, *Histochemistry*, **88**, 267–271.
5. Pedrosa, F., Butler-Browne, G. S., Dhoot, G. K., Fischman, D. A., Thornell, L.-E., 1989, *Histochemistry*, **92**, 185–194.
6. Pedrosa, F., Soukup, T., Thornell, L.-E., 1990, In: *Motorneuronal Plasticity*, p. 82. Bonn, Satellite Symposium of the VII International Congress of Muscular Diseases, Munchen.
7. Rowlerson, A., 1988. In: Hník, P., Soukup, T., Vejsada, R., Zelená, J. (eds), *Mechanoreceptors: Development, Structure and Function*, pp. 45–50. Plenum Press, New York.
8 Soukup, T., Pedrosa, F., Thornell, L.-E., 1990, *Histochemistry*, **94**, 245–256.
9. te Kronnie, G., Donselaar, Y., Soukup, T., van Raamsdonk, W., 1981, *Histochemistry*, **73**, 65–74.
10. te Kronnie, G., Donselaar, Y., Soukup, T., Zelena, J., 1982, *Histochemistry*, **74**, 355–366.
11. Thornell, L.-E., Eriksson, P.-O., Fischman, D. A., Grove, B. K., Butler-Browne, G. S., Virtanen, I., 1988. In: Hník, P., Soukup, T., Vejsada, R., Zelená, J. (eds), *Mechanoreceptors: Development, Structure and Function*, pp.39–44. Plenum Press, New York.
12. Carlson, B. M., 1983, In: Seil F. D. (ed.), *Nerve, Organ and Tissue Regeneration: Research Perspectives*, pp. 431–454. Academic Press, London and New York.
13. Maréchal, G ., 1986, *Biomed . Biochim . Acta*, **45**, 125–130.
14. Studitsky, A. N., 1977, *Transplantatsiya Myshts u Zhivotnykh*, Meditsina Publishers, Moscow. English translation: Carlson, B. M., 1988, *Transplantation of Muscles in Animals*, pp. 1–211 Oxonian Press, New Delhi.
15. Carlson, B. M., Gutmann, E., 1975, *Anat. Rec.*, **183**, 47–62.

16. Dubowitz, V., Brooke, M. H., 1973, In: *Major Problems in Neurology*, W.B. Saunders, London.
17. Rogers, S. L., 1982, *Devl. Biol.*, **94**, 265–283.
18. Rogers, S. L., Carlson, B. M., 1981, *Neuroscience*, **6**, 87–94.
19. Soukup, T., 1988. In: Hník, P., Soukup, T., Vejsada, R., Zelená, J. (eds), *Mechanoreceptors: Development, Structure and Function*, pp. 111–116. Plenum Press, New York.
20. Zelená, J., Sobotková, M., 1971, *Physiol. bohemoslov.*, **20**, 433–439.
21. Cui, F., Walro, J. M., 1989, *Neurosci. Lett.*, **97**, 75–79.
22. Walro, J. M., Kucera, J., Cui, F., Staffeld, C. G., 1989, *Histochemistry*, **92**, 1–13.
23. Morris, C. J., Raybould, J. A., 1971, *J. Neurol. Sci.*, **13**, 181–187.

17
Influence of the Small Bowel Transplantation on Structure and Function of its Muscular Layer

V.P. COULIC

P. Lumumba, Peoples' Friendship University, Moscow 117198, USSR

Introduction

The morphological and functional effect of muscle grafting has been well studied by many authors in the case of skeletal muscle (1–6, and others). At the same time, it is known that grafting interferes with organ motility in such organs as small bowel (7–10). However, very few direct studies of the influence of transplantation on organ and smooth muscle motility have been published. Hence, the aim of the present work is to emphasize different aspects of the influence of the small bowel transplantation not only on the morphology and function of the intestinal smooth muscle, but on the motility of the digestive tract as a whole.

Materials and Methods.

We used two models of grafting: (1) mature (adult) organ transplantation on a vascular pedicle, (2) foetal implantation.The first model was used for the investigation of the peculiarities of the transplanted bowel and their pathogenesis. With the second model we tried to have a glance at regeneration processes in a growing graft in comparison with normal myogenesis.

Muscle and Motility Vol. 2. Proceedings of XIXth European Conference in Brussels
© Intercept Ltd, PO Box 716, Andover, Hampshire, SP10 1YG, UK

Allo- and autogenic grafts of the whole jejuno-ileum or of a 20 cm intestinal segment on a vascular pedicle were undertaken on 211 mongrel dogs (male and female, 10–18 kg) orthotopically, heterotopically in the pelvis and the neck using described techniques (8, 11). Surgical revascularization was immediate (ischaemia always less than 60 min). In 30 cases surgical reinnervation of the graft was performed by para- and epineural suturing opf the visible trunks and ganglions of the mesenteric nervous plexus of the graft with those of mesenteric or lower hypogastric plexus or a branch of nervus vagus (depending on the graft site) of the recipient. In 183 experiments the influence of 50 min ischaemia, neuroreflex isolation (NRI), heterotopic position of the small bowel was investigated.

Implantation of 1 cm segment of foetal bowel (age; 15–20 day) was performed under the ear skin in rats 'August' and 'Fischer', in mice 'Balb' and 'C57B1'.

The bowel smooth muscle function was investigated by balloon kymography in a modification enabling the evacuation speed to be measured in dogs, and in mice and rats by radioscopy; by means of biomicroscopy with a trans-illumination technique. In both groups of animals, histological and histochemical study, some times electronic microscopy were done (coloration of the slices by haematoxylin eosin, Van Gieson, toluidin; schiff reaction; determination of acid and alcaline phosphatases by Gomori, succinate dehydrogenase by Nachlas; Ag impregnation by Gross-Bielshowsky, coloration of neurones by Nissl; uranyl acetate and Pb citrate staining for electron microscopy).

Results

In all cases of bowel grafting in dogs, intestinal motility was affected. In allogenic transplantation the decisive factor of disturbance was the development of a rejection reaction: in parallel with the swelling and the lymphoid infiltration of the graft, a diminution of the bowel contraction amplitude was noted (3–5 days). Tonicity failing and total paralysis of the graft coincide with the beginning of its necrotization (7–9 days). In some cases of total bowel transplantation, an activation of the motility and an acceleration of the transit were to be noted at first in a highly tonic, oedematous graft. This has shown that the rejection process was not the only cause of the motility disturbance in allogenic graft. It was also noted that the myocytes were the last to die together with neurones.

In autografts, the following differences between normal and grafted intestine were noted: during the first 1–2 weeks, inhibition of the segmental rhythmic component and rough increase of tonicity especially in case of mechanical stimulation,hence a tremendous acceleration of intestinal transit, diarrhoea; after 2 weeks, progressive restoration of the rhythmic component but maintenance of the high tonic reactions and transit acceleration, poor condition of the dogs; after 1.5–2 months, reactive modification of the gastric and large bowel evacuatory activity as well as alimentary behaviour of the dogs (slowing of the stomach emptying by evacuation of small sized boli only when the preceding had reached the large bowel, coprophagy of the badly digested stools), improvement of the animal condition which, however, never reached the initial one.

Histological investigations did not point to any substantial alteration of the intestinal smooth muscle morphology or biochemistry. But some alterations of intramural, especially myenteral neurones were observed: swelling, excentric nuclei, fragmentation of Nissl substance and degeneration of some axons. During the whole period of observation; rectoenteral reflex was absent in the grafted bowel and its sensitiveness to humoral agents (metamisyl) was significantly increased.

Studies of the intestinal motility after ischaemia, NRI and heterotopical site for the graft have shown that ischaemia is responsible for motility inhibition during the first 2 weeks p.o. and for some neuronal alterations. Its influence is entirely reversible within 3–4 weeks and never leads to transit acceleration. NRI, though not altering the smooth muscle itself, causes severe disturbances of the bowel motility that have a great similitude with those observed in the grafted bowel 2 weeks p.o. and later; the same may be said about histological pictures. In order to control the hypothesis of the 'nervous' origin of the observed changes of the intestinal graft motility, surgical reinnervation was performed as described. It was proved that when the NRI or grafted intestine is situated in the peritoneal cavity and included in the alimentary tract, normalization of motility and bolus evacuation may be reached within 6–I2 months p.o.(depending on the ortho- or heterotopic localization). (Fig. 1). The reestablishment of rectoenteral reflex and normal sensitivity to metamisyl proved the reinnervation. In case of subcutaneous situation of the graft and absence of food transit through it, a progressive atrophy of the intestinal wall and sclerosis of the muscular sheet were noted, beginning at the 2–3d months independently of reinnervation (which was not successful). Surgical reinnervation of an intestinal allogeneic graft is promissing where immunodepressive therapy is successful, as proved in several of our observations.

The results of foetal bowel implantation are the following. In the case of allogenic implantation, foetal bowel maintains its structure but develops a typical rejection reaction with lymphoid infiltration and necrosis. In the case of syngenic implantation or successful immunotherapy of the rejection process in allogenic one, during the first days p.o. the graft completely loses its structure (Fig. 2a). Later, among the unstructured mass some epithelial, nervous and muscular cells may be seen. At the end of the first p.o. week, a lumen forms lined by epithelial cells. Later, within 10–28 days (depending on the animal strain) the muscular sheet forms (Fig. 2b) and peristalsis may be detected at the biomicroscopic examination: rhythmic contractions; tonic state, extrusion of intestinal secretions. In our experiments with subcutaneous situation of the implant, retention kysts form and peristalsis soon disappears. Sclerosis of the muscular sheet is observed and sometimes a malignant tumor (sarcoma) develops. However a normal smooth muscle layer may be observed within several months after implantation (Fig. 2c). Nevertheless most myocytes are not quite mature (Fig. 2d).

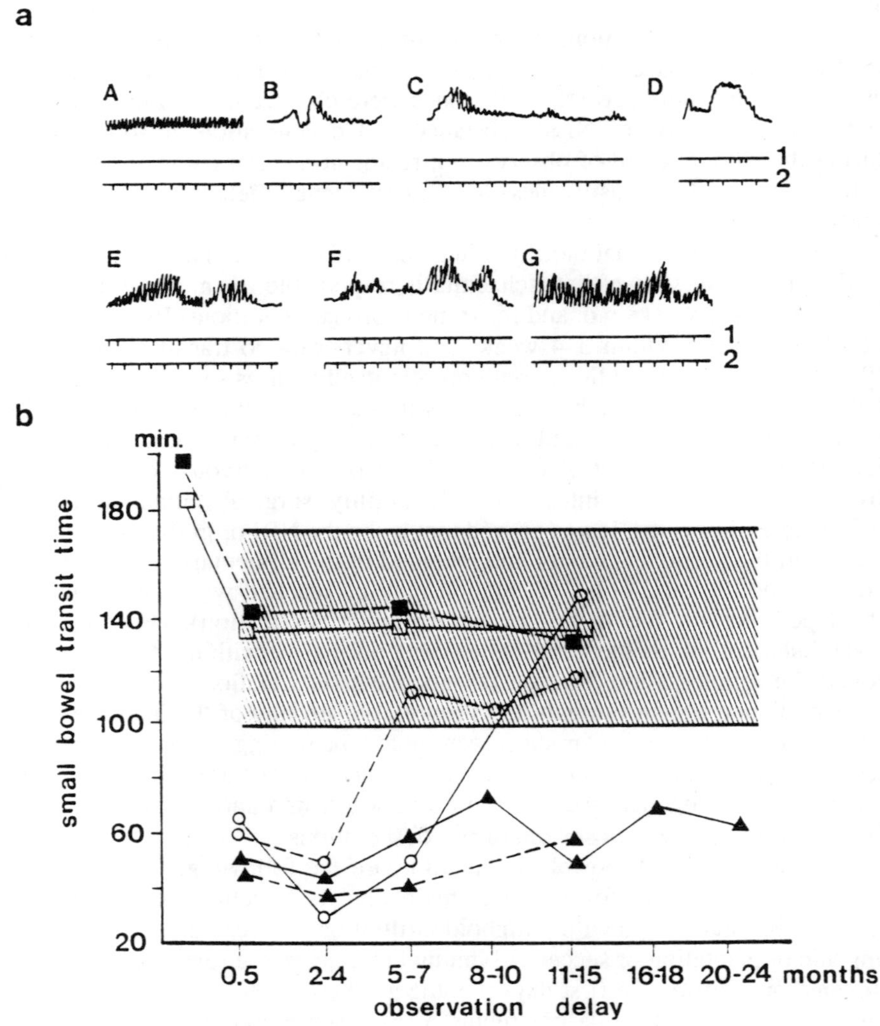

Fig. 1 Motility characteristics of the autotransplanted adult small bowel

(a) Balloon kymographic record of motility:
A, at the end of operation (general anaesthetic effect); B, at the 3d day p.o.(inhibition of the rhythmic component); C, at the 10th day p.o. (beginning of the rhythmic component restoration); D, at the 10th day p.o. (high tonic response to mechanical stimulation); E, at 28 days p.o. (restoration of the rhythmic component but hypertonicity); F, at 6 months p.o. (high tonic response to mechanical stimulation); G, within 2 years p.o. (some hypertonicity remains) 1 record of the evacuation function (I = 1 cm) 2 timing (1 min.)

(b) Evolution of the bolus transit speed through the small bowel after different operations (radiological data) ▨ control; ■— 50 min ischaemia; ◆— sham operation; ▲— autotransplantation; ▲- - - neuroreflex isolation (NRI); ●— autotransplantation with surgical reinnervation; ○— NRI with surgical reinnervation.

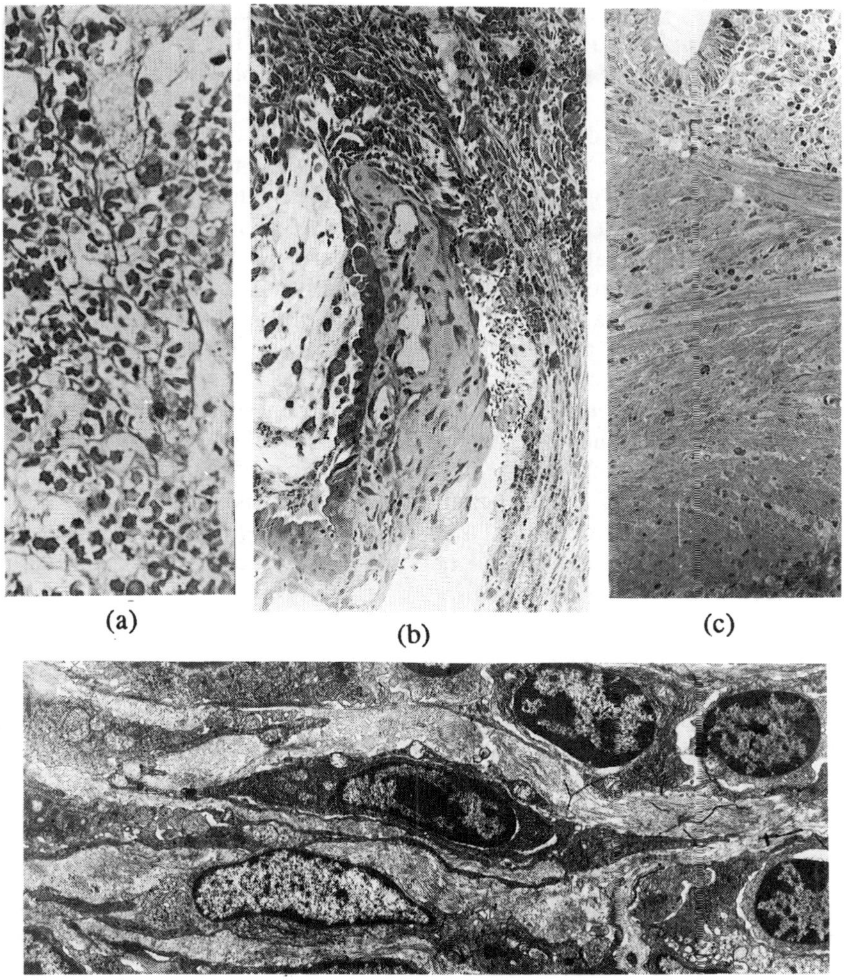

Fig. 2 Morphological characteristics of the muscular layer in foetal small bowel syngenic implantation under the ear skin in rat.
(a) 4 days, complete desintegration of the grafted bowel (G ×64);
(b) 11 days, beginning of reorqanization of the grafted bowel (×25);
(c) 3 months, two layers muscular sheet of the graft ((G ×16);
(d) 3 months, myocytes in the lamina propria mucosae (G ×5000);
(a–c) toluidin blue stains; d: uranyl acetate and Pb citrate.

Discussion

The analysis of obtained data shows that most of the physiological disturbances observed after small bowel transplantation especially at long term, such as diarrhoea, malabsorption, weight loss (7, 10, 12 and others) are mainly a motility problem, as far as secretion increases and the enzymatic features of epithelium remain intact and can ensure a normal digestion in vitro (13), and in spite of the maintenance of normal migration motor complexes (14, 15). Our data show and confirm that Ballinger (7), Sarr (14) and others have correctly identified NRI or extrinsic denervation as the main cause of motor disorders. Our observations also prove that the effect of NRI does not manifest itself only by a discoordination of the grafted bowel motility relative to other parts of the digestive tract, but also by a significant increase of the graft tonicity especially in response to a mechanical stimulation. As Halperin (16) has shown, tonic component of bowel motility is linked with evacuative function. In our case, the tonus increase leads to a tremendous acceleration of the bolus transit. And this causes malabsorption and diarrhoea with subsequent alteration of the physical condition. The mechanism of the tonus increase is not quite clear. It may be the result of the increased sensitivity of denervated structures to humoral factors (17, 18), such as catecholamines, particularly serotonin here, neurotransmitters, hormones (motilin). It was also confirmed that within several weeks after the operation, a compensating process starts and involves the innervated parts of the gastrointestinal tract. It may be thought that it is initiated or, at least, coordinated at the central nervous level, since even the animal feeding behaviour is concerned (manifestation of coprophagy untypical for dogs). Normalization of the grafted bowel motility and the digestive processes after reinnervation controlled by normalization of RER and metamisyl reaction (cf.16) is a complementary argument for NRI factor as a cause of motility disorders in bowel grafting. But it is also a way of treating post-transplantation disturbances, that has to be taken in account, especially by clinicians.

Quite different are the motility problems in foetal bowel implantation. In absence of biological incompatibility a total disintegration of the graft is noted that may be interpreted as 'plastic status' prior to regeneration which is possibly starting from the remaining viable cambial cells: myoblasts, neuroblasts. Our data confirm other authors' (19) in this point that observed myogenesis seems to follow normal patterns, at first. But the delayed observations have shown that complete maturation of the myocytes does not occur, even within months. More, some structural deformations may take place and lead to secondary necrobiosis and even tumor development. This may be due, at least partially, to unphysiological conditions for the foetal bowel growth, i.e. subcutaneous site and absence of food transit through it. This deserves further morphofunctional investigations of the implanted bowel, particularly at subcellular and molecular levels.

Conclusions

1. Small bowel transplantation with minimal ischaemy and out of immunological problems does not significantly alter neither morphology, nor functional pattern (capacity) *per se* of the muscular layer of adult grafts, but it severely modifies the motility regimen and causes hypertonicity, hence transit acceleration with subsequent malabsorption, diarrhoea and impairement of the organism general condition.
2. The main cause of motility disturbances in adult bowel autotransplantation is NRI (extrinsic denervation) of the graft and the means of correction of long-term alterations is the immediate surgical reinnervation of the intestinal graft.
3. In the case of allotransplantation, the myocytes are the last cells to be destroyed (with the nervous ones).
4. Foetal bowel implantation presents a model of neomyogenesis that deserves further attention and opens perspectives to tissue and organ 'culture *in vivo*'.

Acknowledgements

The author thanks A .N. Maximenkova, L. F. Makarova, L. V. Naumetz for their qualified help, I. A. Morozov and Iu. A. Lysikov for electron microscopy, P. Furugian and V. Novikov for technical assistance.

References

1. Carlson, B. M., 1972, '*The regeneration of minced muscles*' (Monografts in *Developmental Biology*, vol. 4). Karger, Basel.
2. Bertrand, C., Playhki, L., Maréchal, G., 1981. In: Freilinger, G., Holle, J., Carlson, B. M. (eds), *Muscle Transplantation*, pp. 39–45. Springer-Verlag, Berlin.
3. Studitsky, A. N., 1964, *Arkh. Anat; Histol. Embryol.*, **46**, 29–49 (in Russian).
4. Zhenevskaya, R. P., 1974, *Neurotrophic Regulation of Plastic Activity of Muscular Tissue*. Moscow (in Russian).
5. Maréchal, G., Schwartz, K., Beckers-Bleukx, G., Ghins, E., 1984, *Eur. J. Biochem.*, **138**, 421–428.
6. Faulkner, J. A., Carlson, B. M., 1985, *Muscle & Nerve*, **8**, 413–418.
7. Ballinger, W. F., Christy B. G., Ashby W. B., 1962, *Surgery*, **52**, 151–163.
8. Coulic, V., 1986, *Transplantation of Digestive Organs*. FPU, Moscow. (in Russian).
9. Lillehei, R. C., 1962, *Surg. Clin. North Amer.*, **42**, 1191-1217; 1963, *Amer. J. Surg.*, **105**, 58–72.
10. Olivier, Cl. *et al.*, 1969, *Presse méd.*, **77**, 1273-1274; *Cirurgie*, **98**, 331–348.
11. Coulic, V., Maximenkova, A. N., 1974, *Function and Structure of Transplanted Bowel*. Nauka, Moscow (in Russian).
12. Lear, P. A. *et al.*, 1987, *Transplant. Proc.*, **19**, 1123–1124.

13. Coulic, V., Maximenkova, A. N., Jesuitova, N. N., Timofeeva, N. M., 1986. In: *Membrane Digestion and Absorption*, pp. 74–75. Zinatie, Riga. 74–75 (in Russian).
14. .Sarr, M. G. *et al.*, 1981, *Gastroenterology*, 81, 303–310.
15. Sarr, M. G. *et al.*, 1990, *S.G.O.*, 170, 338–348.
16. Halperin, Iu. M. *et al.*, 1966, *Bull. Experim. Biol. and Med.*, **9**, 30–32; *Pharmacol. a. Toxicol.*, **6**, 695–698 (in Russian).
17. Cannon, E., Rosenbluth, A., 1951, '*Increasing of the sensitiveness of denervated structures*', Moscow, (in Russian).
18. Youmans, W. B., 1949, *Nervous and Neurohumoral Regulation of Intestinal Motility.* New York.
19. Guttmann, F. M. *et al.*, 1985, *J. Ped. Surg.*, **20**, 747–753.

SECTION THREE

Muscle Plasticity

18
Adaptive Remodelling of the Contractile Apparatus of the Electrostimulated Muscle

A. JAKUBIEC-PUKA* AND U. CARRARO**

*Laboratory of Protein Metabolism, Nencki Institute of Experimental Biology, 02–093 Warszawa, ul. Pasteura 3, Poland
**CNR Unit for Muscle Biology and Physiopathology, University of Padova, Via Trieste 75, 35121 Padova, Italy

Introduction

The length–force relation in the striated muscle has been observed by several authors. There is an optimal sarcomere length at which the force amplitude is maximal (1, 2, 3). It is probably due to this phenomenon that the number of sarcomeres is highly adjustable, being capable of either decreasing or increasing when the muscle is maintained in a shortened or lengthened position (4). Electrical stimulation accelerates this process (5). The length of sarcomeres optimal for function is not clearly defined from the literature; it ranges from about 2.0 μm to 2.9 μm (6, 7). There are several hypotheses for explaining the length–force relationship (7). In the present work the length of sarcomeres was examined in striated muscle maintained in the shortened position and electrically stimulated for 4–45 h.

Methods

Three-month-old female albino rats of Wistar strain were used. Soleus muscle was maintained in shortened position by immobilization of the ankle joint in a plastic tube at an angle of 130–160 degrees. The sciatic nerve was stimulated by an implanted electrode (under phenobarbital and ether anaesthesia) for 4–45 h by pulses of 0.3 ms duration and 20 Hz frequency. Muscles from of (a)

Muscle and Motility Vol. 2. Proceedings of XIXth European Conference in Brussels
© Intercept Ltd, PO Box 716, Andover, Hampshire, SP10 1YG, UK

contralateral legs of experimental animals, (b) from non-operated animals and (c) those stimulated without joint immobilization, served as controls. A minimum of two animals was used in each experiment.

Immediately after decapitation the whole muscle was excised, maintaining the length, measured and attached to a plastic rod. Procedures of fixation, embedding in Epon and staining were performed as described earlier (8, 9).

For measurements of sarcomere length, samples were randomly selected from the central part of the muscles. Sarcomere length was measured on micrographs of the longitudinal section of muscle fibres within regular areas of well preserved contractile structure, as described in detail elsewhere (8). Several fibres from at least four randomly selected samples of each muscle were studied. All samples used for measurements were prepared under analogous conditions.

Results and Discussion

The length of the soleus muscle, electrostimulated in shortened position, was decreased by 10% to 30%. The degree of decrease was dependent on muscle shortening, i.e. on the angle of the ankle joint immobilization.

In the control non-stimulated soleus muscle the mean sarcomere length was about 2.8μm (Table 1). In muscle stimulated in the free position (without joint immobilization), the sarcomere length remained in the same range. In the muscle maintained in the shortened position and stimulated for 4–8 h, a considerable decrease and variability of sarcomere length were observed. The decrease in the mean sarcomere length was proportional to that in muscle length (Table 1). After 45 h stimulation, the length of sarcomeres returned to the range of that in controls, in spite of the considerably decreased length of the stimulated muscle (Table 1). Thus, in order to recover the normal sarcomere length, the contractile structure was rapidly reorganized. During reorganization several anomalies and irregularities appear within it; however necrotic changes are hardly seen (10).

Table 1 Length of Sarcomeres in Muscles Stimulated in shortened position.

Duration of stimulation (h)	Length of muscle % of contralateral	Length of sarcomeres (mean* ± SEM) (μm)
—	100	2.77 ± 0.035
6	100	2.81 ± 0.050
4	88	2.62 ± 0.090
6	90	2.62 ± 0.032
8	81	2.35 ± 0.041
45	68	2.77 ± 0.048

*Minimum 30 fibres from 2 muscles were studied.

These results suggest that the sarcomere length of about 2.8 μm is optimal for function in the continuously contracting muscle. Perhaps such a length of sarcomeres *in vivo* allows the maximal force of muscle to develop.

Acknowledgements

Thanks are due to Mrs.H. Chomontowska, Mr.M. Fabbri, Mr.V. Gobbo and Ms. A. Klesniak for skillful technical assistance and Dr. P. E. Williams for helpful discussions. This work was supported by funds from CPBP 04.01. programme of the Polish Academy of Sciences, from the Italian CNR to the Unit for Muscle Biology and Physiopathology, from the Italian CNR Progetto Finalizzato: Tecnologie Biomediche e Sanitarie and from the Italian Ministero della Pubblica Istruzione 60% n. 12.01.86 021 to Prof. U. Carraro.

References

1. Herzog, W., ter Keurs, H. E. D. J., 1988, *Pflügers Arch.*, **411**, 642–647.
2. Wallinga-de Jonge, W., Boom, H. B. K., Boon, K. L., Griep, P. A. M., Lammeree, G. C., 1980, *Am. J. Physiol.*, **293**, C98–C104.
3. Wohlfart, B., Grimm, A. F., Edman, K. A. P., 1977, *Acta Physiol. Scand.*, **101**, 155–164.
4. Williams, P. E., Goldspink, G., 1973, *J. Anat.*, **116**, 45–55.
5. Tabary, J.-C., Tardieu, C., Tardieu, G., Tabary, C., 1981, *Muscle Nerve*, **4**, 198–203.
6. Edman, K. A. P., Reggiani, C., 1987, *J. Physiol.*, **385**, 709–732.
7. Herring, S. W., Grimm, A. F., Grimm, B. R., 1984, *Muscle Nerve*, **7**, 161–173.
8. Jakubiec-Puka, A., 1985, *J. Muscle Res. Cell Motil.*, **6**, 385–401.
9. Jakubiec-Puka, A., Kulesza-Lipka, D., Krajewski, K., 1981, *Cell Tissue Res.*, **220**, 651–663.
10. Jakubiec-Puka, A., Kordowska, J., Belluco, S., Carraro, U., 1988, In: Carraro, U. (ed.), Sarcomeric and Non-sarcomeric Muscles: Basic and Applied Research Prospects for the 90s, pp. 385–390., Unipress Padova, Padova.

19
Conversion of Cat and Rabbit Soleus Muscle Fibres by Alien Nerves

G. K. DHOOT AND G. VRBOVÁ

The Royal Veterinary and University College, University of London, UK

Introduction

It is now well established that many characteristic properties of mammalian skeletal muscles can be modified by their neuronal inputs. The effect of neuronal influence on contractile characteristics can be readily demonstrated by experiments in which motor nerves from a predominantly fast-contracting muscle are transposed into a slow contracting one and vice versa (1). In contrast to the observations on the rabbit, the transformation of the cat soleus supplied by either FHL (Flexor hallucis longus) or the FDL (Flexor digitorum longus) nerve, however, is often incomplete (2, 3). This could be explained if soleus muscles showed some degree of preference for axons that had previously supplied slow muscle fibres in their parent muscle. Although this may be the case in the early stages of reinnervation, during later stages preferential reinnervation of soleus muscle by slow fibres does not seem to occur (4).

Nevertheless, the composition of the nerve and the proportion of axons of fast and slow motor units may influence the time course of contraction and relaxation for the cat soleus (5). Consistent with this is the finding of Buller (6) that shows a greater change of the cat soleus towards a fast type of muscle, when the nerve from the larger FHL muscle rather than when the nerve from the smaller FDL is used to replace the soleus nerve. However, study of individual motor unit properties shows that no matter whether the soleus muscle is crossinnervated by FHL or FDL nerves (3, 7), the majority of motor units is of an intermediate type. This result has often been interpreted to show

Muscle and Motility Vol. 2. Proceedings of XIXth European Conference in Brussels
© Intercept Ltd, PO Box 716, Andover, Hampshire, SP10 1YG, UK

that the cat soleus muscle fibres cannot be transformed due to inherent properties of their contractile proteins (7).

In the present study an attempt was made to see if the slow muscle fibres in the cat soleus can be converted to fast by FHL nerve in the absence of contamination from the original soleus nerve, and whether in rabbits the conversion of soleus muscle induced by FDL nerve is less complete possibly because the antagonist muscles are left intact and the muscle is being stretched.

Materials and Methods

For methods of surgical procedures, physiological recordings and histochemical and immunocytochemical procedures see references 8–11.

Results

As reported previously (10), eight months after cross-innervation of the cat soleus by the FHL nerve, 22.1% (±5.3%) of the muscle fibres stained for alkali stable myosin ATPase (Fig. 1). These muscle fibres also reacted with antibodies against fast isoforms of troponin I and myosin heavy chains. Two-thirds of these stained muscle fibres also stained with antibodies to slow isoforms of

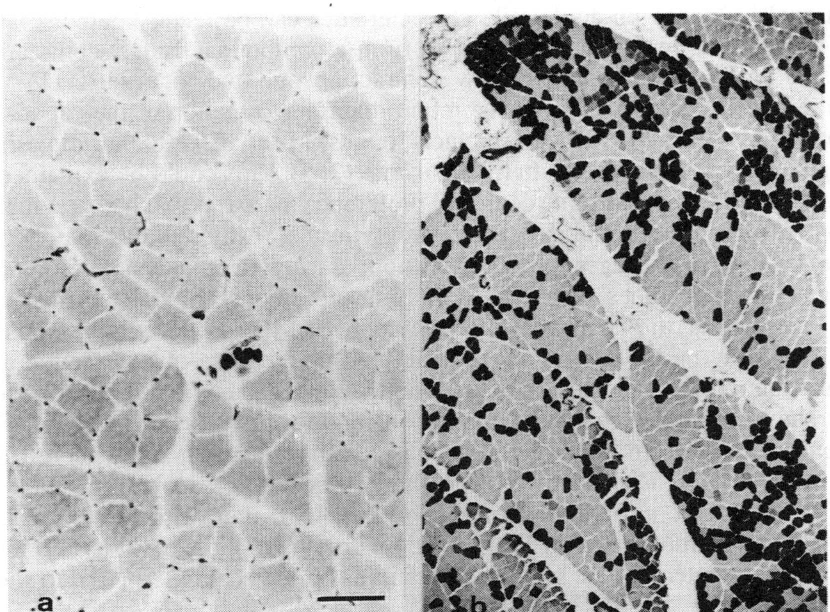

Fig. 1 Cat soleus muscle stained for alkali stable myosin ATPase (pH 9.4). (a) Control, bar = 140 μm; (b) cross-innervated with FHL nerve for eight months, bar = 36 μm.

these proteins. In the remaining one-third of the fibres that had stained with intermediate intensity for fast isoforms almost equally dark staining for slow isoforms could be detected.

For the rabbits, in one group of experiments the peroneal nerve was used to cross-innervate the soleus muscle while, in another group, the FDL nerve was used instead. In control soleus muscles only a few fibres stained for alkali stable myosin ATPase. In those experiments in which the peroneal nerve was used to crossinnervate the soleus muscle, 90% (±2.5%) of the fibres stained for alkali stable myosin ATPase and fast skeletal isoforms of troponin I and myosin heavy chains. Only 1% of the cells stained for both the fast and slow isoforms of these proteins. When the soleus muscle was cross-innervated with the FDL nerve instead of the peroneal nerve, changes in the soleus muscle could still be observed but the degree of conversion, however, was less complete (Fig. 2). Although about 85% (±3.8%) of the cells stained for fast isoforms of troponin I and myosin heavy chains in the muscles cross-innervated with the FDL nerve, 50% (±4.3%) of the cells continued to stain for both fast and slow forms of these proteins. Only 40% of the cells stained for fast isoforms alone. In contrast, slow isoforms of troponin I and myosin heavy chains were suppressed in 90% of the cells in the muscles cross-innervated with the peroneal nerve.

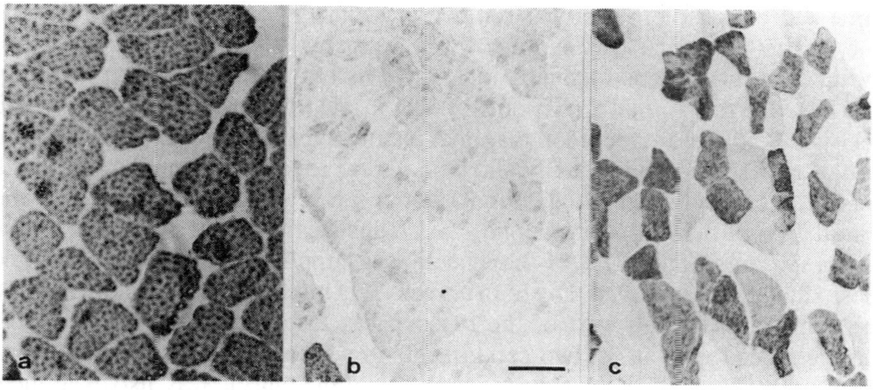

Fig. 2 Expression of slow skeletal myosin heavy chains in rabbit soleus detected by antibody 96J by immunoperoxidase procedure. (a) Control, (b) cross-innervated with peroneal nerve for 28 weeks (c) crossinnervated with FDL nerve for 28 weeks, bar = 102 µm.

In the rabbit, both nerves induced the soleus muscle to become faster contracting but the change appeared to be greater in the soleus muscles cross-innervated by the peroneal nerve (Table 1). In the three experiments shown in Table 1, soleus was reinnervated only by the FDL nerve and there was no self reinnervation by the soleus nerve. In one experiment, both nerves contributed to the innervation of the soleus muscle and in this animal the contraction elicited by the alien nerve had a time to peak of 55 ms and that elicited by the stimulation of the soleus nerve was 80 ms. The relaxation time was also faster

Table 1 Effect of cross innervation by the peroneal or FDL nerve on some characteristics of the rabbit soleus muscle.

Parameter measured	Soleus cross-innervated by peroneal nerve	Control Soleus	Soleus cross-innervated by FDL nerve	Control Soleus
Time to peak (ms)	26.3 ± 1.1 (*n* = 6)	60.5 ± 3.4 (*n* = 6)	33.0 ± 1.6 (*n* = 3)	81.0 ± 4.2 (*n* = 3)
Half relaxation time (ms)	23.0 ± 7.5 (*n* = 6)	72.0 ± 19.0 (*n* = 6)	43.0 ± 20.8 (*n* = 3)	95.0 ± 3.0 (*n* = 3)
Fatigue index	—	—	30 14	90 80

The fatigue index tested in two animals cross-innervated by FDL nerve shows the decline of tension after 3 min of repeated stimulation. Results are given as means ± SD.

when the muscle was stimulated via the alien (FDL) nerve. However, when these values of the incompletely cross-innervated muscle are compared to those of the completely cross-innervated ones, it is apparent that even the contraction elicited by the alien nerve had a slower time course than that produced by the same nerve in a soleus that was reinnervated only by a fast nerve. Since it is known that if a muscle is supplied by a superfluous number of axons some muscle fibres can be permanently innervated by two axons (12), the possibility was tested in this animal too if some of the muscle fibres were supplied by both nerves. We therefore recorded maximal tetanic tension in response to stimulation of either nerve individually and both nerves together. Stimulation of both nerves together elicited 225 g, stimulation of the soleus nerve alone 225 g and stimulation of the FDL nerve alone 150g. Thus all fibres were supplied by the soleus nerve but only 67% of the muscle was supplied by the FDL nerve and these fibres were innervated by both nerves. It is interesting that in spite of this, the contraction of this part of the muscle was faster than that of the whole muscle takėn together. In two experiments in which the FDL nerve was made to innervate the rabbit soleus the fatiguability of the muscle was also examined. The cross-innervated muscle became much more fatiguable than the unoperated control (Table 1).

Discussion

The present results show that the cross-innervation of the cat soleus muscle by the FHL nerve induced the appearance of fast isoforms of troponin I and myosin heavy chains in only 22% of the muscle fibres, and of these one-third of the cells still coexpressed the slow isoforms of these proteins. This observation is in sharp contrast to results obtained in rabbits in which the soleus muscle was reinnervated by the peroneal nerve leading to 90% of the muscle fibres being transformed to fast types. However, here not only was the species different but also the nerve used for cross-innervation. In the present study the rabbit soleus therefore was cross-innervated by the FDL nerve for comparison with the muscles cross-innervated by the peroneal nerve in the same species. Although FDL nerve induced a less complete conversion than the peroneal nerve, still 80–90% of the muscle fibres expressed fast isoforms of troponin I and myosin heavy chains. A large proportion of these cells, however, continued to coexpress the slow isoforms of the proteins investigated. It is clear therefore that although the FDL nerve in the rabbit was less effective in converting the soleus muscle to fast type, it was none the less more effective than the FHL nerve in the cat indicating that the difference in the response between the rabbit and the cat soleus may be partly due to species difference. It appears that in the cat soleus the replacement of slow isoforms of troponin I and myosin heavy chains by fast isoforms is more difficult to achieve than in the rabbit. Examples of such species differences have also been reported for electrically stimulated induced changes in rat skeletal muscles (13).

Nevertheless, when comparing the effects of the two alien nerves on the rabbit soleus, it is apparent from the present results that the peroneal nerve was much more effective in replacing the slow forms of myofibrillar proteins by fast isoforms than the FDL nerve. There are two main possibilities that could account for the different effects of the two nerves on the expression and suppression of isoforms of myofibrillar proteins in the cross-innervated muscle: (1) that the composition of the two nerves with regard to the proportion of the axons from slow motor units is higher in FDL nerve; and (2) that the mechanical situation under which soleus would be contracting in the two experimental situations is different. In the case of cross-innervation with the peroneal nerve, the muscle would not be stretched by its antagonists, for these are paralysed. It would contract in a relatively unloaded fashion, however, when FDL nerve is used the soleus will regularly be stretched by its antagonists during locomotion. If the first possibility was to be correct then there ought to be a considerable difference between the composition of the two nerves. While there are no data available for the rabbits, the results obtained for the cat FDL (3,14) and rat FDL muscle (15) indicate that this is a predominantly fast muscle and contains only a few slow motor units. Indeed, the present reults show that the rabbit soleus reinnervated by FDL nerve had 85–90% muscle fibres that contained fast isoforms of troponin I and myosin heavy chains thus indicating that they had received a fast type innervation. The contractions of the muscle were also fast, and the fatiguability of the muscle increased as expected from a slow to fast converted muscle. Nevertheless, 50% of the fibres continued to express the slow isoforms of these myofibrillar proteins, while muscles reinnervated by the peroneal nerve had only 1% of the converted fibres expressing both types of slow and fast isoforms, and 90% were completely fast.

In view of these results it is unlikely that the lack of complete conversion after reinnervation by the FDL nerve is due to the different composition of the two nerves, and the results are more easily explained by the second suggestion. According to this suggestion the soleus muscle fibres would continue to express the slow isoforms of myofibrillar proteins when cross-innervated by the FDL nerve because in this case the muscle will continue to be stretched during locomotion.

It appears from these findings that the expression of these myofibrillar proteins in the soleus is controlled not only by the activity pattern imposed on the muscle by its motoneurones but also by the mechanical conditions under which the muscle had to contract. Interestingly, other characteristics of the muscle such as its fatigue resistance or even contraction speed, do not seem to be affected to the same extent by the different mechanical conditions.

Acknowledgements

This work was supported by grants from the Medical Research Council and Muscular Dystrophy Group of Great Britain. We wish to thank A. Connold for her assistance during this study.

References

1. Buller, A., Eccles, J. C., Eccles, R. M., 1960, *J. Physiol.*, **150**, 417–439.
2. Buller, A., Mommaerts, W., Seraydarian, K., 1969, *J. Physiol.*, **205**, 581–597.
3. Gauthier, G., Burke, R., Lowey, S., Hobbs, A., 1983, *J. Cell Biol.* **97**, 756–771.
4. Ip, M. C., Vrbova, G. 1983, *Neuroscience*, **10**, 1463–1469.
5. Gillespie, M., Gordon, T., Murphy, P., 1986, *J. Physiol.*, **372**, 485–500.
6. Buller, A. J., 1983, In: Burnstock, G., O'Brien, R., Vrbová, G. (eds) *Somatic and Autonomic Nerve Muscle Interactions*, pp. 277–286. Elsevier, Amsterdam.
7. Chan, A., Edgerton, V., Goslow, G., Kurato, H., Rasmussen, S., Spector, S., 1982, *J. Physiol.*, **332**, 343–361.
8. Vrbová, G., 1963, *J. Physiol.*, **166**, 241–250.
9. Dhoot, G., Perry, S., Vrbova, G., 1981, *Exp. Neurol.*, **72**, 513–530.
10. Eldridge, L., Dhoot, G., Mommaerts, W., 1984, *Exp. Neurol.*, **83**, 328–346.
11. Dhoot, G., 1988, *J. Muscle Res. Cell Motil.*, **9**, 120–131.
12. Ostberg, A., O'Brien, R., Vrbová, G., 1986, *Neuroscience*, **18**, 205–213.
13. Kwong, W., Vrbová, G., 1986, *Pflügers Arch.*, **391**, 200–207.
14. Barany, M., 1967, *J. Gen. Physiol.*, **50** (suppl.), 197–218.
15. Armstrong, R., Laughlin, L., 1985, *J. Exp. Biol.*, **115**, 201–215.

20
Ipsi- and Contralateral Effects after Surgical Overloading of Self- and Foreign-reinnervated Muscles

D.-X. LU, D. RINDERKNECHT AND M. MÜNTENER

Institute of Anatomy, University of Zürich-Irchel, Winterthurerstrasse 190, CH-8057 Zürich, Switzerland

Introduction

It is well known that every muscle displays an individual fibre pattern responding in a plastic manner to a variety of systemic or local stimuli. However, only a decade ago, Rotshenker (1) first showed contralateral changes following unilateral denervation of the cutaneous-pectoris muscle in the frog. Later, contralateral fibre transformations have also been observed in rabbit soleus (SOL) or rat extensor digitorum longus (EDL) muscle after unilateral cross-reinnervation experiments (2–5).

Functional elimination of synergistic muscles (tenotomy or ablation) leads to compensatory hypertrophy of skeletal muscle. This phenomenon has been widely investigated, mainly on SOL and plantaris muscle (6–12). However, contralateral changes have not been studied. Instead, mostly the muscles of the operated side were compared with the corresponding ones of the unoperated contralateral limb (serving as controls).

We studied the effect of functional overloading of (1) normal muscles (response in fast versus slow muscles), (2) transformed muscles (response of

Muscle and Motility Vol. 2. Proceedings of XIXth European Conference in Brussels
© Intercept Ltd, PO Box 716, Andover, Hampshire, SP10 1YG, UK

persisting fibres versus regenerating fibres), and (3) the differential contralateral reaction with respect to the type of overloaded muscle.

Methods

SURGICAL PROCEDURES

Overload of normal muscles

In 5-month-old male rats (strain Zur:SIV) muscular enlargement of the EDL muscle was produced by ablation of the tibialis anterior muscle. Enlargement of the SOL muscle was induced by ablation of the gastrocnemius muscle.

Overload of transplanted muscles

In 4-week-old animals free muscle transplantation of EDL (T-EDL) and SOL (T-SOL) was accomplished as described by Carlson and Gutmann (13). After a consolidation time of 5 months in one group of animals the T-EDL or T-SOL muscles were surgically overloaded as described above. The remaining animals with non-overloaded T-EDL or T-SOL muscles were used for comparison.

Overload of cross-reinnervated muscles

Cross-reinnervation of SOL by the EDL nerve (X-SOL) was performed on 4-week-old animals in the manner previously described (14). After a consolidation time of 4 months in one group of animals the X-SOL muscles were surgically overloaded as described above. Again, the remaining animals with non-overloaded X-SOL muscles were used for comparison.

Controls

Age-matched untreated animals served as controls.

HISTOCHEMISTRY

One, two, and three months, respectively, after the overloading operation the experimental EDL and/or SOL and the contralateral EDL and SOL muscles were removed and frozen in melting isopentane (-160°C). Control muscles were processed in the same fashion. Cryocut cross-sections (12 μm) were reacted for *myofibrillar ATPase* according to Guth and Samaha (15). Consecutive sections (12 μm) were incubated for *cytochrome c oxidase* (E.C. 1.9.3.1.) with catalase added according to the method of Wong-Riley (16).

For *fibre typing and counting* sections stained for myofibrillar ATPase were directly enlarged (by projection) on photographic paper and all fibres counted.

IMMUNOHISTOCHEMISTRY

For the immunolocalization of *parvalbumin* (PV) (17) monoclonal antibodies (antibody 235, ref. 18) were employed. Paraffin sections (5–7 μm) were incubated for 72 h in a moist chamber at 4°C with the primary antisera diluted 1:2000 to 1:5000 in 0.1 M tris-buffered saline (TBS), pH 7.4. After 2 x 5 min washes with buffer the sections were incubated with rabbit anti-mouse IgG/PO (1:500; Bio-Makor, Rehovot). Enzyme activity was demonstrated using H_20_2 and diaminobenzidine/HCl (Polysciences) as substrates.

MORPHOMETRIC PROCEDURES

In every muscle the cross-sectional areas of 80 fibres were assessed by means of planimetry using an HP-86B microcomputer connected to a graphic tablet. Fiber diameters were calculated (for better comparability) by assuming a cylindric form of the muscle fibres.

All the values reported are means ± standard deviations (SD) which were calculated from individual values by standard procedure.

Results

The results of the three-month stages of the different experiments are summarized in Table 1. Surgical overloading of normal muscles led to a fibre transformation from the 'fast-twitch glycolytic' type IIB to the 'fast-twitch oxidative glycolytic' type IIA fibres in the EDL and from the IIA to the 'slow-twitch oxidative' type I fibres in the SOL with a corresponding increase of these fibre populations (Table 1). This process is paralleled by muscle specific changes of the fibre size. In the EDL type IIA fibres and in the SOL type I fibres increased significantly in diameter (Table 1, Fig. 1).

Table 1 Distribution and size of the different fibre types (data for IIC fibres not displayed) in normal (N), crossreinnervated (X) and transplanted (T) muscles with (O) and without overload. All animals are age-matched.

	Distribution (%) of fibre types			Diameter (μm) of fibre types		
	I	IIA	IIB	I	IIA	IIB
N-EDL	4.4 ± 1.1	44.9 ± 2.7	50.7 ± 3.4	33.0 ± 0.7	46.1 ± 4.3	66.5 ± 2.9
O-N-EDL	1.7 ± 0.9*	62.3 ± 3.0**	36.0 ± 3.4**	35.1 ± 1.3	54.4 ± 4.5**	73.2 ± 8.3
N-SOL	95.9 ± 4.9	4.0 ± 4.9	—	70.4 ± 2.8	55.7 ± 9.0	—
O-N-SOL	97.8 ± 1.8	1.4 ± 1.6	—	87.1 ± 11.5***	49.3 ± 1.6	—
X-SOL (~EDL)	22.5 ± 9.8	76.5 ± 9.7	—	54.7 ± 4.2	51.1 ± 3.2	—
O-X-SOL	22.9 ± 11.5	76.4 ± 11.6	—	56.7 ± 0.5	52.9 ± 1.4	—
T-SOL (~EDL)	21.1 ± 5.9	77.9 ± 4.9	—	42.7 ± 6.7	40.5 ± 2.8	—
O-T-SOL	19.7 ± 10.5	79.0 ± 11.2	—	45.6 ± 5.1	46.6 ± 4.2*	—
T-EDL (~SOL)	93.4 ± 4.5	6.6 ± 4.5	—	40.1 ± 2.0	35.1 ± 5.2	—
O-T-EDL	91.3 ± 5.1	8.7 ± 5.1	—	45.8 ± 2.2*	34.7 ± 3.9	—

$***P < 0.001$, $**P < 0.01$, $*P < 0.05$

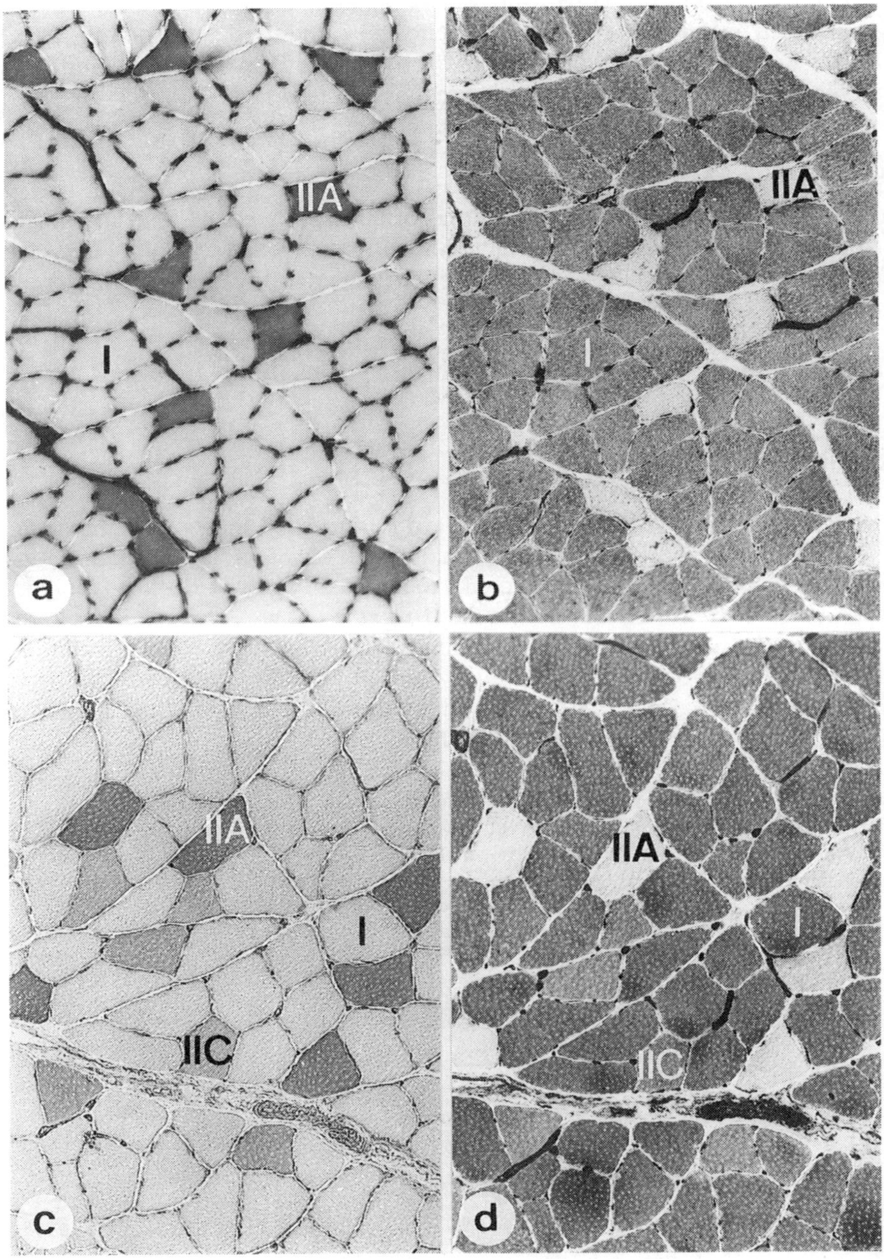

Fig. 1 Consecutive sections of normal (a, b) and overloaded (1 month; c, d) SOL muscles respectively, stained for myofibrillar ATPase after alkaline (a, c) and acid (b, d) pre-incubation. The fibre types are indicated. The transitional IIC fibres exhibit an intermediate staining intensity after alkaline and acid pre-incubation (×100).

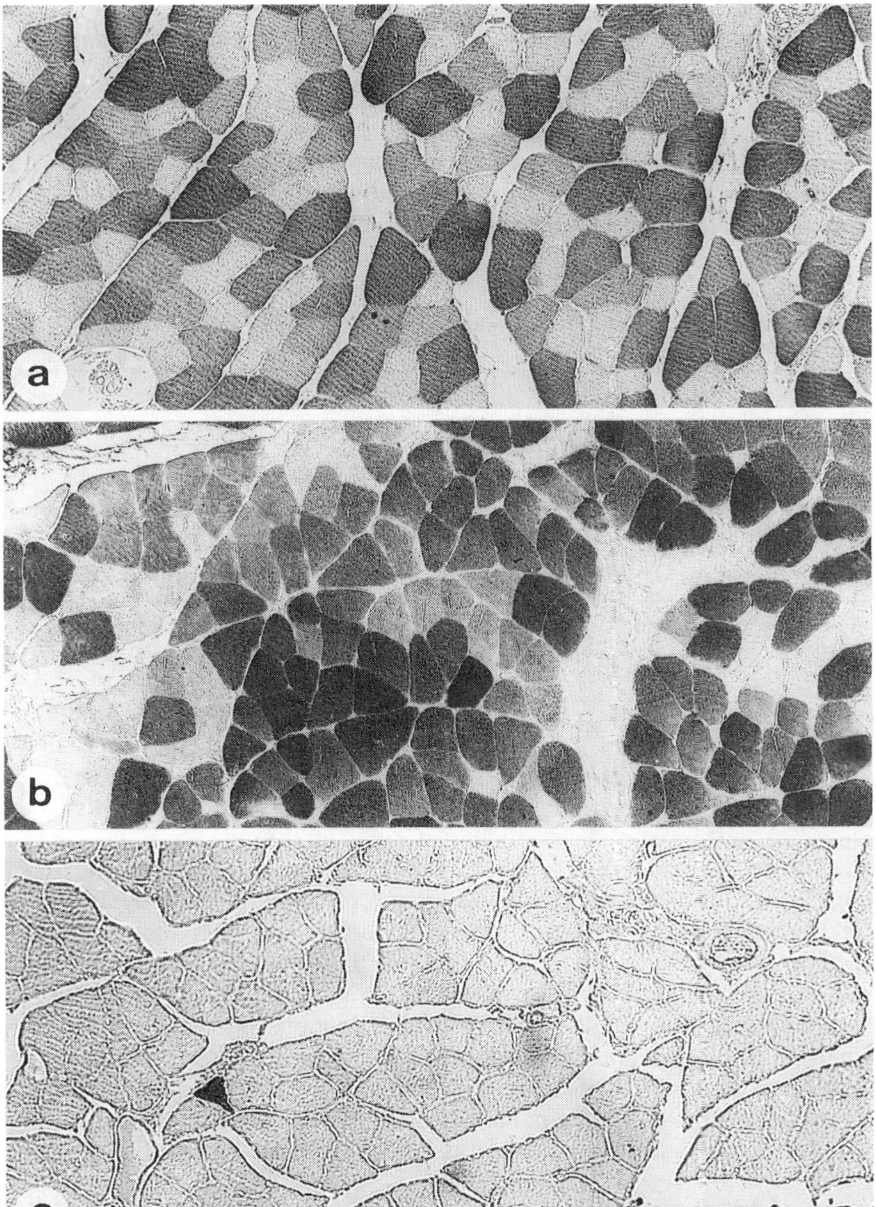

Fig. 2 Cross-sections of normal EDL (a), freely transplanted, non-overloaded SOL (b), and normal SOL (c) immunostained for parvalbumin. According to the staining intensity the transplanted and regenerated SOL (reinnervated by the EDL nerve) displays a nearly complete fibre transformation (×150).

Foreign-reinnervated muscles with persisting (cross-reinnervation) or regenerating (transplantation) fibres showed a fibre distribution essentially corresponding to the new nervous input (Fig. 2). The fibre pattern (irrespective of muscle type) did no more change significantly after overloading and fibre hypertrophy was rather modest (Table 1).

In all experiments the contralateral (i.e. the non-operated) side also exhibited muscular changes. However, they were generally restricted to an alteration of the distribution of fibre types, while the fibre size remained essentially unchanged. In the SOL the increase of the transitional IIC fibres (Fig. 1) indicated a simultaneous fibre transformation on the contralateral side (Table 2).

Table 2 Contralateral changes of distribution (%) of type I, IIA and IIC fibres in overloaded SOL.

Time	Operated side			Contralateral side		
	I	IIA	IIC	I	IIA	IIC
1 month	95.9 ± 4.6	1.4 ± 1.8	2.7 ± 2.9***	92.4 ± 8.0	6.2 ± 7.8	1.4 ± 0.8*
2 month	99.3 ± 0.8	0.4 ± 0.5	0.3 ± 0.4	98.7 ± 0.6	0.8 ± 0.7	0.5 ± 0.4
3 month	97.8 ± 1.8	1.4 ± 1.6	0.8 ± 0.4	96.0 ± 3.3	3.4 ± 3.3	0.6 ± 0.3
Control	94.6 ± 4.9	5.0 ± 4.9	0.4 ± 0.7			

***$P < 0.001$, *$P < 0.05$

Discussion

Functional overloading of a limb muscle with neuronally matching innervation (i.e. normal or transplanted) has two different effects. On the operated side, an increase in performance leads to hypertrophy of the muscle fibres and an altered fibre composition. On the contralateral side, the changes consisting mainly of a shift of the fibre composition are probably induced by an altered efferent activity following alterations of the afferent activity (due to the disturbed locomotion). In the case of neuronally mis-smatched reinnervation (cross-reinnervation) there is an additional 're-programming' in the central nervous system. This could readily explain the more extensive contralateral changes in the crossreinnervation experiments (data not shown). The findings demonstrate that the fibre pattern of a reinnervated muscle — independent of the type of nerve — is rigid and no more changed by increased work. Furthermore, it is evident that contralateral muscles cannot be considered as normal ones.

Acknowledgements

We thank Ms U. Büchter, I. Drescher, I. Holderegger, A. König, C. Meyer, A. Rhyner, S. Varol, and H. Weber for excellent technical assistance. Monoclonal antibodies against parvalbumin were a generous gift of Dr M.R. Celio, Fribourg, Switzerland. This work was supported by the EMDO Stiftung Zürich, the Hartmann Müller-Stiftung, Zürich, the Roche Research Foundation, Basel, and the Stiftung für wissenschaftliche Forschung an der Universität Zürich, Switzerland.

References

1. Rotshenker, S., 1979, *J. Physiol (Lond.)*, **292**, 535–547.
2. Müntener, M., Srihari, T., 1984, *Exp. Neurol.*, **86**, 134–146.
3. Müntener, M., Rowlerson, A. M., Berchtold, M. W., Heizmann, C. W., 1987, *J. Biol. Chem.*, **262**, 465–469.
4. Reichmann, H., Srihari, T., Pette, D., 1983, *Pflügers Arch.*, **397**, 202–208.
5. Srihari, T., Seedorf, U., Pette, D., 1981, *Pflügers Arch.*, **390**, 246–249.
6. Donovan, C. M., Faulkner, J. A., 1986, *J. Appl. Physiol.*, **61**, 288–292.
7. Gutmann, E., Schiaffino, S., Hanzlikova, V, 1971, *Exp. Neurol.*, **31**, 451–464.
8. Ianuzzo, C. D., Gollnick, P. D., Armstrong, R. B., 1976, *Life Sciences*, **19**, 1517–1524.
9. Oakley, C. R., Gollnick, P. D., 1985, *Histochemistry*, **83**, 555–560.
10. Pearson, J. K., Sickles, D. W., 1987, *J. Appl. Physiol.*, **63**, 2301–2308.
11. Schiaffino, S., Pierobon Bormioli, S., 1973, *Exp. Neurol.*, **40**,126–137.
12. Williams, P. E., Goldspink, G., 1981, *Cell Tissue Res.*, **221**, 465–470.
13. Carlson, B. M., Gutmann, E., 1974, *Experientia*, **30**, 1292–1294.
14. Müntener, M, Berchtold, M.W., Heizmann, C.W., 1985, *Muscle & Nerve*, **8**, 132–137.
15. Guth, L., Samaha, F. J., 1970, *Exp. Neurol.*, **28**, 365–367.
16. Wong-Riley, M., 1979, *Brain Res.*, **171**, 11–28.
17. Heizmann, C. W., Celio, M .R., 1987, *Meth. Enzymol.*, **139**, 552–570.
18. Celio, M. R., Baier, W., Schärer, L., De Viragh, P. A., Gerday, Ch., 1990, *Cell Calcium*, **9**, 81–86.

21
Influence of Changes in Innervation of Striated Muscle on the Conformational State of F-actin

J. SZCZEPANOWSKA*, YU. S. BOROVIKOV** AND A. JAKUBIEC-PUKA*

**Laboratory of Protein Metabolism, Nencki Institute of Experimental Biology, 02-093 Warsaw, Pasteur str.3, Poland*
***Group of Cell Motility, Institute of Cytology, Academy of Sciences of the USSR, 194064 Leningrad, Tikhoretskii str.4, USSR*

Introduction

Actin is one of two major proteins composing the contractile system of muscle. It is conservative protein and only a few of its isoforms are known. The numerous functions of actin are probably connected with the changes in its conformation state (1, 2, 3). Conformational changes of F-actin are involved in the process of active tension development in the muscle fibres (4).

Following denervation considerable changes were observed in the conformation state of F-actin and in the flexibility of the actin filament (5). The present study was performed to explain whether lack of function or lack of intact muscle-nerve connection are the reason of the changes in the conformational state of F-actin. The structure of F-actin was studied by the method of polarized fluorimetry (6, 7, 8) in the following models of atrophying muscles: (a) in denervated muscle; (b) in muscle with depressed activity, owing to the reduction of the activity of the afferent system; (c) in muscle immobilized, normal innervation being preserved.

Muscle and Motility Vol. 2. Proceedings of XIXth European Conference in Brussels
© Intercept Ltd, PO Box 716, Andover, Hampshire, SP10 1YG, UK

Materials and Methods

The soleus muscle of female 2.5–3-month-old Wistar rats was studied. The muscle used as models for atrophy (14 days after cutting the sciatic nerve, 4 days after immobilization of the ankle joint between aluminium rods, 4 days after section of the Achilles tendon), as well as intact muscles, were glycerinated according to Rome (9). In each experiment several animals were studied in parallel. Single fibres were isolated from glycerinated muscle and incubated in 800 mM KC1, 1 mM $MgC1_2$, 10 mM ATP and 67 mM phosphate buffer (pH 7.0) as described previously (10). Fibres prepared in such a way are called 'ghost fibres'. In the 'ghost fibres' actin remains in filamentous form while other contractile proteins are extracted. The remaining amount of these proteins is negligible and their fluorophores are randomly arranged.

For the study of the F-actin conformation state, polarized fluorescence was measured. Changes in fluorescence polarization correspond to changes in the fluorophore order which reflect changes in the protein molecule arrangement (7, 8, 10).

Tryptophan, a component of actin, is a very convenient intrinsic fluorophore. Polarized fluorescence of the tryptophan residues in actin was measured prior to and after the addition of 5 mM ATP. Intrinsic tryptophane fluorescence of actin was excited with polarized light at 303 ± 5 nm. The emitted light was recorded in the region between 320 and 380 nm.

Four intensities of polarized fluorescence ${}_{\perp}I_{\parallel}$, ${}_{\parallel}I_{\perp}$, ${}_{\perp}I_{\perp}$, ${}_{\parallel}I_{\parallel}$ were measured. They indicate the perpendicular ($\perp$) and parallel ($\parallel$) direction of the light, relative to the fibre axis: incident light on the left side and emitted light on the right side. These four components were used for computer calculation of fluorescence polarization ($P_{\perp}$ and $P_{\parallel}$ anisotrophy (A) index (11).

Computer analysis of the experimental data enables the angle between the filament long axis and the fibre axis (θ) to be calculated. $\sin^2\theta$ characterizes the flexibility in thin filaments. The theoretical basis for the calculations is given in (7, 8, 11).

Results and Discussion

The polarization parameters $P_{\perp}$ and $P_{\parallel}$ and anisotrophy are noticeably different for intact, denervated, immobilized and tenotomized muscles. We observed changes in the degree of polarization of F-actin intrinsic tryptophan fluorescence (Table 1): $P_{\perp}$ denervated > $P_{\perp}$ tenotomized > $P_{\perp}$ immobilized > $P_{\perp}$ intact muscle and $P_{\parallel}$ denervated < $P_{\parallel}$ intact < $P_{\parallel}$ immobilized < $P_{\parallel}$ tenotomized muscle. Differences in anisotropy were also found between muscles studied: denervated > intact > immobilized > tenotomized (Table 1). Thus orientation of the tryptophan residues of F-actin was different for all studied muscles.

Computer analysis of the data gave some information on the flexibility of actin filaments. Differences were found in the value of $\sin^2\theta$ between the studied muscles: denervated < intact = immobilized ≤ tenotomized. The value of N was constant. Evaluation of these values allows to conclude that the actin

Table 1 Influence of ATP on polarization of tryptophan fluorescence

Type of experiment	$P_{\perp}$	$P_{\parallel}$	A
	F-Actin in the absence of ATP		
Intact	0.194 ± 0.001	0.158 ± 0.001	1.23
Immobilized	0.199 ± 0.001	0.174 ± 0.001	1.14
Tenotomized	0.205 ± 0.001	0.185 ± 0.001	1.11
Denervated	0.210 ± 0.001	0.144 ± 0.001	1.46
	F-actin in the presence of ATP		
Intact	0.199 ± 0.001	0.154 ± 0.001	1.29
Immobilized	0.202 ± 0.001	0.175 ± 0.001	1.15
Tenotomized	0.205 ± 0.002	0.188 ± 0.001	1.09
Denervated	0.200 ± 0.001	0.150 ± 0.001	1.33

P, polarization of fluorescence; ||, ⊥, parallel and perpendicular to the fibre axis direction of polarization; *A* anisotropy of fluorescence ($P_{\perp}/P_{\parallel}$).
Mean values ± SEM from at least 50 measurements are presented.

filaments of denervated muscle are more rigid than those of other muscles studied while the flexibility of intact, immobilized and tenotomized muscles is very similar (Fig. 1).

ATP was used in order to study its influence on the arrangement of intrinsic fluorophores. As known, besides the high affinity-binding site for ATP, the latter binds in polymeric actin at another site with low affinity inducing conformation changes in the monomeric actin molecule in F-actin filaments (14).

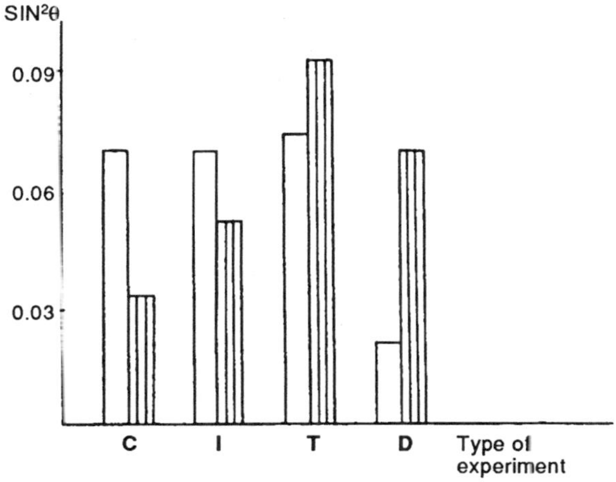

Figure 1 The Effect of ATP on Structural Changes of F-Actin. C, control, intact muscle; I, immobilized muscle; T, tenotomized muscle; D, denervated muscle. The value $\sin^2\theta$ characterizes the flexibility of thin filaments, calculated as described in Materials and Methods. Average values of at least 50 measurements are presented ☐ value of $\sin^2\theta$ in the absence of ATP; ▥ value of $\sin^2\theta$ in the presence of ATP.

In the presence of ATP, polarization of tryptophan fluorescence is changed in comparison with that obtained in the absence of ATP (Table 1): in intact fibres, $P_\perp$ increased, $P_\parallel$ decreased and A increased, whereas, in denervated fibres $P_\perp$ decreased, $P_\parallel$ increased and A decreased.

The data suggest ATP-induced changes in the conformational state of F-actin. These changes are different in character in denervated and intact muscle. In the tenotomized muscle they are similar to those in the denervated, while in the immobilized to those in the intact muscle. In other words, ATP decreased the flexibility of the F-actin filaments in normal and immobilized muscle while increasing it in the denervated and tenotomized muscles (Fig. 1).

The results show that the differences of all parameters studied were larger between denervated muscle and other ones than between normal muscles and immobilized or tentomized ones. It suggests that lack of intact muscle–nerve connection, is the main reason of the changes in the conformational state of F-actin, while decreased activity has much less influence.

Acknowledgements

Thanks are due to Ms. A. Kleśniak for skillful technical assistance. This work was supported by funds from CPBP 04.01. programme of the Polish Academy of Sciences.

References

1. Yanagida, T., Oosawa, F., 1980, *J. Mol. Biol.*, **140**, 313–320.
2. Borovikov, Yu. S., Levitski, D. I., Kirillina, V. P., Poglazov, B. F., 1982, *Eur. J. Biochem.*, **125**, 342–347.
3. Dobrowolski, Z., Borovikov, Yu. S., Nowak, E., Galazkiewicz, B., Dabrowska, R., 1988, *Biochim. Biophys. Acta*, **956**, 140–150.
4. Prochniewicz-Nakayama, E., Yanagida, T., Oosawa, F., 1983, *J. Cell Biol.*, **97**, 1663–1667.
5. Szczepanowska, J., Borovikov, Yu. S., Jakubiec-Puka, A., 1987, *Eur. J. Cell Biol.*, **43**, 394–402.
6. Rozanov, Yu. M., Chernogryadskaya, N. A., Barsky, I. Yu., Borovikov, Yu. S., Shudel, M. S., 1971, *Tsitologiya*, **13**, 190–200.
7. Yanagida, T., Oosawa, F., 1978, *J. Mol. Biol.*, **126**, 507–524.
8. Wilson, M. G. A., Mendelson, R. A., 1983, *J. Muscle Res. Cell Motil.*, **4**, 671–693.
9. Rome, E., 1972, *Cold Spring Harbor Symp. Quant. Biol.*, **37**, 331–339.
10. Borovikov, Yu. S., Gusev, N. B., 1983, *Eur. J. Biochem.*, **136**, 363–369.
11. Kakol, I., Borovikov, Yu. S., Szczesna, D., Kirillina, V. P., Levitsky, D. I., 1987, *Biochim. Biophys. Acta*, **913**, 1–9.

22
Effect of Fibre Size on Proteinase Activity in the Stretched Skeletal Muscles

I. SOHÁR, J. ROHÁN AND F. GUBA

Department of Biochemistry, A. Szent-Györgyi Medical University, 6701 Szeged, P. O. Box 415, Hungary

Introduction

In previous works we found that the lysosomal proteinase activities increased in skeletal muscles immobilized in shortened or lengthened position (4). Observing these results, we were interested in the causes inducing the high proteinase activities. Today we can say that redox state, ATP concentration, glucocorticoids, and prostaglandins play an important role in the faster protein degradation, but insulin, testosterone and endogenous proteinase inhibitors have not proved to be main factors in this phenomenon. Metallo ions like Ca^{2+}, Zn^{2+}, vitamins or lack of vitamins (E, A, C, K), pH, thyroids may also have effect on protein degradation (6).

Muscle fibre size was decreased by starvation or immobilization of skeletal muscle in the shortened position, but fibre size was increased by immobilization in lengthened position. In this work the effect of fibre size on proteinase activity in stretched skeletal muscle has been studied.

Methods

The right hind leg of New Zealand rabbits weighing 2300 ± 100g was fixed by plaster cast in full flexion while soleus muscle was in lengthened position for 4 or 7 days. In another group (12 animals) rabbits were starved for 4 or 6 days. In the third group the right hind leg was fixed in flexion (7 days) after starvation

Muscle and Motility Vol. 2. Proceedings of XIXth European Conference in Brussels
© Intercept Ltd, PO Box 716, Andover, Hampshire, SP10 1YG, UK

for 6 days. Control values were measured in soleus muscle from untreated animals.

To measure enzyme activity, the soleus muscle was homogenized in a solution KCl 1% + TritonX-100 0.1%. After centrifugation at 20 000 g for 20 min, supernatant was used for protein and protease activity measurement. Activity of cathepsin D was determined by using hemoglobin substrate at pH 3.5. 5 μM Z-Arg-Arg-amino-methyl-coumarine (AMC) for cathepsin B and 5 μM Suc-Ala-Ala-Pro-Phe-AMC for MMP-7ase as substrates were used (2, 5). The amount of liberated aminomethyl-coumarine was measured with Hitachi 650-10 spectrofluorimeter (Ex: 360 nm, Em: 460 nm). The protein content was determined by the method of Goa (1).

Results

Table 1 demonstrates the results of this study. Activity of cathepsin D was augmented (28%) after 7 days in stretch position, but this increase in activity did not happen when stretch was applied at animals having previously been starved.

Table 1 Effect of starvation on stretch induced proteinase activation.

	CD ng Tyr x min^{-1} / mg	%	MMP-7ase μU/mg	%	CB μU/mg	%
Control	43 ± 3	100	298 ± 16	100	31 ± 2	100
Stretch for 4 days	53 ± 5	123	490 ± 35	164	75 ± 5	242
Stretch for 7 days	55 ± 3	128	330 ± 28	111	62 ± 4	200
Starvation for 4 days	50 ± 4	116	179 ± 11	60	28 ± 2	90
Starvation for 6 days	30 ± 2	73	324 ± 23	109	15 ± 1	47
Starvation for 6 days + Stretch for 7 days	44 ± 3	102	368 ± 25	123	50 ± 4	161

CD, cathepsin D; CB, cathepsin B; MMP-7ase, muscle-metalloproteinase-7; U = μmol amino-methyl-coumarine min^{-1}.

Similar results were observed at activities of MMP-7ase and cathepsin B. The highest elevation of activity in the stretched muscle was found for cathepsin B activity.

Discussion

In previous experiments we found that the diameter of muscle fibre was increased by stretching, but it did not change by stretching if the muscle had been immobilized in shortened state before stretching (3). In the present study we observed that starvation similarly to immobilization in shortened position could inhibit the enlargement of the muscle cell diameter. This phenomon was correlated to lysosomal (cathepsin B and D) and extralysosomal (MMP-7ase) proteinase activities in stretched skeletal muscles. In other words, after starvation and immobilization in the shortened position proteinase activities were not elevated by stretching. In our hypothesis the proteinase enzymes are activated in low oxygen tension (decreased redox potential). When the diameter of the muscle cell is increased, the oxygen transfer way is longer to the middle of the cell. Afterwards the metabolism of the cell is faster during stretch, but the oxygen supply is slower and cannot keep the redox potential in the cell. After starvation or immobilization in the shortened position, the cell size is smaller and the oxygen transfer way is shorter therefore oxygen supply is sufficient for the faster metabolism. These hypotheses might explain the protective effect of starvation on proteinase activation.

There is one question: why are the proteinases in activated form in skeletal muscle immobilized in the shortened or lengthened state? Our hypothesis is shown in Fig. 1. During reduced redox-potential state the cysteine proteinases and thiol-activated enzymes are activated by opening the disulphide-bridge. The activated cysteine proteinases can activate the procathepsin D to cathepsin D by limited proteolysis. Cathepsin D can inactivate the inhibitor of cysteine proteinases by a feedback activation mechanism. Cysteine proteinases are also activated by a metalloproteinase in a limited proteolysis. We proposed that MMP-7ase is that metalloproteinase (5) activated by —SH compound. The low oxygen tension can activate a proteinase activating xanthine dehydrogenase to xanthine oxidase that catalyses the xanthine-hypoxanthine conversion when oxygen free radicals are generated. The OH free radicals can also convert —S—S— to —SH groups.

This hypothesis might explain the higher proteinase activities in immobilized or exercised skeletal muscles.

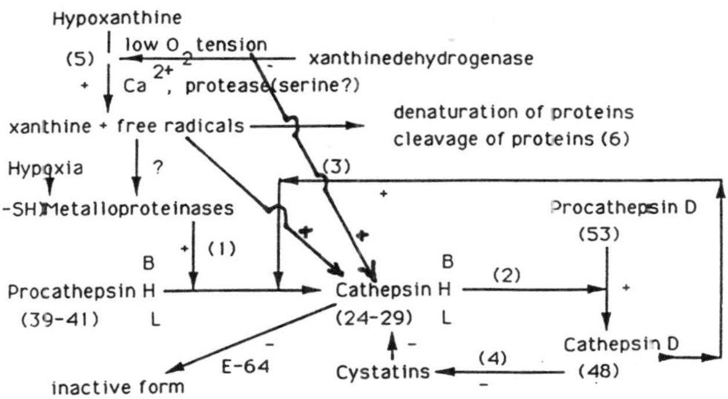

(1) K. Hara et al. 1988 - cultured rat peritoneal macrophages
(2) A.M. Samarel et al. 1986 - rabbit cardiac muscle
(3) Y. Nishimura et al. 1988 - primary cultures of rat hepatocytes
(4) B. Lenarcic et al. 1988 - in vitro
(5) D.A. Parks et al. 1983 - digestive tract
(6) R. Dean 1987

Fig. 1 Activation of proteinases in skeletal muscles; +, activation; -, inactivation.

References

1. Goa, J., 1953, *Scand. J. Clin. Lab. Invest.*, **5**, 218–222.
2. Kirschke, H., Wood, L., Roisen, F. J., Bird, J.W.C., 1983, *Biochem. J.*, **214**, 871–877.
3. Rohán, J., Hideg, J., Guba, F., 1984, *J. Muscle. Res. Cell Motility*, **5**, 218.
4. Sohár, I., Nagy, I., Takács, Ö., Kovacs, Z., Guba, F., 1981, *Adv. Physiol. Sci.*, **24**, 213–220.
5. Sohár, I., Fekete, E., Yorke, G., Cosentino, B., Roisen, F. J. Bird, J. W. C., 1987, *Biomed. Biochim. Acta*, **7**, 571–579.
6. Sohár, I., 1991, *Acta Biol. Acad. Sci. Hung.*, in press.

23
Recovery of Mechanical Properties of the Soleus Muscle after Suspension Hypokinesia of Young Rats

M. VOGTMANN AND G. ASMUSSEN

Carl-Ludwig-Institute of Physiology, University of Leipzig, Germany

Introduction

Suspension hypokinesia (SH) as an animal model of weightlessness leads to an atrophy especially in slow-twitch muscles with postural functions such as the soleus muscle (SOL). Accompanying this, the SOL adopts some histochemical, biochemical and mechanical properties characteristic to fast-twitch muscles (for references see 1). Recent studies have shown that the action of SH is more pronounced in SOL of young developing than in adult rats. In young animals the changes induced by SH are developed within 3 weeks (1) but up to now their reversibility has not been investigated systematically. In this study, the time course and the extent of the recovery of contractile properties of the SOL was determined 1–14 days after 3 weeks of SH performed in 3-week-old rats.

Methods

After weaning (3 weeks of age) 46 Wistar rats were randomly divided into 2 groups. One group (11 female, 7 male) served as control; the other (11 female, 17 male) was used for SH. Three weeks later the animal were released from SH, and the contractile parameters of the SOL were recorded *in vitro* after 1, 3, 5, 7 and 14 days of recovery. The performance of the suspension, the animal conditions during SH, the recording technique, and the parameters determined were described in details elsewhere.

Muscle and Motility Vol. 2. Proceedings of XIXth European Conference in Brussels
© Intercept Ltd, PO Box 716, Andover, Hampshire, SP10 1YG, UK

Results and Discussion

The experimental animals showed no marked differences in total body growth and development as compared to the controls. Immediately after releasing from SH the weight-bearing function of the hindlimbs was disturbed visually but normalized within a few days of recovery. Table 1 summarizes some changes of the SOL observed after SH and during the recovery. SH is followed by an atrophy expressed by a strong decrease in the diameter of type I fibres, whereas the growth and development of the type IIa fibres is relatively unchanged (2).

During recovery the wet weight increases by about 20% on the first day, it was almost doubled on the fifth one, and almost normal after a fortnight. This time course corresponds to that observed in adult rats SH (5) or after plaster cast immobilization (3). In Fig. 1 typical mechanical responses of the SOL are

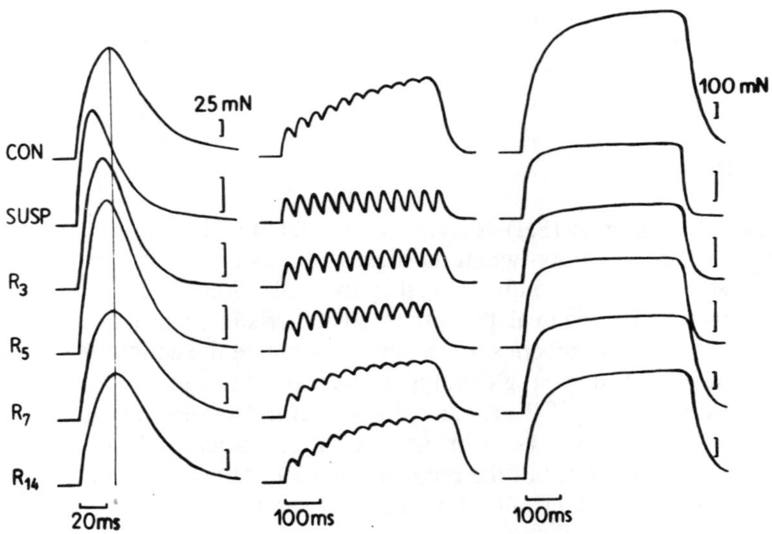

Fig. 1 Typical single twitches (left), unfused (stimulation frequency 35 Hz, middle) and fused (stimulation frequency 200 Hz, right) tetanic contractions of rat soleus muscles. The changes after a suspension hypokinesia lasting for 3 weeks (SUSP) and the recovery after 3, 5, 7 and 14 days (R) are shown. The thin line marks the time to peak of a normal muscle (CON).

depicted. It is clearly shown, that SH changes the contractile properties of the SOL from a slow-twitch muscle to a more fast-twitch one. The contraction (CT) and half-relaxation (HRT) times of single twitch decreases, the same is true for the time parameters of the tetanus. The fusion frequency (FF) rises from 50 to about 100 Hz, and the SOL exhibits a post-tetanic potentiation (PTP), a property only observed in fast-twitch muscles. The CT of 17 ms is in accordance with the CT of fast fatigue-resistant motor units of rats (4) and corresponds to the observation that in growing rats the normal conversion of type IIa fibres to type I fibres is arrested (2). Fig. 1 also demonstrates the fast recovery after SH. The CT increases about 1 ms per day and after 7 days no significant differences to controls were found. A comparable time course was also observed for the recovery of the HRT, the summation properties, the FF, and the PTP. Such a fast recovery of the time parameters after SH was not observed in muscles of older rats (5) or after long-term-SH (7 months, Elder, pers. comm) of young rats. SH decreases the tension output as well of the twitch as of the tetanus (Table 1). However, the decrease of the twitch is smaller

Table 1 Changes of some mechanical properties of the soleus muscle of young rats after suspension hypokinesia(SH) lasting 3 weeks and the recovery after 5 (R_5) and 14 days (R_{14}) in comparison to controls (CON). Data are presented as mean ±SD, the number of observations is given in parenthesis.

Parameter	CON (36)	SH (10)	R5 (10)	R14 (8)
Muscle weight (mg)	80.8 ± 19.7	27.5 ± 5.1 –66%,*	51.9 ± 7.5 –36%*	75.5 ± 9.3 –7%
Mean muscle area (mm²)	4.75 ± 0.91	1.47 ± 0.14 –69%,*	2.59 ± 0.54 –45%*	3.83 ± 0.57 –19%**
Contraction time (ms)	27.7 ± 2.3	17.0 ± 0.7 –39%*	22.4 ± 1.7 –19%*	26.8 ± 2.3 –3%
Twitch tension (mN)	180 ± 39	69 ± 14 –62%*	119 ± 21 –34%*	144 ± 23 –20%**
Tetanus tension (Po, mN)	886 ± 114	216 ± 48 –76%*	313 ± 43 –65%*	720 ± 21 –19%, &
Po/muscle area (N/cm²)	18.7 ± 2.2	14.7 ± 4.3 –21%*	12.1 ± 2.4 –35%*	18.9 ± 1.3 +1%
Twitch-tetanus-ratio	0.2 ± 0.04	0.32 ± 0.03 +60%,*	0.38 ± 0.04 +90%,*	0.2 ± 0.03 ± 0%

* $P < 0.01$, ** $P < 0.05$ in comparison to controls.

than that of the tetanus and consequently the twitch-tetanus-ratio (TTR) rises. The recovery of twitch tension is faster than that of the tetanic one, and therefore the TTR is enhanced during the first week of recovery, but normal after two weeks. The restitution occurs during the second week. A possible explanation could be, that the contractile proteins produced during recovery are initially not fully functional.

References

1. Asmussen, G., Miersch, H., Soukup, T., 1989, *Biomed. Biochim. Acta*, **48**, 426–431.
2. Asmussen, G., Soukup, T., 1987, *Histochem. J.*, **19**, 613.
3. Booth, F. W., Seider, M. J., 1979, *J. Appl. Physiol.*, **47**, 435–439.
4. Close, R., 1967, *J. Physiol.*, **193**, 45–55.
5. Templeton, G. H., Padalino, M., Manton, J., Glasberg, M., Silver, C. J., Silver, P., Demartino, G., Leconey, T., Klug, H., Hagler, H., Sutko, J. L., 1984, *J. Appl. Physiol.*, **56**, 278–286.

24
A New Way of Muscle Activity Regulation: Thiamine Participation in Neuromuscular Transmission

A. V. ROMANENKO

A. V. Palladin Institute of Biochemistry Academy of Sciences of the Ukrainian SSR, 252030, Kiev, Leontovich str. 9, USSR

Thiamine plays an important role in nerve and muscular systems function. Thiamine deficit in humans and animals provokes disturbances of movement coordination, reflexes, alimentary tract motility, paralysis and atrophy of muscles. The mechanism of these disturbances and the specific link between them and thiamine, as well as the role of thiamine in the regulation of muscle contraction, were unknown till recently.

It is impossible to explain the specific role of thiamine in nerve and muscular systems function only on the basis that vitamine B_1 coenzyme form — thiaminediphosphate — participates in some enzymatic reactions because:

1. In B_1-hypovitaminosis animals neurological and muscular disturbances occur earlier than the changes in thiaminediphosphate-dependent enzymes activity.
2. Injections of vitamine B_1 antagonist pyrithiamine lead to myoneural disturbances in animals without expressed changes in thiaminediphosphate-dependent enzymes activity.
3 Pyrithiamine does not influence thiaminediphosphate-dependent enzymes activity in isolated nerve fibres.
4. In animals vitamine B_1 antagonist oxythiamine injections provoke a decrease of thiaminediphosphatedependent enzymes activity without development of the myoneural symptoms of B_1-hypovitaminosis (1, 2, 3).

We have compared some biochemical and electromyographical characteristics in normal and B_1-hypovitaminosis rats (4). It was found that because of feed on low-vitamin B_1 synthetic diet (0.1 mg of vitamin B_1 per 1 kg of diet) during 28–31 days, there were 1.9, 2.0 and 6.8 times decrease of vitamine B_1 level (including thiamine and thiaminephosphates) in blood, brain and skeletal muscles respectively. These data indicate B_1-hypovitaminosis development in experi-

Muscle and Motility Vol. 2. Proceedings of XIXth European Conference in Brussels
© Intercept Ltd, PO Box 716, Andover, Hampshire, SP10 1YG, UK

mental rats. Although there were differences neither in blood pyruvate level, which is an indicator of thiaminediphosphate-dependent enzymes activity level, nor in electromyographically measured nerve impulse velocity in tail ventral nerve trunk of control and B_1-hypovitaminosis rats, amplitude of skeletal muscle action potentials, which appear in lateral muscle of 8–10 tail segments in response to tail ventral nerve trunk supramaximal electrical stimulations, were in B_1-hypovitaminosis rats 3.1 ± 0.5 mV ($n = 22$), in control rats 4.1 ± 0.4 mV ($n = 75$) and in rats, which received synthetic diet with vitamine B_1 supply of 5 mg per 1 kg of diet, 4.7 ± 0.6 mV ($n = 33$). So in B_1-hypovitaminosis rats in response to nerve stimulation fewer muscle fibres in skeletal muscle are simultaneously involved in activity than in control animals. The reason for this is a disturbance of neurotransmitter release from nerve endings in B_1-hypovitaminosis animals.

When nerve was stimulated by pair of supramaximal impulses with 10 ms interval between these two impulses, the amplitude of the second skeletal muscle action potential was greater than the first one in the group of B_1-hypovitaminosis rats ($n = 22$) in 32% and in control group ($n = 64$) in only 9% of experiments. Thiamine intraperitoneal injection (50 mg/kg) led to rapid neuromuscular transmission and neurotransmitter release from nerve endings normalization and first skeletal muscle action potential amplitude increased some minutes after injection.

It is possible to conclude that thiamine is a factor, which is necessary for normal neuromuscular transmission. This point of view is supported by our results obtained by microelectrode and 'sucrose-gap' techniques, which indicate that thiamine can facilitate neuromuscular transmission by increasing quantal neurotransmitter release from nerve endings in isolated frog n.ischiadicus–m.sartorius preparations, cray-fish m.abductor dactilopodit, guinea-pig, taenia coli. Thiamine can also evoke increase in guinea-pig taenia coli smooth muscle cells electrical and mechanical activity.

Vitamine B_1 participation in neuromuscular transmission does not link with its coenzyme properties. According to our data (5) both vitamine B_1 coenzyme form—thiaminediphosphate—and vitamin B_1 antagonist oxythiamine, which provokes thiaminediphosphate-dependent enzymes block in animals, do not influence neuromuscular transmission. Moreover, thiamine derivatives 3-benzyl-4-methyl-5-β-hydroxyethylthiazolium chloride and 3-benzyl-4-methylthiazolium chloride, which imitate in some chemical reactions thiaminediphosphate coenzyme action in enzymatic reactions (6), block neuromuscular transmission and quantal neurotransmitter release from nerve endings in frog skeletal muscle (7).

Another vitamin B_1 derivative, its antagonist pyrithiamine, also effectively blocks neuromuscular transmission (8, 9). Pyrithiamine intraperitoneal injections (50–100 mg/kg) provoke in mice the development of typical clinical myoneural symptoms of B_1-hypovitaminosis including convulsions and sometimes ophtalmoplegia. Electromyographical recordings after pyrithiamine intraperitoneal injection show the decrement of amplitude of skeletal muscle action potentials, which appear in mice m.gastrocnemilus in response to n. ischiadicus supramaximal electrical stimulation with 3 Hz frequency. According to the generally accepted point of view, it reflects the disturbance of neuromuscular transmission. In some experiments the decrease of amplitude of first skeletal muscle action potential is also observed. So, after pyrithiamine intraperitoneal injection, fewer mice skeletal muscle fibres are activated in response to nerve stimulation than in control. Such pyrithiamine action can be easily explained. In our experiments, pyrithiamine decreased amplitude and quantal content of end-plate potentials, frequency of miniature end-plate potentials in frog n. ischiadicus–m. sartorius isolated preparations. So pyrithiamine is an effective blocker of quantal neurotransmitter release from nerve endings.

In conclusion the mechanism of vitamine B_1 involvement in neuromuscular transmission can be proposed. Vitamine B_1 in the form of thiaminephosphates is localized in excitable, particularly, in synaptic membranes (10, 11) and nerve excitation provokes these thiaminephosphates dephosphorylation and thiamine releases from membranes (3, 12). Thiamine interacts with specific presynaptic receptors and that provokes a facilitation of neurotransmitter release from nerve endings. Probably, pyrithiamine, 3-benzyl-4-methylthiazolium chloride and 3-benzyl-4-methyl-5-β-hydroxyethyl-thiazolium chloride block this process.

Thiamine involvement in synaptic transmission allows to explain the specific myoneural vitamine B_1 role in the organism and practically all myoneural disturbances, which occur in the organism as a result of B_1-hypovitaminosis or pyrithiamine action and also the absence of typical myoneural disturbances in the organism after oxythiamine injection.

References

1. Butterworth, R. F., Giguere, J.-P., Besnard, A. M., 1985, *Neurochem. Res.*, **10**, 1417–1428.
2. Cooper, J. R., 1968, *Biochim. biophys. acta*, **156**, 368–373.
3. Itokawa, Y., 1977, *J. Appl. Nutr.*, **29**, 5–16.
4. Romanenko, A. V., Fenik, V. B., 1990, *Dokl. Acad. Sci. Ukr. SSR. (Ser. B)*, **N9**.
5. Romanenko, A. V., 1988, *Dokl. Acad. Sci. Ukr.SSR. (Ser. B)*, **N2**, 76–79.
6. Breslow, R., 1958, *J. Amer. Chem. Soc.*, **80**, 3719–3726.
7. Romanenko, A.V., Vovk, A.I., 1988, *Dokl. Acad. Sci. Ukr. SSR. (Ser. B)*, **N11**, 75–77.
8. Romanenko, A. V., 1989, *Dokl. Acad. Sci. Ukr. SSR. (Ser. B)*, **N1**, 80–82.
9. Romanenko, A. V., Zinov'eva, M. L., Kokshareva, N. V., 1989; *Dokl. Acad. Sci. Ukr. SSR. (Ser. B)*, **N11**, 74–77.
10. Matsuda, T., Cooper, J. R., 1981, *Anal. Biochem.*, **117**, 203–207.
11. Schoffeniels, E., Dandrifosse, G., Bettendorf, L., 1984, *J Neurochem.*, **43**, 269–271.
12. Muralt., A., 1962, *Ann. N.Y. Acad. Sci.*, **98**, 499–507.

SECTION FOUR

Cardiovascular Function

25
The Use of Transplanted Skeletal Muscle for Cardiac Assistance

U. CARRARO, C. CATANI, C. RIZZI, S. BELLUCO AND L. DALLA LIBERA
CNR Unit for Muscle Biology and Physiopathology, Institute of General Pathology, University of Padova, Italy

D. DANIELI-BETTO
Institute of Human Physiology, University of Padova, Italy

G. MIRACOLI
Sorin Biomedica, Cardiovascular Division, Saluggia, Italy

G. ARPESELLA, P. MIKUS, M. CIRILLO, M. PARLAPIANO, M. SENNI AND A. PIERANGELI
Chirurgia del Cuore e dei Grossi Vasi, University of Bologna, Italy

Introduction

Skeletal muscle can be used in a variety of ways to repair or assist the heart. Its use has been suggested in revascularization of the heart, creation of heart valves, repair of cardiac and great vessel walls and as an energy source to power cardiac assist devices. Skeletal muscle-powered cardiac assistance (SM-P/CA) takes advantage of the use of skeletal muscle to dynamically reinforce or substitute the heart or great vessel walls and to actuate cardiac assist devices. Functional electrostimulation of transplanted skeletal muscle used in Dynamic Cardiomyoplasty or skeletal muscle-powered cardiac assist devices (SM-P/CAD) promises to suitably reinforce aneurysmatic ventricles, correct right heart's congenital defects, treat terminal congestive heart failure.

End-stage cardiac failure is a common and difficult problem for a large segment of world's population (1): 30 000 heart transplant candidates in USA, 25 000 in Europe (2500 in Italy), 20 000 in East Europe, 5000 in Japan and

Muscle and Motility Vol. 2. Proceedings of XIXth European Conference in Brussels
© Intercept Ltd, PO Box 716, Andover, Hampshire, SP10 1YG, UK

Australia, several thousands in Latin America (Chagas' cardiomyopathy) and developing countries die per year due to shortage of donors. Furthermore ten times more patients with congestive heart failure associated with diabetes or slow-progressive tumours, etc. (patients not suitable for heart transplant), or patients with acute cardiogenic shock (heart infarction) and congenital heart defects need cardiac assistance (1–4).

Actual managements of pharmacologically intractable congestive heart failure are at the stage of accepted clinical therapy or clinical experimentation, namely:

1. *homologous heart transplant* — this therapy covers only 1% of the needs due to problems related to patients' selection and donor shortage;
2. *transcutaneously powered electromechanical or fluidic ventricular assist devices* (VAD) (e.g. intraortic balloon pump, Jarvik's artificial ventricle, Novocor) — but maximal survival with these devices is only of months; therefore they are used as bridge to transplant;
3. *dynamic cardiomyoplasty* — shows good clinical benefit, but low haemodynamic improvement demonstrable by non-invasive or invasive haemodynamic analyses.

At the stage of animal experimentation are: Xenotransplants, total artificial heart (TAH) and Skeletal muscle-powered cardiac assist devices (SM-P/CAD).

The concept of using skeletal muscle energy to assist the failing heart was first reported in 1959 (5). The primary impediment to the use of skeletal muscle as cardiac assist has been skeletal muscle fatigue. Chronic electrical stimulation of fast skeletal muscle induces an adaptive transformation, which includes a marked increase in resistance to fatigue (6, 7). This has reopened the possibility of diverting skeletal muscle from its normal function to augment the function of a failing heart. Three main approaches to the use of skeletal muscle for cardiac assistance have been considered:

1. wrapping a sheet of muscle around the heart or the aorta;
2. shaping the muscle as an auxiliary ventricle;
3. using a skeletal muscle as the power source for a mechanical assist device.

Components of a device for skeletal muscle-powered cardiac assistance are:

1. drive unit — the skeletal muscle;
2. transmission — either full biologic as in dynamic cardiomyoplasty or aortomyoplasty, or biomechanic as in the case of skeletal muscle ventricle (SMV), skeletal muscle tube (SMT), or dual chamber SM-p/CAD;
3. control unit — commercially available (8) or experimental (9) demand cardiomyoneuroelectrostimulator.

Three steps can be recognized for the development of a Skeletal Muscle Engine:

1. engine construction optimizing the configuration of the selected muscle (latissimus dorsi flap) for dynamic cardiomyoplasty or different muscle options for one of the several SM-p/CAD;

2. muscle conditioning by chronic electrostimulation of the skeletal muscle to achieve its indefatigability at moderate power output;
3. muscle training — supplementing the continuous stimulation with daily sessions of muscle electrostimulation at near-maximum exercise to increase through muscle mass its maximum sustenable external power output.

Experimental configurations of SM-p/CAD can be subdivided according to types of transmission chosen:

1. Fluidic transmission as in skeletal muscle ventricle, (SMV) (10), skeletal muscle tube, (SMT) (11), and extraortic dual chambered pump for implantable counterpulsation assist system (9);
2. Linear configuration of the muscle with mechanical transmission, but this is an open field for which new ideas are urgently needed.

Indeed the major open problem in biomechanical configurations is the induced thromboembolism at the device/blood interface. Actual options are those offered by full biologic transmission (e.g., dynamic cardiomyoplasty (8) and aortomyoplasty (12), or autologous lined SMV or SMT (13). Future developments could arise from improvements of material and design of actual devices or from new approaches. Though not fully optimized, solutions are available for the problem of device control: the actual cardiomyoneurostimulator is composed of a sensing electrode and can be operated at the conditioning setting (continuous or cyclic stimulation mode) and at the exercise mode delivering burst of stimuli either during systole or diastole . Different designs of stimulating electrodes can be employed: intramuscular, perineural (cuff or multichannel), intraneural (14).

In the following sections we describe our contribution to the international efforts in research and development of SM-p/CAD. In summary the Italian group settled a muscle conditioning approach with a cyclic accelerated protocol (CAP), which allows muscle conditioning and a sustainable external power output (SEPO) from the end of the first week of conditioning. Furthermore we analysed SEPO of normal and conditioned LD of adult sheep and performed a feasibility study of linear configurations of SM-p/CAD. In this paper we present biochemical data confirming muscle transformation after the cyclic accelerated protocol.

Methods

Experiments were performed on sheep. Under general anaesthesia, a cutaneous incision was performed in the subscapular dorsal region with anterolateral direction toward the posterior axillary line. Subcutaneous muscles were incised gaining the superior plane of latissimus dorsi muscle (LDM). The muscle was implanted with modified Sorin electrodes and either an Orion 60 pacemaker (Sorin Biomedica, Italy) or an Itrel Medtronic programmable neurostimulator model 7420, essentially according to Chachques *et al.* (8), as described in (15). The anode was placed close to the nerve entrances into the LDM, the cathode was placed distally in the muscle. LDM electrostimulation began after one day of healing of the skin suture with either the 2 Hz continuous or the cyclic stimulation protocol (Table 1).

Table 1 Protocols of Muscle Conditioning: cyclic accelerated protocol (CAP)

Days	Hz	S/B	Mode/time (h/day)		ON - OFF (s)	Tetani/min	Tetani/day
1	healing						
2 - 3	20	2	cyclic	24	0.125 - 10	6	8640
4 - 7	20	2	"	"	0.125 - 4	15	21600
8 - 11	20	2	"	"	0.125 - 1	54	77760
12 - x	43	5	"	"	0.125 - 1	54	77760

Balanced cathodic stimuli 0.125 ms long and of 2–5 V amplitude. S/B, Stimuli per burst. Continuous Protocol: Stimulation at 2 Hz with balanced cathodic stimuli, 0.125 msec long and of 2–5 V amplitude.

Histochemical and biochemical analyses were performed on muscle biopsies taken at day 1, 4 and 8 weeks. Transverse cryostat sections were used to localize SDH and ATPase activities (15).

Purification of myosin, two-dimensional analysis of myosin light chains (MLC), one-dimensional analysis in sodium dodecyl sulphate-polyacrylamide gel electrophoresis (SDS PAGE) of myosin heavy chains (MHC) from muscle fragments were performed essentially as described in Carraro *et al.* (10–19); electrophoresis under nondissociating conditions (PPi PAGE) of myosin isoforms was performed essentially as described in (20): MLC, MHC and isomyosin analyses were performed following the slight, but with essential modifications introduced in 1989 and 1990. The updated methods are described in detail in (21).

Results

Fig. 1A shows the results of the analysis by electrophoresis under non-dissociating conditions (PPi PAGE) of isomyosins from contralateral LDM and after four and eight weeks of cyclic electrostimulation. While after four weeks the slow isomyosin content of the stimulated muscle is not higher than that of contralateral LDM, the eight-week biopsy taken from the centrodistal portion of the stimulated LDM contains slow isomyosin only. Coanalysis of myosins from contralateral and stimulated biopsies confirms the adult slow nature of the isomyosin present in eightweek stimulated muscle.

SDS 6% PAGE of MHC confirms and extends these results. Indeed, Fig. 1B shows that the percentage content of MHC 1 in the studied biopsies of LDMs of sheep Bo 12-90 correlates with the isomyosin pattern in non-dissociating conditions. Furthermore, Fig. 1B shows that biopsies taken from different

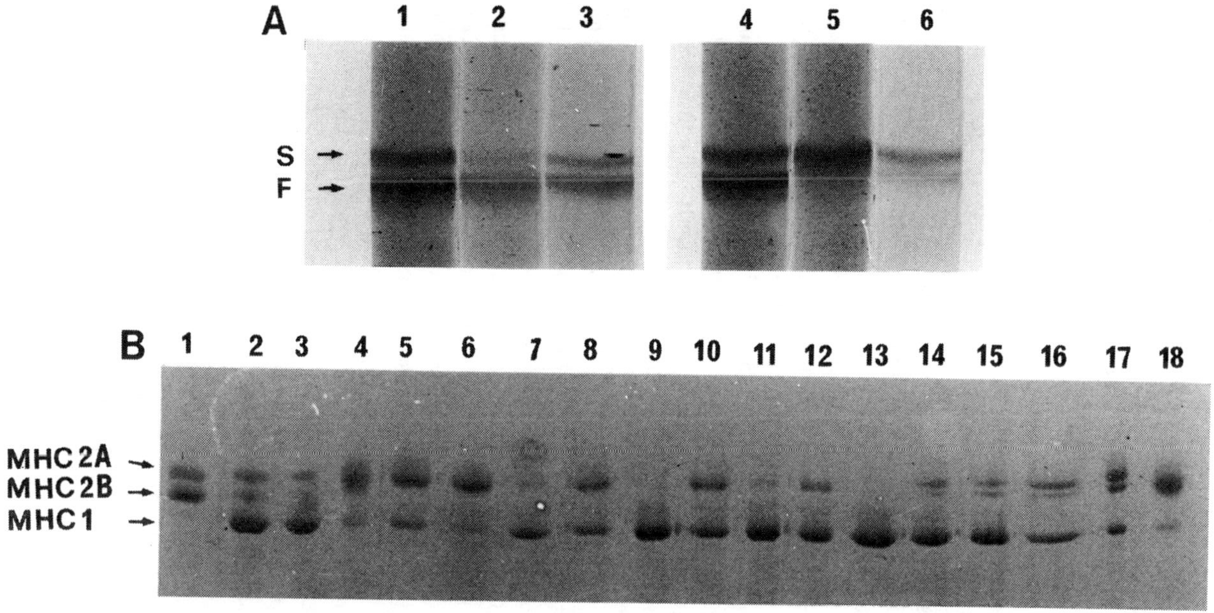

Fig. 1 Electrophoretic analyses of isomyosins and myosin heavy chains from electrostimulated LDM (sheep Bol2).
(A) PPi PAGE of myosin purified from: contralateral normal LDM (lanes 1 and 4); LDM after four-week cyclic electrostimulation (lane 2); coelectrophoresis of 1 + 2 (lane 3); LDM after eight-week cyclic electrostimulation (lane 5); coelectrophoresis of 4 + 5 (lane 6). S, isomyosin peculiar of slow myofibers; F, isomyosins peculiar of fast myofibres .

(B) SDS 6% PAGE of myosin heavy chains from: rat EDL (lane l); rat soleus + EDL coelectrophoresis (lane 2); rat soleus (lane 3); normal LDM of sheep 202 (lane 4); normal LDM of sheep Bol2 (lane 5); LDM of sheep Bol2 after four-week cyclic electrostimulation (lane 6); sample A from LDM of sheep Bol2, after eight-week of cyclic electrostimulation (lane 7); sample B from LDM of sheep Bol2 after eight-week of cyclic electrostimulation (lane 8); sheep soleus (lane 9); 9 + 6 (lane 10); 9 + 7 (lane 11); 9 + 8 (lane 12); sheep ventricle (lane 13); 9 + 16 (lane 14); 9 + 17 (lane 15); human LDM, case 1 (lane 16); human LDM, case 2 (lane 17); normal LDM of sheep 202 (lane 18). MHC 1, myosin heavy chain peculiar to slow myofibres; MHC 2A, myosin heavy chain peculiar to fast oxidative myofibres; MHC 2B, myosin heavy chain peculiar of fast glycolytic myofibres.

portions of the stimulated LDM show different degrees of muscle transformation: while the centrodistal biopsy contains MHC 1 only, the costolateral biopsy shows a pattern composed of equal amounts of fast and slow MHC. This result may be the consequence of differential mechanical effects of cyclic stimulation on different portions of the LDM.

Fig. 1B also confirms that MHC 1 from different animal species (i.e. sheep and human) comigrate, confirming that MHC analysis is an easy, reliable and sensitive method for typifying fatigue-resistant muscle using as starting material few cryostat sections of bioptic tissues (21).

Discussion

Gel electrophoretic analysis of the native myosin molecule shows that LDM of normal adult sheep is a mixed muscle in which fast isomyosins largely prevail. Present results confirm that after few weeks of uninterrupted electrostimulation there is a substantial increase in the proportion of slow isomyosin (up to 100%), compared to the 10–20% in the control pre-stimulated LDM. The effects of the continuous 2 Hz pattern (24 h per day, i.e. 165.600 twitches per day) or of cyclic but uninterrupted protocol (24 h per day of 125 m long 43 Hz trains followed by 1 s rest, i.e. 230.000 0.2 ms long electric currents, producing 76.800 fused tetani per day) are comparable: after two months of conditioning the contractile proteins of LDM are transformed in those characteristic of the slow, fatigue-resistant muscle (15, 18, 21). It is worth mentioning that in individual sheep the full transformation is achieved in one month (21). The analyses of native myosin and of its light and heavy subunits reveal that the stimulated LDM is turned into a typical slow type skeletal muscle, i.e. the tissue specific lS myosin light chain is expressed (18, 21). Recently it has been claimed that a 'cardiac' myosin is expressed during early myogenesis (22). This finding reopened the contention that eventual accumulation of ventricle myosin heavy chain (MHC 1 is a gene expressed in both slow myofibre and ventricle myocardiocytes) in conditioned LDM could be interpreted as a demonstration of regenerative events during muscle transformation. Our results showing that a typical skeletal slow myosin is expressed in the conditioned muscle negate this interpretation and confirm that muscle damage is not a necessary step of muscle conditioning to fatigue resistance (18).

During this study we accumulated evidence that by means of functional electrostimulation a fast-fatiguable muscle not only changes its gene expression according to the functional requirements, but become able to sustain continuous (24 h per day) tetanic/cardiac-like work during the second week of muscle conditioning. Therefore, it can be easily predicted that with an escalating cyclic protocol early after surgical construction of a muscle-powered cardiac assist device it will be possible to obtain a low, but significant amount of work, which could be used both for cardiac support and to drive changes of muscle characteristics.

The results obtained in clinical experimentation by Chachques *et al.* (8, 12) demonstrate that dynamic cardiomyoplasty is able to prolong and improve the quality of life of patients suffering from severe chronic and irreversible dysfunction (8). Animal experimentation shows that a SM-p/CAD can function in circulation chronically: an autogenously lined skeletal muscle ventricle demonstrates that canine skeletal muscle can provide aortic diastolic counterpulsation for up to 15 months without clinically apparent thromboembolic complications (11, and pers. comm.).

In conclusion on the sound ground of our experimental knowledge we believe that skeletal muscles can be made tireless so that they could activate biomechanical circulation support devices and that such devices will have in the near future an increasing role in reducing disability and death from heart diseases.

Acknowledgements

Supported by funds from the Italian CNR to the Unit for Muscle Biology and Physiopathology, to the Progetto Finalizzato: Tecnologie Biomediche e Sanitarie (TBMS) and from the Italian Ministero della Pubblica Istruzione 60% to Prof. U. Carraro.

References

1. Cardiovascular Disease, The office of Disease Prevention and Health Promotion, U.S. Public Health Service, US Department of Health and Human Services. 1988, In: *Disease Prevention/Health Promotion: The facts,*. pp 242–252. Bull Pub. Palo Alto.
2. Franciosa, J. A., Wilen, M., Ziesche, S., Cohn, J. N., 1983, *Am. J. Cardiol.*, **51**, 831–841.
3. Heck, C. F., Shumway, S. J., Kaye, M.P. 1989. *J. Heart Transpl.*, **8**, 271–276.
4. Pae, W. E., Miller, C. A., Pierce, W. S., 1988., *J. Heart Transpl.*, **8**, 277–280.
5. Kantrowitz, A., Mc Kinnon, W. M. P., 1959, *Surg. Forum*, **9**, 266–268.
6. Pette, D., Vrbová, G., 1985, *Muscle & Nerve*, **8**, 676–689.
7. Salmons, S., Henriksson, J., 1981, *Muscle & Nerve*, **4**, 94–105.
8. Chachques, J. C., Grandjean, P. A., Pfeffer, I. A., Perier, P., Dreyfus, G., Jebara, V., Acar, C., Levy, M., Bourgeois, I., Fabiani, J -N., Deloche, A., Carpentier, A., 1990, *J. Heart Transplant*, **9**, 239–251.
9. Li, C. M., Hill, A., Desrosiers, C., Grandjean, P., Chiu, R. C., 1989, *ASAIO Trans.*, **35**, 405–407.
10. Acker, M. A., Hammond, R. L., Mannion, J. D., Salmons, S., Stephenson, L. W. 1987, *Science*, **236**, 324–327.
11. Anderson, W. A., Andersen, J. S., Bridges, C .R., Hammond, R. L., DiMeo, F., Frisch, E. E., Salmons, S., Stephenson, L. W., 1988, *ASAIO Trans*, **34**, 241–246.
12. Chachques, J. C., Grandjean, P. A., Fischer, I., *et al.* 1990, *Ann. Thorac. Surg.*, **49**, 224–230.

13. Anderson, D. R., Pochettino, A., Hammond, R. L., Hohenhaus, E., Spanta, A. D., 1990. *J. Thorac. Cardiovasc. Surg.*, in press.
14. Agnew, W. F., McCreery, D. B. (eds), 1990, *Neural Prostheses: Fundamental Studies*. Prentice Hall, Englewood Cliffs, NJ.
15. Carraro, U. (ed), 1988, *Sarcomeric and Non-Sarcomeric Muscles: Basic and Applied Research Prospects for the 90s*. Unipress, Padova.
16 Carraro, U., Catani, C., Belluco, S., Cantini, M., Marchioro, L., 1986, *Exp. Neurol.*, **94**, 537–553.
17. Carraro, U., Catani, C., Saggin, L., Zrunek, M., Szabolcs, M., Gruber, H., Streinzer,W., Mayr, W., Thoma, H., 1988, *Muscle & Nerve*, **11**,1026–1038.
18. Carraro, U., Dalla Libera, L., Catani, C., Rizzi, C., 1990. In: Freilinger, E., (ed), *Vienna Muscle Symposium*, in press.
19. Carraro, U., Morale, D., Mussini, I., Lucke, S., Cantini, M., Betto, R., Catani, C., Dalla Libera, L., Danieli-Betto, D., Noventa, D., 1985, *J. Cell Biol.*, **100**, 161–174.
20. Pauletto, P., Piccolo, D., Belluco, S., Dalla Libera, L. 1989, In: Carraro, U., Dalla Libera L., (eds) *Elettroforesi e HPLC di Biopolimeri e loro Frammenti*. pp. 101–103. CLEUP Editore, Padova.
21. Carraro, U., 1991 In: *Seminars in Thoracic and Cardiovascular Surgery*, Magovern, G. J., (ed.),W. B. Saunders, in press.
22. Wick, M., Bandman, E. 1989, *J. Cell Biol.*, **109**, (4) part 2: 262a.

26
Analysis of Gene Expression at Protein Level in Skeletal Muscle used for Cardiac Assistance

D. V. E. CUMMING, G. A. O'BRIEN, H. A. WILLIAMSON, M. J. DUNN, C. W. PATTISON AND M. H. YACOUB

Department of Cardiac Surgery, National Heart and Lung Institute and the Royal Veterinary College, London, UK

Introduction

The use of skeletal muscle for cardiac assistance is a rapidly expanding area of research. Mobilized latissimus dorsi (LDM) may be wrapped around the heart and stimulated to contract in systole (cardiomyoplasty) (1). Alternatively LDM may be used to create auxilliary skeletal ventricles acting as counterpulsation devices (2). Our approach differs in that we have created an entirely autologous counterpulsation system with no foreign material/blood surface interface (3).

Electrical stimulation is necessary not only for appropriate muscular contraction during the cardiac cycle, but also to initiate the fibre type transformation from fast to slow, fundamental for the LDM to perform long term work (for a review see 4). We have analysed this muscular adaptation using one- and two-dimensional electrophoresis.

Methods

SURGERY AND ELECTRICAL STIMULATION

We have operated on 25 normal Welsh mountain ewes under the guideline laid

Muscle and Motility Vol. 2. Proceedings of XIXth European Conference in Brussels
© Intercept Ltd, PO Box 716, Andover, Hampshire, SP10 1YG, UK

down by the Home Office for Great Britain and Northern Ireland. The left latissimus dorsi was mobilized on an intact neurovascular pedicle and wrapped around the proximal descending thoracic aorta. Stimulation leads were positioned at the proximal end of the muscle to achieve neurovascular stimulation and a sensing electrode placed on the left ventricle to allow synchronization with the native cardiac cycle.

A purpose built external pulse train stimulator was designed and built by the Department of Biomedical Engineering, at the Royal Brompton and National Heart Hospital (5). A standard stimulation regime for the LDM was adopted for the assessment of fibre type transformation. Stimulation was initiated two days after the operation, every fourth cardiac cycle, at 35 Hz, 3 V and with a 240 μs pulse width.

SAMPLE PREPARATION

Muscle samples were taken from the LDM and the left ventricle and stored in liquid nitrogen. Small samples (ca. 100 mg) of muscle tissue were pulverized while still frozen into a powder. Proteins were extracted twice by homogenization in 1% SDS followed by spinning at 12 000 g for 10 min at 4°C.

GEL ELECTROPHORESIS

To analyse changes in myosin heavy chain composition one-dimensional polyacrylamide gel electrophoresis was performed using a modified method of Carraro *et al.* (6). 37% v/v glycerol was used in both the stacking and separating gels. The separating gel was 4% w/v acrylamide. Two-dimensional gel electrophoresis using the technique of Anderson (7), with 15% acrylamide in the second dimension was used to analyse changes in myosin light chain (MLC) profile. Gels were silver stained using (8) or for the two-dimensional gels the Daiichi 2D staining kit (Koch Light).

Results

Fig. 1 demonstrates myosin heavy chain transformation in LDM stimulated using the standard regime for up to 4 months. In ventricular muscle (V) only the slow isoform is detected. In control unstimulated LDM (O) the fast myosin heavy chain isoform is the major species although the slow isoform is detected. This demonstrates the typical mixed fibre type compostion of normal LDM. By 28 days of stimulation there is a marked increase in the slow myosin heavy chain isoform and by 120 days this is the major component.

The myosin light chain profile found in LDM stimulated up to 6 months is shown in Fig. 2. During the first month of stimulation there is little alteration from the pattern shown for 10 days. By 3 months an increase in $MLC2_S$ is observed and at 6 months this isoform is the predominant species for MLC2. Transformation of the alkali light chains is not detected by this time point.

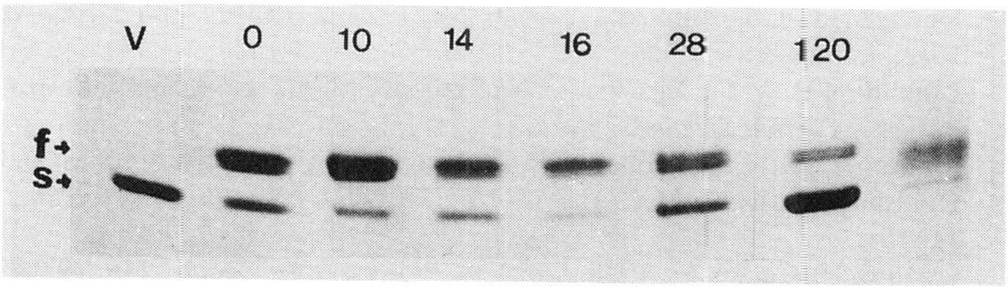

Fig. 1 Myosin heavy chain fast and slow isoforms separated using one-dimensional electrophoresis as described in methods. f, fast isoform, s, slow isoform, V, ventricle, 0–120 days of electrical stimulation to LDM used for counterpulsation.

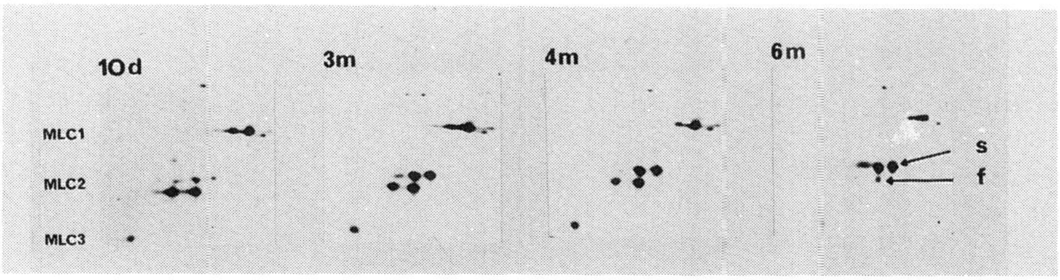

Fig. 2 Myosin light chain analysis using two-dimensional gel electrophoresis as described in methods. Sample 10d = 10 day stimulated LDM; 3, 4 or 6m = 3, 4 or 6 month stimulated LDM, all with standard regime.

Conclusion

Fibre type transformation of LDM from a predominantly fast composition toward a slow, fatigue-resistant phenotype can occur whilst the muscle is being used as an autologous counterpulsation device. We have analysed myosin heavy and light chain transformation in LDM used for circulatory support using one- and two- dimensional gel electrophoresis. This has shown that LDM stimulated with our standard regime exhibits transformation of myosin heavy chains (detected at protein level) before myosin light chains. Analysis of *in vitro* translation products over this time course will allow us to relate these changes to mRNA levels.

References

1. Hagege, A. A., Desnos, M., Chachques, J. C., Carpentier, A., Fernandez, F., Fontaliran, F., Guerot, C., 1990, *The Lancet*, **335**, 1122–1124.
2. Kochamba, G., Desrosiers, C., Dewar, M., Chiu, R. C. J., 1988, *Ann. Thorac. Surg.*, **45**, 620–625.
3. Pattison, C. W., Cumming, D. V. E., Williamson, A., Clayton Jones, D. G.,

Dunn, M. J., Goldspink, G., Yacoub, M., 1990, *J. Thorac. and Cardiovasc. Surg.*, in press.

4. Cumming, D. V. E., Pattison, C. W., Yacoub, M. H., 1989, *Annual Cardiac Surgery, Current Science*, 75–82.
5. Pattison, C. W., Logan Sinclair, R., Cumming, D. V. E., Clayton Jones, D. G., Goldspink, G., Dunn, M. J., Yacoub, M. H., 1990, *J. Med. Engin. and Tech.*, **14**, 11–15.
6. Carraro, U., Catani, C., 1988, *Biochem. Biophys. Res. Comm.*, **116**, 793–802.
7. Anderson, L., 1988, *Two Dimensional Electrophoresis*. Large Scale Biology Press, Washington DC.
8. Patel, K., Easty, D. J., Dunn, M. J., 1988. In: Walker, J. M. (ed.), *Methods in Molecular Biology (Humana)*, **3**, 159–168.

27
Morphometric Analysis of Human Latissimus Dorsi Muscle in Middle-aged Subjects: an Autopsy Study

R. SCELSI*, P. POGGI** AND G. AZAN†

**Department of Human Pathology, **Institute of Histology and †Department of Neurology, University of Pavia, Via Forlanini 16, 27100 Pavia, Italy*

Introduction

The use of the electrostimulated autologous latissimus dorsi muscle (LDM) was recently proposed for substitution and/or reinforcement of the cardiac wall in patients with chronic irreversible myocardial disorders and for construction of experimental pumping chambers (1–7). Morphometric analysis of the enzyme-histochemical profile of fibres from normal and electrostimulated LDM have been performed on animals (4, 8, 9), but rare studies on normal human LDM are available (10). The present study is a qualitative and quantitative analysis of the fibre types and of the microvasculature of normal LDM from autopsy of 20 sedentary male subjects aged 40–65 years who died without signs of neuromuscular pathology. The selection of these subjects was performed because the higher incidence of chronic myocardial failure in respect to others of different ages.

Methods

Twenty sedentary male subjects aged 40–65 years, without neuromuscular diseases, who died suddenly from head injury or for acute ventricular failure,

Muscle and Motility Vol. 2. Proceedings of XIXth European Conference in Brussels
© Intercept Ltd, PO Box 716, Andover, Hampshire, SP10 1YG, UK

were selected for this investigation. Muscle specimens were removed from the superficial and deep parts of the LDM in the subscapular region. Blocks measuring approximately 1 cm^3 were frozen in isopentane cooled in liquid nitrogen at -170°C for enzyme-histochemistry, Other blocks were fixed in 10% formalin for morphology. Serial transverse cryostat sections were treated for myofibrillar ATPase at pH 9.6, for myofibrillar ATPase EDTA reversal at pH 4.35 and for ATPase at pH 4.35; for DPNH-d and for cytochrome *c* oxidase. According the staining properties for acid myofibrillar ATPase, the fibres were classified into type I, type IIA and IIB (11). The fibre type count and size were made on transverse cryostat sections stained for acid ATPase activities on three micrographs for each subject at the final magnification ×400 (12). Samples of muscle fixed in formalin were embedded in epoxy resin and transverse semithick sections were stained with Dominici's stain and with Gomori's silver impregnation for reticulin: the following values were determined for each of the specimens: capillaries per fibre (C/F); capillary density per square millimetre of muscle fibre (CD) and the number of capillaries surrounding a single fibre (CAF). The C/F ratio was determined by counting the muscle fibre and capillaries on a total of 10 micrographs from four semithin sections (final magnification ×900 (13). Quantitative evaluation of the fibre types and diameters, and of capillaries were performed with an Automatic Interactive Image Analysis System (IBAS I and II, Kontron; Munich,FRG). Data were stored on a floppy disc for the statistical evaluations by mean of the IBAS I system (12, 13).

Results

MUSCLE MORPHOLOGY

The human LDM is composed by large fascicles of polygonal fibres with mean fibre size 40 ± 3.0 μm. A discrete variation in the fibre size is generally observed ,but no structural fibre changes are seen. In the superficial parts of the muscle samples, there is a more marked variation in the fibre diameters. Occasional central nuclei are seen. The muscle is divided by large perymisial septa containing peripheral nerves and vessels; narrow endomysial reticular tissue between muscle fibres is evident. Frequently, grouped fat cells are seen between muscle fibres and in the perymisium.The mean percentage of the fat tissue is 19% of the total cross-sectional area (range 12–25%).

FIBRE TYPE COMPOSITION

The mean fibre diameter and the percentages of type I, IIA and IIB fibres in the 20 autopsy samples taken from the deep portion of the LDM are reported in Table 1. The type I fibres are the largest (range 38–48 μm) and the LDM contains 50% of slow type I fibres (range 48.2–68.2); the smallest type IIA and IIB fibres are 32% and 18%, respectively. In the caudal and superficial parts of the muscle, a higher content of type II fibres is generally observed.

Table 1 Morphometric results on LDM fibre types (mean ±SD)

No. cases	Mean age	Fibre types %			Fibre type diameter µm			Fat tissue*
		I	IIA	IIB	I	IIA	IIB	
20	52	50	32	18	42.2 ±6.4	40.0 ±7.2	38.0 ±4.8	19.0 Range 12–25

*% of the total cross-sectional area

DISTRIBUTION OF CAPILLARIES

The distribution of capillaries in the LDM is reported in Table 2. The capillary measures (C/F; CD; CAF) are similar to those seen in different muscles studied by others (12, 13) and there is an approximate linear relationship between the capillary number and the fibre diameter. In fact, with few exceptions the largest type I fibres are surrounded by a greater number of capillaries than type II fibres.

Table 2 Overall capillary morphometry (C/F; CD; CAF) and mean of capillaries surrounding the type I and II fibres (±SD).

C/F	CD	CAF	Diameter 38–48 Type I	Diameter 28–38 Type II
1.20 ± 0.08	220 ± 6.8	3.2 ± 0.10	3.36*	2.10*

*based on minimum of 100 total fibres (each fibre diameter group).

Discussion

The LDM is an adductor muscle and in normal sedentary people is predominantly used in the inspiratory mouvements. Data on the fibre type distribution of LDM are available in an autopsy study from six male normal subjects aged between 17 and 30 years: in these cases the fibre type I percentage range 35.5–65, with a significant variability of the type I proportions between the samples (10).

In the present study,we selected the LDM specimens from subjects aged 40–65 years because the higher incidence of the irreversible heart failure necessitating of cardiac transplantation or cardiomyoplasty. The main morphological findings in the present study are:

1. The presence of abundant fat tissue seen in intrafascicular and perifascicular areas; this might be a negative parameter in the use of the LDM for cardiomyoplasty and electrostimulation.
2. the high variability in the fibre type composition in different subjects, with frequent type I fibre predominance, characterizing the slow muscles. In the

normal subjects,type I predominance is said to occur when more than 55% of fibres are type I (11). In the present study, four cases have shown a true type I predominance. In half of the cases, the muscle surface has shown a higher proportion of type II fibres. However, it is known that differences in the fibre type composition are detected in muscle samples taken from the superficial and deep parts of muscles (10). The fibre type composition of the human LDM differs from that in animals,in which the muscle shows the properties of a fast and fatiguing muscle containing 13–23% of type I fibres (8, 9). The rare studies on the fibre type composition of the human LDM revealed 50% of type I fibres, but no differentiation between type IIA and IIB fibres was performed (10, 15).

3. In the past, studies on human muscle microvasculature have been performed on normal sedentary subjects and on athletes. These studies revealed a relationship among fibre diameter,type and capillary number. Typically, the greatest number of capillaries surround the type I fibres with the largest diameter (12, 13). In the LDM, the high number of capillaries per fibre and the elevated capillary density reflects the numerical predominance of the type I muscle fibres with the highest levels of oxidative enzymes.

The sequential and progressive muscle electrostimulation following cardiomyoplasty results in the transformation of the fast glycolytic type II fibres into slow, oxidative type I fibres, having the enzymehistochemical characteristic of the myocardium (6-8-9). The results obtained in the present study emphasize that the LDM part usually utilized for cardiomyoplasty shows predominant characteristic of a slow, fatigue resistant muscle, with an high capillary number per fibre. These parameters indicate that the LDM is an adequate muscle for the application of functional electrostimulation.

References

1. Acker, M. A., Anderson, W. A., Hammond, R. L., Chin, A. S., Buchanan, J. W. Morse, C. C., Kelly, A. M., Stephenson, L. W., 1987, *J. Thorac. Cardiovasc. Surg.*, **94**, 163–174.
2. Acker, M. A., Hammond, R. L., Mannion, J. D., Salmons, S., Stephenson, L.W.,1987 *Science*, **236**, 324–327.
3. Carpentier, A., Chachques, J. C., 1985, *Lancet*, **1**, 8440.
4. Carraro, U., Catani, C., Dell'Antone, P., Danieli-Betto, D., Arpesella,A., Parlapiano, M., Cirillo, M., Albanese, S., Pierangeli, A.1988. In: Carraro, U. (ed.), *Sarcomeric and Non-Sarcomeric Muscles: Basic and Applied Research Prospects for the 90s*, pp. 459–470. Unipress, Padova.
5. Chachques, J C., Grandjean, P. A., Vasseur, B., Hero, M., Perier, P., Bourgeois J., Eardeau, M., Carpentier, A., 1986. In: Nose', Y. (ed.) *Progress in Artificial Organs*. pp. 409–412. ISAO Press, Cleveland.
6. Chachques, J. C., Grandjean, P. A., Tommasi, J. J., Perier, P., Chavaud, S., Bourgeois, I., Carpentier, A., 1987, *Life -Support Systems*, **5**, 323–327.
7. Chachques, J. C., Grandjean, P. A., Schwartz, K., Mihaileanu, S., Fardeau, M., Swynghedauw, B., Fontaliran, F., Romero, N., Wisniewsky, C., Perier, P.,

Chavaud, S., Bourgeois, I., Carpentier, A., 1988, *Circulation*, **78** (suppl III), 203–216.

8. Chachques, J .C., Grandjean, P. A., Schwartz, K., Pfeffer, T. A., Fardeau, M., Nataf, P., Carpentier, A., 1988. In: Carraro, U., (ed.), *Sarcomeric and Non-Sarcomeric Muscles: Basic and Applied Research Prospects for the 90s*, pp. 429-447. Unipress, Padova.
9. Gruber, H., Frey, M., Happak, W., Burggasser, G., Mayr, W., Losert, U., Deutinger, M., Windberger, M., Thoma, H., In: Carraro, U. (ed.), *Sarcomeric and Non-Sarcomeric Muscles: Basic and Applied Research Prospects for the 90s*, pp. 471476. Unipress, Padova.
10. Johnson, M. A., Polgar, J., Weightman, D., Appleton, D., 1973, *J. Neurol. Sci.*,**18**, 111–129.
11. Dubowitz, V., *Muscle Biopsy: a Practical Approach.* 1985, Baillière Tindall, London.
12. Scelsi, R., Marchetti, C., Poggi, P., 1980, *Acta Neuropath.(Berl.)*., **51**, 99–105.
13. Carry, M. R., Ringel, S. P., Starcevich, J. M., 1986, *Muscle and Nerve*, **9**, 445–454.
14. Poggi, P., Marchetti, C., Scelsi, R., 1987, *Anatomical Record*, **217**, 30–34.
15. Zumstein, A., Mathieu, O., Howald, H., Hoppeler, H., 1983, *Pflügers Arch.*, **397**, 277–283.

28
In situ Hybridization Study on Myosin Heavy Chain Expression in Human Heart Failure

T. BRAND, J. SCHAPER, B. MÜNKEL, H. S. SHARMA, N. BLEESE* AND W. SCHAPER

*Max-Planck Institute, Department of Experimental Cardiology and *Kerckhoff-Clinic, Department of Thoracic and Cardiovascular Surgery, Benekestrasse 2, 6350 Bad Nauheim, Germany*

Introduction

Extensive research over the last few years has revealed a complex regulation of myosin heavy chain (MHC) isoform expression in heart. In rodents, different MHC isoforms are expressed under normal and pathophysiological conditions (1). However, in humans, ventricular myocytes express mainly β-MHC, and no change in isoform expression has been detected in diseased heart (2). In contrast, measurements of myofibrillar ATPase activity in congestive heart failure revealed a 20–30% decrease in enzymatic activity (3), of which the molecular basis is not known. Previous ultrastructural studies of human failing heart revealed degenerative changes in myocytes, characterized by loss of contractile material, increased nuclear size, and proliferation of T-tubules (4). By immunohistochemistry decreased staining for myosin and actin was detected, cross striation was diminished and was missing in many areas of the cell (5). To clarify whether the pathological changes in ultrastructure and protein expression are due to an altered pattern of myocyte gene expression, we studied by *in situ* hybridization MHC gene expression in human heart failure.

Muscle and Motility Vol. 2. Proceedings of XIXth European Conference in Brussels
© Intercept Ltd, PO Box 716, Andover, Hampshire, SP10 1YG, UK

Material and Methods

Tissue samples from human hearts with end-stage heart failure were excised from left ventricle during transplantation surgery. For comparison, normal papillary muscle from a patient undergoing coronary bypass surgery as well as heart, soleus, extensor digitorum longus (EDL) and liver from adult rat (Wistar; -200 g) were used. Samples were snap frozen in liquid N_2 and were stored at -70°C. Frozen sections (5 μm) were cut on a Cryostat and fixed in 4% paraformaldehyde, washed in distilled water and dehydrated through graded ethanols. Without further pretreatment 50 μl hybridization buffer (0.3 M NaCl, 10 mM NaH_2PO_4, pH 6.8, 10 mM Tris-Cl, pH 7.5, 5 mM EDTA, pH 8.0, 0.02% BSA, 0.02% Ficoll, 0.02% polyvinylpyrrolidone, 50% formamide, 100 μg/ml yeast RNA, 10 mM DTT, 10% dextran sulphate) containing 100 ng/ml [^{35}S]-UTP labelled cRNA probe, was applied to the sections and hybridized over night at 50°C. Sections were washed three times in wash buffer (0.3 M NaCl, 10 mM NaH_2PO_4, pH 6.8, 10 mM Tris-Cl, pH 7.5, 5 mM EDTA, pH 8.0, 10 mM DTT and 50% formamide) at 50°C for 1 h. The sections were treated with RNAse A for 30 min at 37°C and finally washed twice in 2 x SSC at 65°C for 30 min. For autoradiography, sections were coated with LM-1 emulsion (Amersham) and were exposed for 12 days at 4°C. For the synthesis of cRNA, a cDNA clone (pBSa/b) (6) containing a 528 bp insert common in sequence to α- and β-MHC mRNA was used. Tissues used for *in situ* hybridization were also analysed by northern blotting. Total RNA was isolated (7) and size-fractionated by formaldehyde-agarose-gelelectrophoresis. RNA was vaccuum transferred to Hybond-N membrane (Amersham) and hybridized under stringent conditions (8) with MHC cDNA and subsequently probed with GAPDH cDNA (9).

Results

Film autoradiographs of *in situ* hybridization with antisense and sense probe for MHC showed that our hybridization conditions were specific. With antisense probe only, heart and soleus were labelled, whereas with sense probe a faint

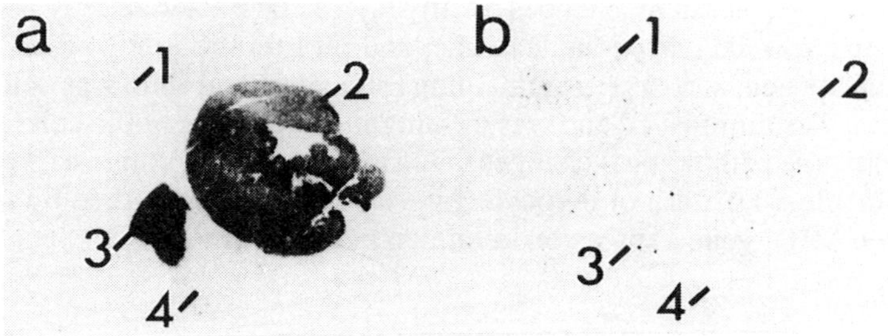

Fig. 1 Film autoradiography of rat tissues hybridized with (a) antisense MHC cRNA probe, (b) with sense probe. 1–liver; 2–ventricle; 3–soleus muscle; 4–EDL muscle.

background labelling of all tissues was seen (Fig. 1). Microscopic examination of autoradiographs of MHC antisense probe hybridized to soleus muscle revealed a checkerboard pattern of labelled and unlabelled fibres (Fig. 2). A

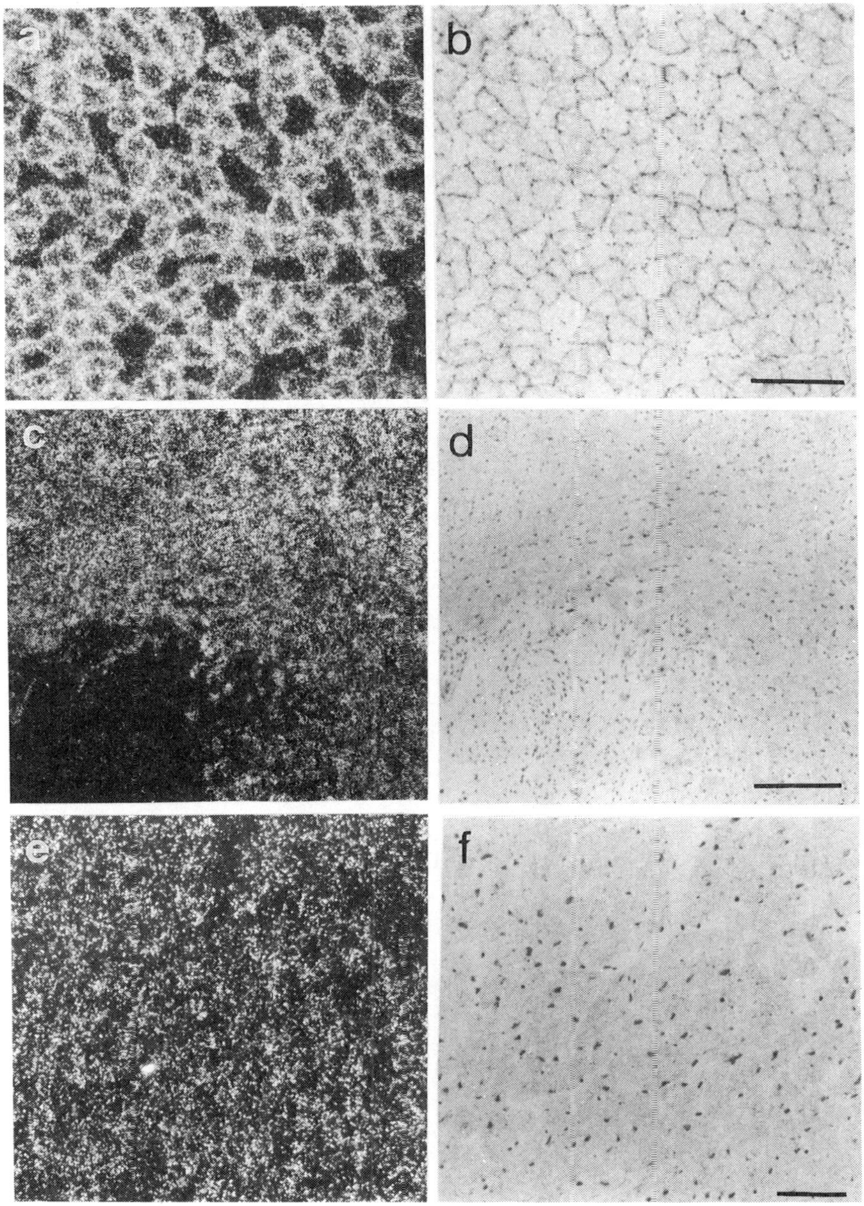

Fig. 2 *In situ* hybridization of rat and human tissues hybridized with antisense MHC probe: (a, c, e) dark field; (b, d, f) light field; (a, b) soleus muscle; (c, d) rat ventricle; (e, f) normal human papillary muscle. Bar represents 200 µm in a, b; 100 µm in c.

nonhomogeneous distribution of MHC mRNA was noticed, labelling was strongest near the sarcolemma. In rat heart, myocytes were uniformly labelled,

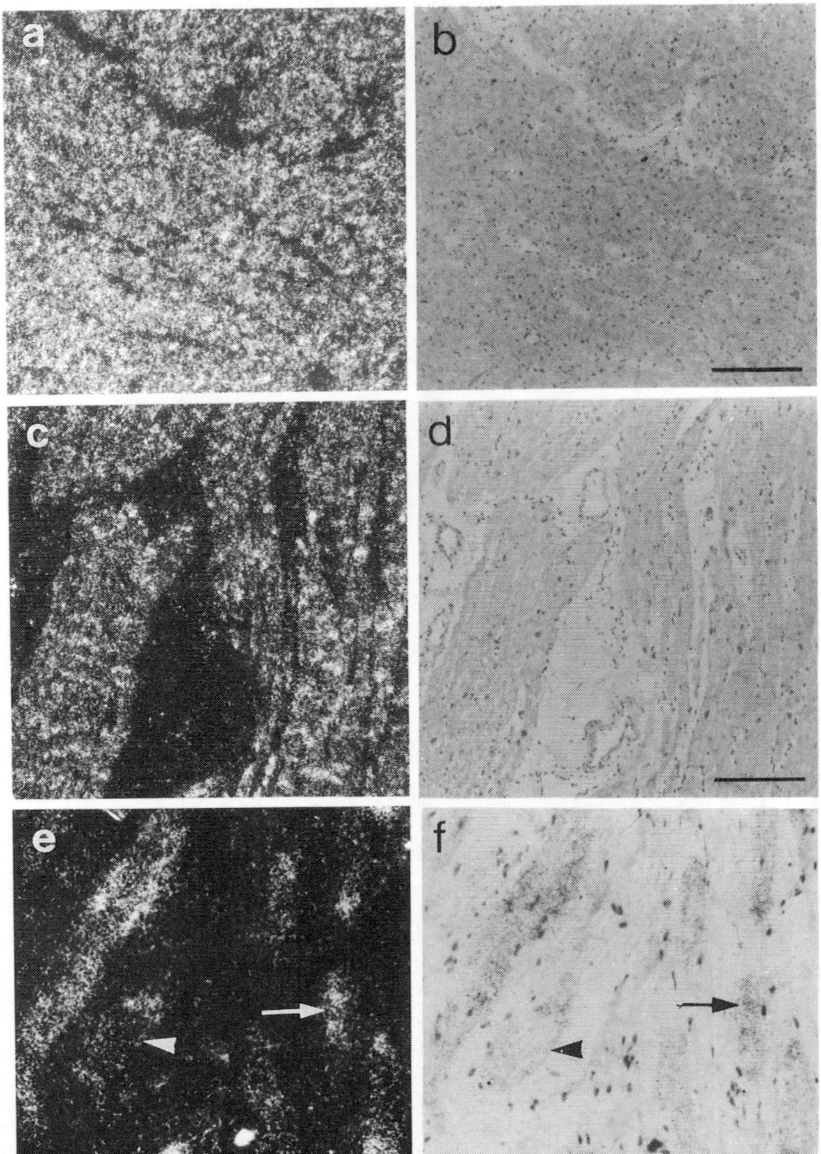

Fig. 3 MHC *in situ* hybridization of human heart failure specimens. (a, c, e) dark field; (b, d, f) light field; (a, b) normal appearing specimen; (c, d) biopsy with high amount of nonhybridizing fibrotic tissue; (e, f) heterogenous labelling of individual myocytes — unlabelled myocyte (arrowhead); normal labelled myocyte (arrow). Bar represents 500 μm in a, b; 100 μm in c.

whereas extracellular matrix and capillaries were unlabelled (Fig. 2). Normal human heart muscle showed a similar labeling pattern compared to rat heart (Fig. 2). However, in diseased human hearts a different picture emerged. Large areas of tissue were unlabelled, indicating a high degree of fibrosis (Fig. 3). However, strongly labelled strands of myocytes were also observed in these specimens. In some patients, we observed myocytes with variable grain densities; some myocytes were nearly unlabelled (Fig. 3). In rare cases, specimens did not show large amounts of fibrotic tissue and labeling intensity was similar to normal human heart (Fig. 3). Northern blot hybridization of tissues used for *in situ* hybridization confirmed our histological results. In all specimens of human heart failure a lower MHC/GAPDH ratio, compared to control heart, was observed. However, the extent of reduction was quite variable and ranged between 20–80% of control values (Fig. 4).

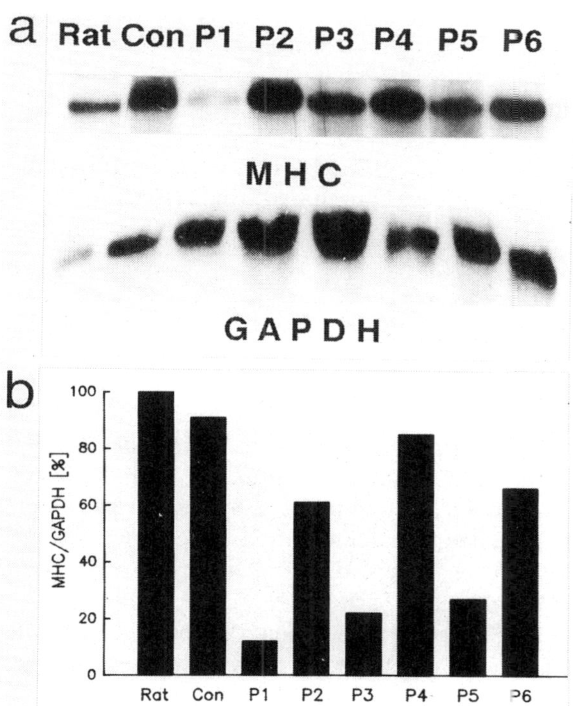

Fig. 4 Northern blot analysis of mRNA content for MHC: (a) hybridization with MHC- and GAPDH- cDNA probe; (b) quantification by densitometric analysis. Values are expressed as percentage of MHC/GAPDH ratio of rat left ventricle.

Discussion

In situ hybridization is a powerful technique to visualize gene expression in morphological intact tissue. In skeletal muscle, mRNA for MHC was mainly localized beneath the sarcolemma, this was also reported by Dix (10) and may be due to the subcellular distribution of polysomes in skeletal muscle which are mainly located subsarcolemmal (11). Some fibres in soleus muscle were unreactive with our probe; these fibres presumably express type II MHC isoform (12). We demonstrated, that myocytes of human heart failure express MHC gene in an apparently normal way. Even myocytes surrounded by a huge amount of fibrotic tissue retain the capability for MHC expression. Only in very rare cases we observed unlabelled myocytes. This result was unexpected as with immunohistochemistry a reduced amount of MHC protein was detected (5). With northern blotting, a reduced MHC mRNA concentration in failing human heart was observed, but reduction was quite variable and may represent the extent of fibrosis in each biopsy. We therefore conclude that in human heart failure a normal MHC gene expression occurs, irrespective of diminished amounts of MHC protein and altered myofibrillar organization. It is therefore very likely that defects at the translational level and/or defects of myofibrillogenesis are important factors in this disease.

Acknowledgements

We thank Dr K. Schwartz, INSERM U127, Paris for providing us the pBSa/b clone. The GAPDH probe was purchased from ATTC.

References

1. Swynghedauw, B., 1986. *Physiol. Rev.*, **66**, 710–771.
2. Mercadier, J. J., Bouveret, P., Gorza, L., Schiaffino, S., Clark, W. A., Zak, R., Swynghedauw, B., Schwartz K., 1983. *Circ. Res.*, **53**, 52–63.
3. Alpert, N. R., Gordon, M. S., 1962. *Am. J. Physiol.*, **202**, 940–946.
4. Schaper, J., Froede, R., Buck, A., Bleese, N., 1988. In: Schultheib H.-P. (ed.), *New Concepts in Viral Heart Disease*. Springer, Heidelberg, 295–302.
5. Schaper, J., Bernotat-Danielowski, S., Froede, R., Hein, S., Konertz, W., Bleese, N., 1990. In: Lewis, H.P., Kinchi, A. (eds), *Heart Failure — Mechanisms and Management*. Springer, New York, in press.
6. Schiaffino, S., Samuel, J. L., Sassoon, D., Lompre, A. M., Garner, I., Marotte F., Buckingham, M., Rappaport, L., Schwartz, K., 1989. *Circ. Res.*, **64**, 937–948.
7. Chomzynski, P., Sacchi, N., 1985. *Anal. Biochem.*, **1987**, 162, 156–159.
8. Sambrook, J., Fritsch, E. F., Maniatis, T., 1989. *Molecular Cloning: A Laboratory Manual*. Cold Spring Harbor, New York.
9. Tso, Y. J., Sun, X. H., Kao, T., R. Wu, 1985. *Nucl. Acid Res.*, **13**, 2485–2501.
10. Dix, D., Eisenberg, B., 1988. *J. Histochem. Cytochem.*, **36**, 1519–1526.
11. Horne, Z., Hesketh, J., 1990. *Biochem. J.*, **268**, 231–236.
12. Butler-Browne, G. S., Whalen, R. G., 1984. *Dev. Biol.*, **102**, 324–334.

29
Is Atrial Light Chain 1 a Marker of End-stage Failure in Dilated Cardiomyopathy?

T. TRAHAIR*, T. YEOH*, T. CARTMILL**, V. CHANG†, P. GUNNING†† AND C. DOS REMEDIOS*

*Muscle Research Unit, Department of Anatomy, The University of Sydney, ††Muscle Genetics Unit, Children's Medical Research Foundation, **Department of Paediatric Surgery, Royal Alexandra Hospital for Children, Camperdown, †National Heart Transplantation Program, St. Vincent's Hospital, Darlinghurst, Australia

Introduction

The contractile apparatus consists chiefly of two proteins, actin and myosin, both of which are essential for the development of force and movement. Molecules of actin (G-actin) polymerize to form filaments (F-actin). Specific myosin isoforms are expressed according to tissue type and developmental stage (1). Ventricular myosin consists of six subunits, two myosin heavy chains (MHCs), two alkali light chains (VLC1) and two regulatory light chains (VLC2). Atrial light chain one (ALC1) is a major mammalian developmental isoform expressed in embryonic atria, ventricle and skeletal muscle (12). Postnatally ALC1 found in embryonic ventricle is replaced by VLC1 (3, 4) (i.e. ALC1 is not found in normal adult ventricles). ALC1 is normally expressed only in adult atria, perhaps reflecting the more primitive nature of this chamber (5). There are recent reports that an ALC1-like protein is expressed in diseased ventricles (6–9). High levels of this protein are encountered in cases of dilated cardiomyopathy (CM) (7, 8) and congenital heart disease (CHD) (6, 9). The presence of this atrial isoform appears to be related to wall stress in the ventricle (7, 8) and could be a compensatory mechanism for cardiac overload. Preliminary data found no evidence for significant expression of ALC1 in

Muscle and Motility Vol. 2. Proceedings of XIXth European Conference in Brussels
© Intercept Ltd, PO Box 716, Andover, Hampshire, SP10 1YG, UK

dilated CM, which is in apparent contrast to others. Therefore we have investigated expression of ALC1 in both CM and CHD.

Methods

Cardiomyopathic (CM) samples were obtained during heart transplantation. Control left atrial specimens came from donor hearts. Control left ventricular samples came from an atrial septal defect and from one rheumatic heart disease. One atrial and five ventricular specimens were obtained from children during repair of congenital heart disease. All specimens were obtained within minutes of cardiopulmonary bypass and stored in liquid nitrogen. Total cellular RNA was extracted using the acid guanidinium-phenol-chloroform method of Chomczynski *et al.* (10). The final RNA pellet was dissolved in sterile water. Yield was determined from the OD at 260 nm, purity by running 0.5–1 μg of RNA in an agarose mini-gel. Northern blots and hybridization were carried out as before (11). The ALC1 probe was labelled with α^{32}PdCTP using the Bresatec OLK-C random priming kit, except a specific primer was used instead of the hexanucleotide buffer provided. Crude myosin was prepared according to Price *et al* (4), except that the final pellet was dissolved in 5 mM phosphate buffer with 0.1 mM EDTA, 1% SDS and 0.1% β-mercaptoethanol. Extracted protein was stored at -80°C. Crude myosin was electrophoresed in SDS PAGE, using a 5% stack with either a 10% or 14% separating gel (12), and stained with Coomassie brilliant blue or electroblotted (BioRad Transblot) onto nitrocellulose (BioRad). Total protein was stained 0.45% amido black to visualize protein bands, unstained membrane was probed using the 2h2 monoclonal antibody (13).

Results

Examination of ventricular myosin extracts separated in a 10% stacking gel revealed a band migrating close to the position of ALC1 (result not shown). When the same samples were run in a 14% gel this band migrated significantly slower than ALC1 (Fig.1). Lane 1 contains crude myosin from a CM atrial sample with the position of ALC1 and ALC2 bands marked. Lane 2 contains ventricular crude myosin (from the same patient), notice the band higher than ALC1 which migrated at the position of ALC1 in a 10% gel. Clearly ALC1 is not present in this sample. Both VLC1 and VLC2 are marked. Lane 3 contains both atrial and ventricular samples to demonstrate that these bands are distinct. This pattern is characteristic of the CM tissue examined. Western blots were then probed with the 2h2 monoclonal antibody which binds specifically to myosin light chain isoforms (13). The 2h2 antibody binds to VLC1 (W. Boey, pers. comm.) and these results shows that 2h2 binds ALC1, VLC2 but not ALC2. Sequence analysis of ALC1 (14) and VLC2 (15) shows that both contain the proposed 2h2 epitope (W. Boey, pers. comm.). Therefore if ALC1 is expressed in CM ventricles there should be three distinct bands corresponding

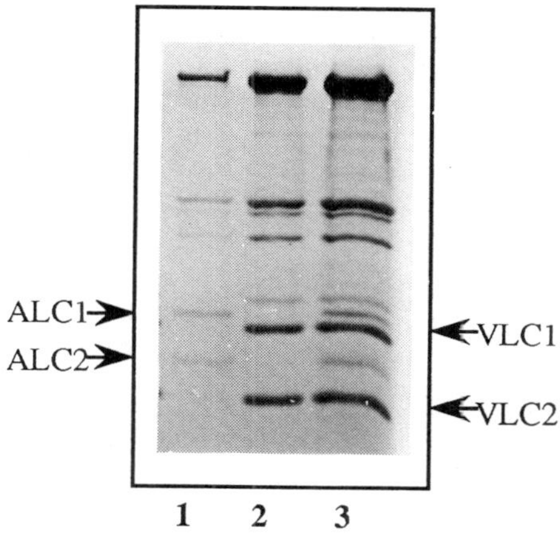

Fig. 1 SDS PAGE of CM crude myosin extracts. Lane 1, CM atrial crude myosin showing ALC1, ALC2. Lane 2, CM ventricular crude myosin with VLC1 and VLC2 marked. Note the band migrating above ALC1 in this sample. Lane 3 shows comigration of atrial and ventricular myosins demonstrating that bands are distinct.

to ALC1, VLC1 and VLC2 in that order. Part of the membrane was stained with 0.45 % amido black to identify protein bands (Fig. 2, panel A). Lanes 1 and 2 contain control ventricular and atrial crude myosin extracts respectively. Fig. 2, panel B shows the membrane probed with the 2h2 antibody. Lane 1 contains a control atrial crude myosin extract while lane 2 contains control ventricular crude myosin. Fig. 2, panel C contains different CM ventricular crude myosins (Lanes 1, 2, 3 CM ventricular crude myosins). Two distinct bands in the ventricular extracts (corresponding to VLC1 and VLC2) can be seen as well as 2 bands in the control atrial lane (corresponding to ALC1 and VLC2). Expression of VLC2 in normal adult atria has been shown to occur. Western blot analysis shows that ALC1 is not significantly expressed in these CM samples.

Northern blots of CM and CHD samples were used to examine expression of ALC1 mRNA. The ALC1 cDNA probe is specific for the ALC1 mRNA 3′ untranslated region (3′ UTR) and does not bind to other mRNAs. Northern blot analysis shows that ALC1 mRNA is clearly expressed in CHD but not in CM. Fig. 3, panel A shows the results for CHD. Lanes 1 to 5 contain CHD ventricular RNA. Lane 6 contains CHD atrial RNA. Lanes 7 and 8 contain control ventricular and atrial RNA respectively. The ALC1 band can clearly be seen in all the CHD ventricular samples as well as in both atria. Panel B shows a northern blot analysis of CM ventricles compared to one atrial sample. Lanes 1, 3, 4, 7, 8 contain CM ventricular RNA, lane 2 has CM atrial RNA (note the clear band), lanes 5 & 6 contain control ventricular RNA. Clearly ALC1 mRNA is not expressed in CM ventricles.

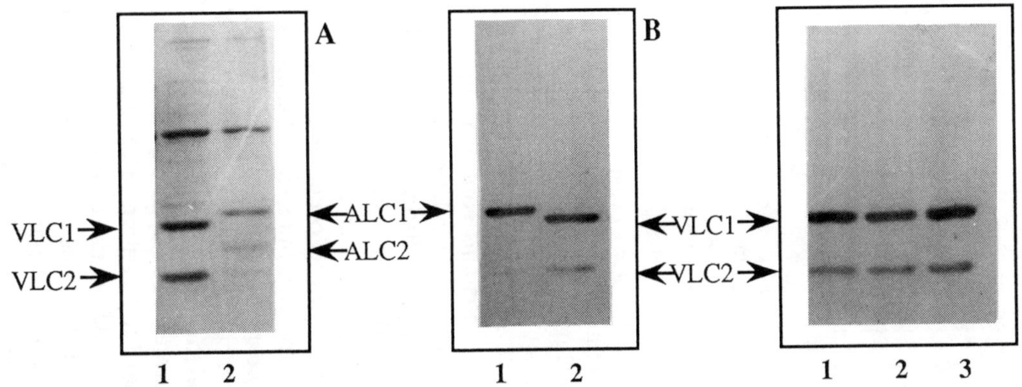

Fig.2 Western blot using 2h2 antibody.
Panel A. Total Protein stained with 0.45% amido black. Lane 1, control ventricular crude myosin with VLC1 and VLC2 marked. Lane 2, control atrial crude myosin showing ALC1 and ALC2.
Panel B Control atria and ventricle probed with 2h2 antibody. Lane 1, control atria showing ALC1 band. Lane 2 control ventricle with clear bands corresponding to VLC1 and VLC2, note the absence of ALC1.
Panel C CM ventricles probed with 2h2 antibody. Lanes 1, 2, 3 contain different CM ventricle. In all samples note the distinct bands corresponding to VLC1 and VLC2. ALC1 is clearly not expressed in CM tissue.

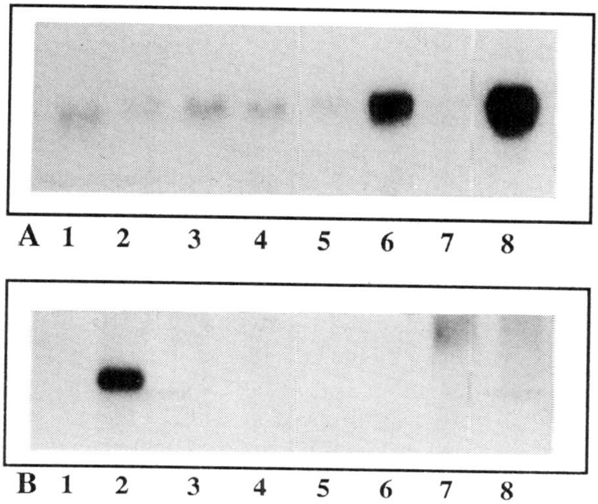

Fig. 3 Northern Blot Analysis.
Panel A: CHD RNA probed for ALC mRNA. Lanes 1–5, CHD right ventricular RNA. Lane 6, CHD right atrial RNA. Lane 7, control ventricular RNA. Lane 8, control atrial RNA. 5 µg RNA per lane. *Panel B*: CM RNA probed for ALC1 mRNA. Lanes 1, 3, 4, 7, 8, CM ventricular RNA. Lane 2, CM atrial RNA. Lanes 5, 6, control ventricular RNA. 5 µg RNA per lane.

Discussion

ALC1 is expressed in foetal ventricle but is replaced post-natally by the ventricular isoform (VLC1) (3, 4). ALC1 is not found in normal ventricle but has been clearly demonstrated in CHD, using two-dimensional gels (6). It appears that ALC1 actually replaces VLC1 (6, 7). Our data on ALC1 mRNA expression in CHD confirm this. Hirzel *et al* found a significant correlation between the expression of ALC1 and average peak circumferential wall stress in overloaded ventricle (7, 8). The reappearance of ALC1 may reflect the plasticity (adaptive capability) of the overloaded myocardium.

Our CM results showing a clear lack of ALC1 expression, might seem to conflict with those of Schaub and Hirzel (8) and Hirzel *et al.* (7), who found significant expression of ALC1 in CM and a tendency for decreasing expression with disease progression. However, their tissue samples came from myocardial biopsies and they stated 'none of the patients was in clinically apparent heart failure' (7). Symptoms displayed by this group could be classified as class 1 or 2 according to the New York Heart Association (NYHA) criteria for cardiac functional capacity (i.e. slight limitation of physical activity). Our patients were heart transplant recipients who displayed clinical signs and symptoms of advanced cardiac failure (orthopnoea, hepatosplenomegaly, increased jugular venous pressure and ankle oedema) and were classified as NYHA class 4 or worse (i.e. unable to carry on any physical activity with signs of cardiac insufficiency at rest). Thus the overwhelming clinical picture of our patient population is end stage cardiac failure due to CM. The difference between these two patient populations is cardiac failure. The deficit of ALC1 expression in CM tissue we examined could be a function of cardiac failure. This underlies the concept of disease progression in CM. There is clear evidence for ALC1 expression early in the disease, before end-stage failure, and a clear deficit of expression late in CM when heart failure supervenes.

Timing of surgical repair in CHD is a compromise between the age of the patient and the degree of cardiac compensation for the defect. Usually definitive repair is left for as long as the patient's condition will allow, without risking irreversible myocardial damage and generally occurs before the onset of end-stage heart failure to ensure a successful outcome. In CHD the heart is still able to adapt to overload, evidenced by the lack of end stage failure. However, in the CM group, end-stage cardiac failure (i.e. failure of cardiac adaptation) dominates the clinical picture. Therefore differences in ALC1 expression between these two groups could relate to the progress of cardiac adaptation to overload. Hirzel *et al.* (7) speculated that ALC1 expression could be lost when the heart loses its capacity to compensate overload. Our data appears to support this interpretation. While the heart is still able to cope with overload ALC1 is expressed (as seen in CHD), but when adaptive capability has been exhausted (end-stage CM) ALC1 is no longer expressed.

This leads us to the hypothesis that the loss of ALC1 expression in CM is a reflection of end-stage heart failure and the loss of adaptive capacity. The corollary is that expression of ALC1 earlier in the disease could be a measure of cardiac plasticity, i.e. continuing ability to respond to stress. Therefore it is important to examine the time course of ALC1 expression in CM and its possible relation to progression of clinical symptoms of heart failure. Expression of ALC1 could be a marker of continuing cardiac adaptation in early CM and the inability to express ALC1 might be a sign of end-stage CM and irreversible cardiac failure.

Acknowledgements

One of us was supported by Beckman Instruments, Australia. This research was supported by grants from the National Health & Medical Research Council of Australia.

References

1. Swynghedauw, B., 1986, *Physiol. Rev.*, **66**, 710–771.
2. Barton, P., Buckingham, M., 1985, *Biochem. J.*, **231**, 249–261.
3. Cummins, P., Price, K., Littler, A., 1980, *J. Muscle Res. Cell. Motil.*, **1**, 357–366.
4. Price, K., Littler, A., Cummins, P., 1980, *Biochem. J.*, 191, 571–580.
5. Cummins, P. 1983. In: Alpert, N. (ed.), *Perspectives in Cardiovascular Research*, **7**, pp. 417–423. Raven Press, New York.

6. Auckland, L., Lambert, S., Cummins, P., 1986, *Cardiovasc. Res.*, **20**, 828–836.
7. Hirzel, H., Tuchschmid, C., Schneider, J., Krayenbuehl, H., Schaub, M. 1985, *Circ. Res.*, **57**, 729–740.
8. Schaub, M., Hirzel, H., 1987, *Basic Res. Cardiol.*, **82**, (Suppl 2), 357–367.
9. Tuchschmid, C., Srihari, T., Hirzel, H., Schaub, M., 1983. In: Jacob, R., Gulch, R., Kissling, G. (eds), *Cardiac Adaptation to Hemodynamic Overload, Training and Stress*, pp. 123–128. Steinkopff, Darmstadt.
10. Chomczynski, P., Sacchi, N. 1987, *Anal. Biochem.*, **162**, 156–159.
11. Gunning, P., Hardeman, E., Wade, R., Ponte, P., Bains, W., Blau, H., Kedes, L., 1987, *Mol. Cell Biol.*, **7**, 4100–4114.
12. Laemmli, U., 1970, *Nature*, **227**, 131–135.
13. Noakes, P., Everett, A., Bennett, M., 1986, *J. Comp. Neurol.*, **248**, 245–256.
14. Kurabayashi, M., Komuro, I., Tsuchimochi, H., Takaki, F., Yazaki, Y. 1988, *J. Biol. Chem.*, **263**, 1393–13936.
15. Dalla Libera, L., Hoffman, E., Floroff, M., Jackowski, G., 1989, *Nucleic Acids Res.*, **17**, 2360.
16. Cummins, P., Russell, G. 1986, *Comp. Biochem. Physiol.*, **84B**, 343–348.

SECTION FIVE

Intracellular Regulations

30
The Regulation of Catch in Molluscan Muscles

R.K. ACHAZI, R. HAUCK, G. HELLWIG, M. RUPP-DAHLEM AND M. SAILER

Institut für Tierphysiologie, Freie Universität Berlin, Grunewaldstraße 34, D-1000 Berlin 41, Germany

Introduction

Bivalve skeletal muscles, like shell adductor muscles, display different ultrastructural features according to their physiological function. The fast, phasic fibres of these adductors may be either obliquely or cross-striated, while the slow, tonic fibres are smooth (1). The regulatory system of the fast fibres is myosin bound, at least in scallops, and does not involve the phosphorylation of myosin light chains (2). In the smooth, tonic fibres phosphorylation of contractile proteins seems to be important for regulation (3–6). The details of this mechanism are still unknown.

Muscle and Motility Vol. 2. Proceedings of XIXth European Conference in Brussels
© Intercept Ltd, PO Box 716, Andover, Hampshire, SP10 1YG, UK

Methods

UTRASTRUCTURAL INVESTIGATIONS

Unskinned muscle fibres were mounted into a transducer and immersed in unbuffered, artificial sea water of 15°C. The different states of contraction were induced with 10^{-5} mol of FMRFamide and 5-Ht. The preparation for electron microscopy was performed according to Hauck *et al.*(7).

InsPn ANALYSIS

ABRM fibres were incubated in 500 nmol/l ^{3}H-inositol for 20 h (10°) till isotopic equilibrium, washed three times and incubated with 10^{-5} mol/l ACh or 5-Ht. InsPn were analysed according to Berridge (8).

PROTEIN KINASES

Isolation and characterization of the enzymes were performed according to Sailer *et al.*(9). The subunits of kinase A were separated by cAMP-agarose chromatography.

IN SITU PHOSPHORYLATION

In situ phosphorylation of contractile proteins with ^{32}P-γ-ATP was performed with saponin skinned fibres pre-incubated with ^{32}P-γ-ATP. Active contraction was induced by 10^{-5} mol/l Ca^{2+}, catch by reduction of free Ca^{2+} to $<10^{-7}$ mol/l and relaxation by 5-Ht. The fibre bundles were freeze clamped at the desired functional state and analysed.

Results

THE FUSING OF THICK FILAMENTS IN CATCH

The paramyosin hypothesis of catch (10) is supported by ultrastructural results, showing the melting of thick filaments in this stage (7, 11, 12). So far conclusive results have been obtained only with skinned ABRM fibres. These investigations were performed on unskinned fibres of the posterior shell adductor muscles (PAM) of *Mytilus edulis* and *Mya arenaria* with FMRFamide and 5-Ht. The PAM of *M. edulis* contains smooth fibres only, while the one of *M. arenaria* contains two fibre types, present in a 1:3 ratio. Type I fibres can be classified as smooth, displaying thick filaments of an extraordinary large cross-sectional area, whereas type II fibres resembles obliquely striated fibres , because of the size and the arrangement of the thick filaments.

In both muscles removal of FRMFamide resulted in the fusion of thick filaments of the smooth fibres, while the thick filaments of juxtaposed type II fibres in the *Mya* adductor did not fuse (Fig.1). Addition of 5-Ht resulted in the segregation of the fused thick filaments and in instant relaxation.

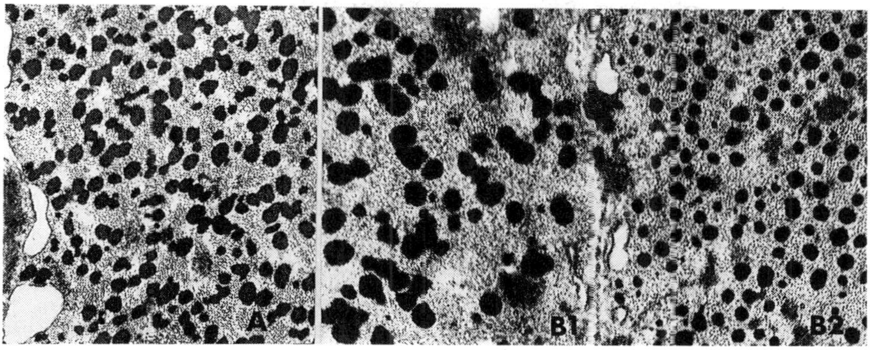

Fig. 1 Cross-section of posterior molluscan shell adductor muscles (PAM) in catch. A: PAM of *Mytilus edulis*; b: PAM *Mya arenaria*, B1: type I fibres (smooth) B2: type II fibre (obliquely striated); ×33 500

THE PHARMACOMECHANICAL COUPLING IN THE ABRM

Stimulation of muscarinic ACh- as well as 5-Ht$_2$-receptors induce physiological changes by the second messenger InsP$_3$. Pharmacological evidences suggest that these two receptor subtypes are present in the ABRM of *M. edulis* besides the nicotinic ACh- and 5-Ht$_1$-receptors (13, 14). The involvement of InsP$_n$ in the contraction process was tested with ABRM fibres equilibrated with ^{3}H-inositol. Incubation with ACh increased the cytosolic concentration of ^{3}H-InsP$_3$ and -InsP$_2$ already during the first 10 s. In presence of 5-Ht their accumulation was significant only after 25 s (Table 1). The cytosolic ^{3}H-InsP concentration did not change significantly during this period.

Table 1 The time dependent effect of 10^{-5} mol/l ACh and 5-Ht on the concentration of ^{3}H-InsP$_3$ and ^{3}H-InsP$_2$ (mean ± SD) in the ABRM of *M.edulis* (n = 3).

	ACh		5-Ht	
Incub.time (s)	^{3}H-InsP$_3$ (fmol/mg)	^{3}H-InsP$_2$ (fmol/mg)	^{3}H-InsP$_3$ (fmol/mg)	^{3}H-InsP$_2$ (fmol/mg)
0	56 ± 22	19 ± 3	44 ± 3	25 ± 1
10	103 ± 16	28 ± 3	50 ± 16	28 ± 9
25	122 ± 13	34 ± 6	72 ± 6	44 ± 7
60	116 ± 16	38 ± 7	78 ± 16	44 ± 9

PROTEIN KINASES IN *MYTILUS* CATCH MUSCLE

In the catch muscles of *M. edulis* the protein kinases A, C, G and CaM dependent protein kinase should be present according to the second messenger systems demonstrated. So far protein kinase A and CaM dependent kinase are characterized (Table 2). The monomeric CaM dependent kinase displays a narrow protein specificity. Its favourite substrate is paramyosin, whose

Table 2 Characterization of the Ca^{2+}-CaM dependent (CaM Kin.) and the cAMP-dependent (Kin. A) protein kinases of *M. edulis* catch muscle.

	Molecular weight Holoenz. (kD)	Subunit (kD)	Activator (mol/l)	Max. ^{32}P-incorp. in Pm mol/mole	Kinase substrates
CaM Kin.	36 ± 2	—	Ca^{2+} 10^{-5}	3.9 ± 0.8	Pm (MHC, 42kD)
Kin. A	145 ± 10	r:32 ± 2 c:41 ± 3	cAMP 10^{-5}	0.9 ± 0.4	Pm, casein histon IIa CaM kinase (MHC, 42kD)

r: regulatory subunit, c: catalytic subunit, MHC: myosin heavy chain, Pm: paramyosin.

phosphorylation is completed in 2 min. The kinase A is tetrameric and has a broad substrate specificity. Paramyosin phosphorylation is complete only after 40 min. The CaM dependent kinase seems to be substrate of the kinase A.

PHOSPHORYLATION OF CONTRACTILE PROTEINS OF SKINNED ABRM FIBRES IN DIFFERENT FUNCTIONAL STATES

Skinned ABRM fibres incorporated during active contraction and 5-Ht-induced relaxation about two times the amount of ^{32}P/mg protein as during resting and catch state (Table 3). SDS-PAGE followed by autoradiography disclosed that following proteins of the thick filaments were labelled in all four functional states: 300 kDa, 205 kDa (myosin heavy chain), 100 kDa (paramyosin), 66 kDa and 18 kDa (regulatory MLC). Of thin filament proteins 86 kDa-, 42 kDa-, 39 kDa- (tropomyosin) and 31 kDa-protein were phosphorylated.

Table 3 *In situ* incorporation of ^{32}P-γ-ATP into the proteins of saponin skinned ABRM fibres during a contraction–catch–relaxation cycle ($n = 3$–4)

	Resting	Contraction	Catch	Relaxation
^{32}P-incorporation (pmol/mg protein)	113.9 ±29.8	237.8 ±23.7	111.82 ±21.0	192.1 ±12.8

Discussion

The crucial point in all theories about catch is the question whether the stretch resistance is the result of a slowing down of the crossbridge cycle or an interaction of thick filaments with one another without involvement of actomyosin cross bridges. The ultrastructural results of different groups (7, 11, 12) favour the latter alternative. The fusion of the thick filaments in catch can not be classified as an artifact, as the ones of juxtaposed obliquely striated fibres do not fuse during catch.

The other dispute concerns the regulation of the contraction–catch–relaxation cycle of catch muscles. The results of *in vitro* and *in situ* phosphorylation of contractile proteins (3–6, 9) of catch muscles implicate that the contraction–catch–relaxation cycle of smooth bivalve muscles is controlled by Ca^{2+}- and cAMP dependent phosphorylation of contractile proteins. The importance of paramyosin is emphasized by the fact that only paramyosin of catch muscles can be phosphorylated (15).

In the ABRM a multitude of proteins are phosphorylated during the contraction–catch–relaxation cycle. The doubling of phosphorylation during contraction and relaxation is obviously the result of an activation of the paramyosin specific CaM kinase and the unspecific kinase A respectively. Kinases C and G may be active during both processes, though ACh as well as 5-Ht activate the InsP system. The phosphorylation of the different proteins of the thick filaments during contraction and relaxation may induce a conformational change which favors the stabilization of paramyosin-myosin rod interaction. During catch this interaction is destabilized and thus thick filament melting becomes possible.

References

1. Bowden, J., 1958, *Int. Rev. Cytol.*, **7**, 295–335.
2. Kendrick-Jones, J., Lehman, W., Szent-Györgyi, A. G., 1970, *J. Mol. Biol.*, **54**, 313–326.
3. Achazi, R. K., 1979, *Pflügers Arch.*, **379**, 197–201.
4. Achazi, R. K., Baché, E., Stiller, D., 1984, *J. Mus. Res. Cell Mot.*, **5**, 197
5. Castellani, L., Cohen, C., 1987, *Proc. Natl. Acad. Sci. USA*, **84**, 4058–4062.
6. Sohma, H., Morita, F.,1986, *J.Biochem*, **100**, 1155–1163.
7. Hauck, R., Achazi, R. K., 1987, *Eur. J. Cell Biol.*, **45**, 30–35.
8. Berridge, M. J., 1983, *Bioch. J.*, **212**, 849–858.
9. Sailer, M., Reuzel-Selke, A., Achazi, R. K, 1990, *Comp. Bioch. Physiol.*, **96B**, 533–541.
10. Johnson, W. H., Kahn, J. S., Szent-Györgyi, A. G., 1959, *Science*, **130**, 160–161.
11. Gilloteaux, J., Baguet, F., 1977, *Cytobiol.*, **15**, 192–220.
12. Heumann, H.-G., Zebe, E, 1968, *Z. Zellforsch.*, **85**, 534–551.
13. Murakami, H., Ishikawa, T., Watanabe H., 1984, *Comp. Biochem. Physiol.*, **77C**, 273–277.
14. Murakami, H., Sano, M., Tsukimura, T., Yamazaki, A., 1989 *Comp. Biochem. Physiol.*, **93C**, 341–344.
15. Watabe, S., Sri Kantha S., Hashimoto, K., Kagawa H.,1990, *Comp. Biochem. Physiol.*, **96B**, 81–8.

31
Increased Thermic and Fura-2-Fluorescent Signals during Stationary States of Increased Calcium Cycling in Non-contracting Muscles

V. VAN DER BENT*, A. DECROUY AND A. CHINET

University Departments of Physiology, Faculty of Science, and Faculty of Medicine CMU, 1 rue Michel-Servet, 1211 Geneva 4, Switzerland*

Introduction

Two types of measurements have been used extensively to describe calcium mobilization in a wide variety of mammalian cell types. These are calcium currents and isotopic fluxes across the plasma membrane on one hand, and fluorimetric indices of intracellular Ca activity (aCa_i) on the other. The latter yielded abundant illustration of the characteristic biphasic pattern of aCa_i changes during cell stimulation (i.e. spike phase and a subsequent plateau phase), and provided the first direct evidence for stimulated influx of extracellular divalent cations into the cytosol of dye-loaded cells (1). ^{45}Ca uptake — sometimes combined with fura-2-fluorescence — measurements, and Ca current measurements were used to demonstrate calcium entry into cells through receptor-operated (2, 3) and voltage-operated (4) channels. Under extreme conditions of cellular Ca mobilization (severe hyperosmotic stress, exposure to large concentrations of caffeine), ^{45}Ca effluxes were observed during the sustained phase of stimulation and taken to reflect increases in aCa_i and consecutive stimulation of the active, regulatory limb of a transplasmalemmal Ca recirculation. Such increased Ca cycling was also found to occur under more physiological conditions, where aCa_i during the plateau phase may hardly dif-

Muscle and Motility Vol. 2. Proceedings of XIXth European Conference in Brussels
© Intercept Ltd, PO Box 716, Andover, Hampshire, SP10 1YG, UK

fer from the original, prestimulation value, at least when carbonic acid/bicarbonate-buffered media are used (5).

We present here a complementary approach to the description and analysis of Ca mobilizations: the microcalorimetric determination of the rate of energy expenditure devoted to intracellular Ca homeostasis, which we think is a very sensitive index of Ca cycling across cellular membranes. The present work has two aims: (a) to bring evidence that intracellular membranes, namely that of sarcoplasmic reticulum in muscle (SR membrane), do partake in Ca cycling during steady states where the rate of calcium cycling across plasma membrane is increased, and (b) to compare the relative steady-state changes of aCa_i (more precisely a fura-2-fluorimetric index of aCa_i) and of the rate of cell calcium cycling (more precisely a calorimetric index of cell Ca cycling). We show that a significant part of muscle-cell energy expenditure in the absence of contraction is suppressed under the effect of 2,3-butanedione monoxime (BDM), a reversible inhibitor of passive calcium transfer through the SR Ca channel (6, 7), and that this calcium-dependent energy expenditure undergoes very significant steady-state increases under conditions where steady-state increases of aCa_i are small or even undetectable.

Methods

Skeletal and smooth muscle tissue-cell preparations were the intact *extensor digitorum longus* (EDL) muscle isolated from decapitated C57-BL10 mice (own breed), and an unfolded, cut-open segment of the anterior left-descending branch of the pig coronary artery (whose endothelium had been removed by gentle rubbing with a cotton tip) taken from freshly killed animals at the local slaughterhouse, respectively. Both were mounted on thin stainless steel frames, which maintained them slightly stretched at constant length, and were superfused with a Krebs-Ringer solution, bicarbonate-buffered at pH 7.4 by equilibration at 30°C with a 5% CO_2-95% O_2 (calorimetric experiments) or a 5% CO_2-20% O_2-75% N_2 (fluorimetric experiments) gaseous mixture. Heat production rate ($\dot{E}$) was measured as the voltage difference between two series of six thermal gradient layers surrounding test and control superfusion chambers of a twin calorimeter (8). The proportionality constant between $\dot{E}$ and voltage difference was determined at the chamber perfusion flow rate of 2 ml/min using a calibrated field-effect transistor as the heat source. Results are expressed in microwatts (total heat flux) or in mW/g (normalized for muscle wet weight measured at the end of the experiment, after removal of tendons from EDL, and adventitia from coronary artery preparations, respectively). Loading of the preparations with fura-2 AM (10^{-5} M) was at 37°C in the presence of pluronic F-127 (3.10^{-5} M) in a HEPES (20 mM) buffered solution for 45 min. The fluorescence emitted at 510 nm by the surface cell layers of the preparations superfused in narrow chambers with the normal, bicarbonate-buffered solution was then observed through a quartz lamella with a Nikon inverted microscope (Diaphot) equipped with the P1 photometer. Alternate measurements at two excitation wavelengths (340 and 380 nm filters changed manually about every minute) were recorded and the results are presented as such (raw signals) on Figs 1, 3 and 4, or as the 340/380 signal ratio, calculated at every change of the filters and taken as an index of aCa_i (9) on Figs 2, 5 and 6. The advantage of the dual excitation ratioing is that it corrects for instrument artefacts and for reduction of fluorescence from photobleaching.

Results

We first compared thermic and fluorescent signals under conditions expected or known to produce definite increases of aCa_i. In skeletal muscle this was obtained by exposure to 4 mM caffeine in the presence of a 70-mosM hyperosmotic stress created by a 25% increase of NaCl and $NaHCO_3$ concentrations in the buffer (which also increased extracellular Na and pH). It is known that exposure

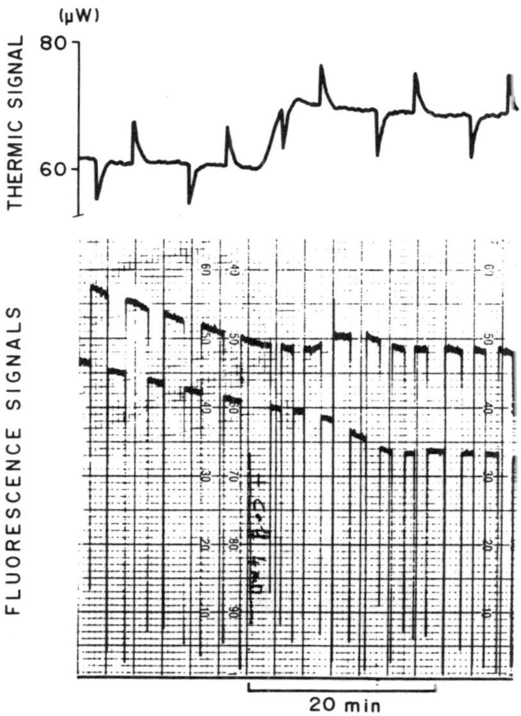

Fig. 1 Raw thermic (upper trace) and fura-2-fluorescent (lower traces) signals recorded as a function of time in skeletal muscle (EDL) preparations superfused with a hyperosmotic solution (see text). Under these conditions, exposure to 4 mM caffeine (starting at the beginning of the time calibration bar) induced clear-cut, sustained increases of the 340/380 signal ratio and total heat production rate. Note that the sharp, alternate perturbations of the raw thermic signal are technical artefacts (due to switches of the calorimeter chamber effluents by means of a fluid distributor in front of a pO_2 cathode) that are omitted in the following figures.

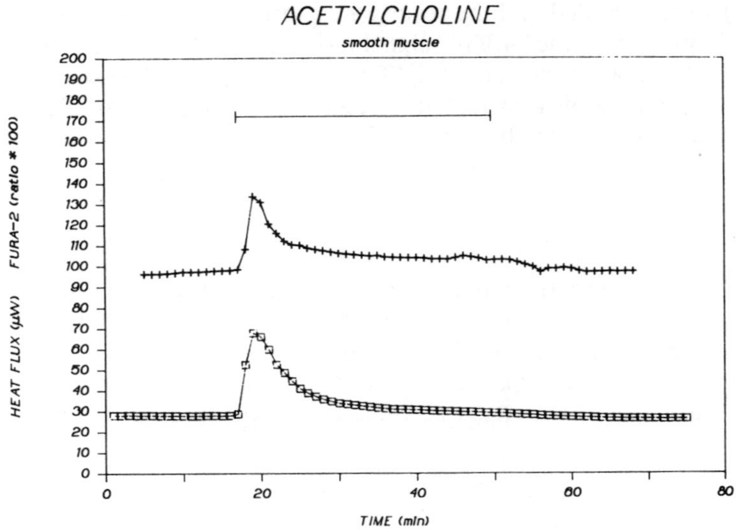

Fig. 2 Total rate of heat production (squares for heat flux) and fura-2 ratio (crosses) as a function of time in smooth muscle (pig coronary artery) preparations exposed for 33 min to 10^{-5} M acetylcholine. Relative increases (with respect to extrapolated basal values) of heat flux and fura-2 ratio are 115% and 36% at the spike, respectively, and 20% and 5% after 15 min. The specific rate of heat production ($\dot{E}$) at 30°C for pig coronary artery (normalized for wet weight of the media) was 0.93 mW/g ±0.06 SE (n = 16). This is significantly lower than $\dot{E}$ at the same temperature for mouse EDL muscle, which was 2.39 mW/g ± 0.11 SE (n = 7).

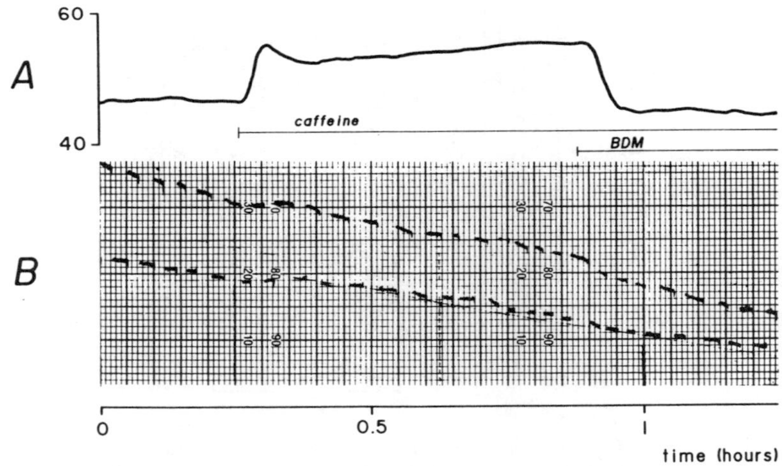

Fig. 3 Redrawn thermic signal (A), and raw fura-2 signals (B: upper trace at 340 nm, lower trace at 380 nm excitation wavelength) as a function of time in EDL preparations superfused with a normal, bicarbonate-buffered Krebs-Ringer solution. The effect of 4 mM caffeine on the fura-2 ratio was undetectable and that of BDM hardly detectable (see text), whereas heat flux was clearly increased by caffeine, and cut down to less than basal by BDM.

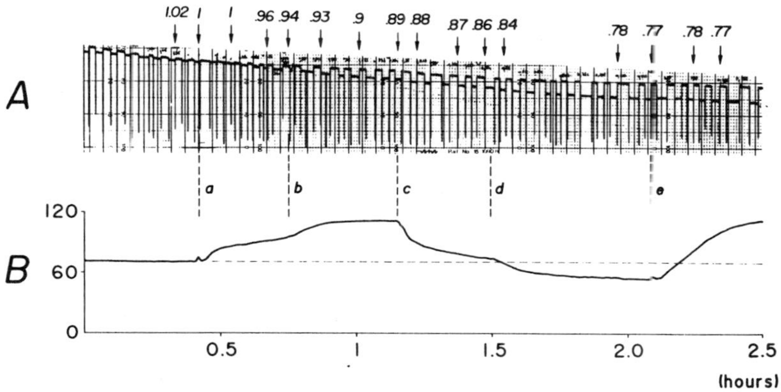

Fig. 4 Raw fura-2 signals (A) and heat flux (B) as a function of time in EDL muscle preparations superfused with the normal, bicarbonate-buffered solution with the following successive additions or subtractions: (a) + sucrose (80 mM), (b) + caffeine (4 mM), (c) – sucrose, – caffeine, (d) + BDM (10 mM), (e) – BDM, + sucrose, + caffeine. Modifications of the fura-2 ratio, here superimposed upon a downward trend, are only transitory and very discrete whereas, as in Fig. 3, modifications of the heat flux are large and sustained.

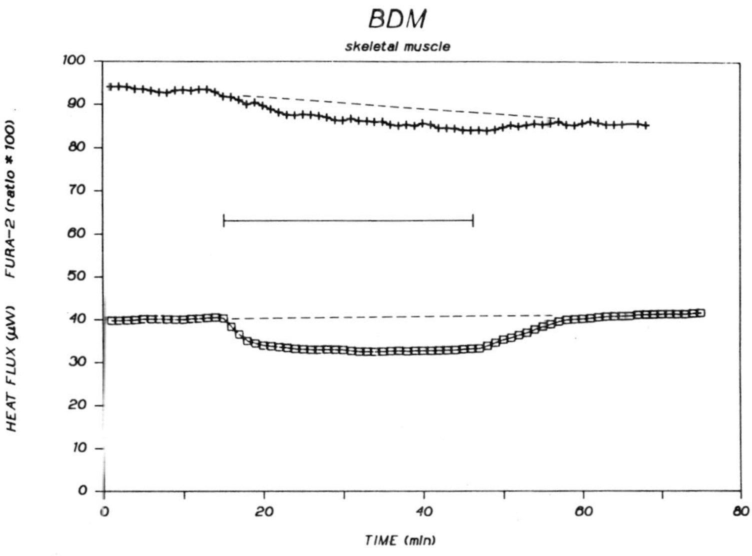

Fig. 5 Reversible effects, on fura-2 ratio (times 100, upper graph) and heat flux, of a 30-min exposure to BDM in EDL muscle preparations. The relative downward shift is 19% for the graph of heat flux versus time; it is only 4% for that of the fura-2 ratio versus time.

to hyperosmotic media elevates intracellular pH in a sodium-dependent way (10) and should therefore facilitate Ca release from intracellular stores (11). Fig.1 shows raw-signal recordings before and after exposure of EDL to caf-

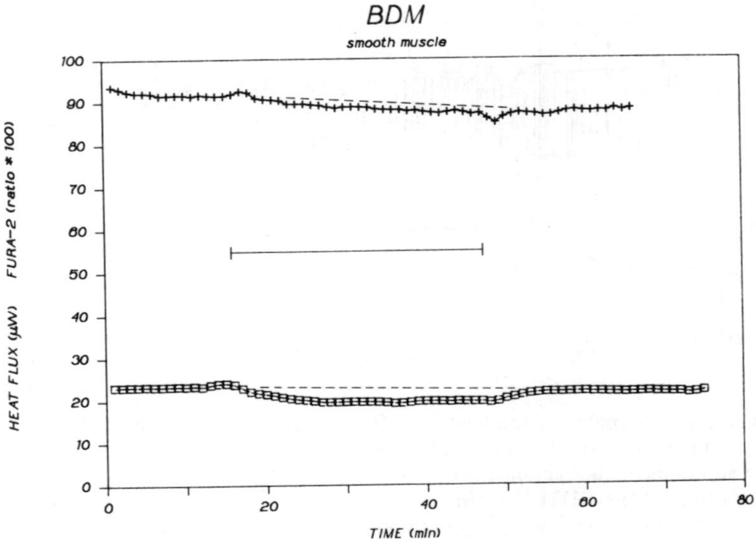

Fig. 6 Reversible effects of BDM on smooth muscle (pig coronary artery) preparations. Same protocol and format as in Fig. 5. Relative downward shifts are 14% and 2% for the heat flux and fura-2 graphs, respectively.

feine in such an experiment. In smooth muscle, exposure to 10^{-5} M acetylcholine (Fig. 2) induced the well-known increase of aCa_i whose time course is similar to that of tension, and a concomitant increase in $\dot{E}$ which, as clearly visible from the graphs whose common ordinate starts at zero, is relatively larger than the change of the 340/380 signal ratio (fura-2 ratio).

When skeletal muscle was exposed to the same (4 mM) concentration of caffeine as in Fig. 1, but in the absence of the hyperosmotic stress, $\dot{E}$ increased by about the same amount. However, there was no detectable change of the fura-2 ratio because the two emitted signals (Fig. 3B, upper one under 340 nm and lower one under 380 nm excitation wavelengths) increased by the same relative amount. Upon subsequent exposure to 10 mM BDM, however, when $\dot{E}$ was sharply cut down to somewhere below the initial basal value in spite of the continued presence of caffeine, *some* decrease of the fura-2 ratio could be suspected but the actual computation (not shown) was disappointing due to a low signal-to-noise ratio. In another experimental protocol, only transient and very slight changes (superimposed on a monotonous decrease) of the fura-2 ratio could be detected (Fig. 4A), whereas $\dot{E}$ underwent marked and sustained changes (Fig. 4B).

In both Figs. 3 and 4, inhibition of Ca release from sarcoplasmic reticulum by BDM reduced $\dot{E}$ to less than its initial, basal value, which indicates that part of basal $\dot{E}$ is BDM sensitive. This is better illustrated in Figs. 5 and 6, for skeletal and smooth muscle preparations respectively. Lower graphs show sustained and reversible 19% and 14% decreases of $\dot{E}$, concomitant with 4% and 2% downward shifts of the fura-2 ratio (upper graphs).

Discussion

Our calorimetric approach is based on the generally admitted hypothesis that energy metabolism is tightly controlled by the rate of overall ATPase processes. Among these only one, the Na-K ATPase, is responsible for the maintenance of the Na electrochemical gradient across plasma membrane. This permits direct observation of the immediate energetic consequences of specific inhibition of that particular ATPase, and therefore gives an idea of the overall energy cost of the dependent ion transports. In contrast, several ATPase processes are responsible for the maintenance of aCa_i: the calmodulin-dependent Ca ATPase of plasma membrane, the Mg-Ca ATPase of the endo(sarco)plasmic membrane and probably a third one which, in the same membrane, allows calcium ions to be exchanged for protons (12). As a consequence, direct estimation of the overall energy cost of aCa_i homeostasis by specific inhibition of these ATPases proved complicated because all three processes had to be inhibited together (using the anticalmodulin W-7, 2,5-di(*tert*butyl)-1,4-benzohydroquinone and N-ethylmaleimide as inhibitors, respectively). Indeed, inhibition of one single process appears to immediately increase energy dissipation by the others (unpublished observations from this laboratory). Another difficulty arose, in muscle, from the increase of aCa_i consecutive to inhibition of all three processes: it quickly stimulated myosin ATPase and induced contracture. We therefore chose *indirect* inhibition of the active processes by cutting down passive Ca entries into the sarcoplasm, with the implicit assumption that any consecutive decrease of aCa_i would immediately decrease active Ca reuptake. We found that specific inhibition of one of these entries, the one which borrows the SR Ca channel, with sodium dantrolene in noncontracting skeletal muscles from mice immediately suppressed a significant part of steady-state energy dissipation at rest, and all of the excess $\dot{E}$ induced by subcontracture K depolarization (13). BDM was then used, instead of dantrolene, as a more convenient inhibitor of Ca discharge from internal stores (6). Contrary to the effect of dantrolene, the effect of BDM is indeed fully reversible (see Figs. 5 and 6).

What is shown here is that very small, and sometimes undetectable, changes of aCa_i can induce significant changes of Ca cycling to which passive Ca release from SR appears to contribute. Such a contrast between relative steady-state changes of aCa_i and Ca cycling rates is in keeping with the prediction that Ca clearance processes are governed by a high power (fourth power) of aCa_i (14).

To conclude, the microcalorimetric approach provides strong evidence that aCa_i is maintained at physiological value, in noncontracting muscles, thanks to energy-dissipating calcium re-uptake processes which continuously counterbalance a BDM-sensitive influx of calcium from internal stores. The present results therefore suggest that at least one intracellular membrane (the SR membrane) partakes in cellular Ca cycling, and that it does so not only during plateau phases of stimulated Ca mobilization but also under basal conditions. Finally, the finding that aCa_i during plateau phases of stimulated Ca mobilization may hardly differ from its pre-stimulation value, as juged from fura-2-fluorescence measurements, suggests that the steady-state Ca signal is subject to powerful feedback mechanisms that limit its amplitude. This is at the cost of significant amounts of extra energy dissipation, as reflected by BDM-sensitive excesses of heat production with respect to basal rate.

Acknowledgements

This work was supported by grants nos 3.053-0.87 and 3.203-0.88 from the Swiss National Science Foundation.

References

1. Hallam, T. J., Rink, T. J., 1985. *FEBS Lett.*, **186**, 175–179.
2. Muallem, S., Schoeffield, M. S., Fimmel, C. J., Pandol, S. J., 1988. *Am. J. Physiol.*, **255**, G229–G235.
3. Schilling, W. P., Rajan, L., Strobl-Jager, E., 1989. *J. Biol. Chem.*, **264**, 12838–12848.
4. Donaldson, P. L., Beam, K. G., 1983. *J. Gen. Physiol.*, **82**, 449–468.
5. Rassmussen, H., 1990. *Biol. Chem. Hoppe-Seyler*, **371**, 191206.
6. Decrouy, A., Stucker, M., Chinet, A., 1989. *Arch. Int. Physiol. Bioch.*, **97**, A127.
7. Fryer, M. W., Gage P. W., Neering, I. R., Dulhunty, A. F., Lamb, G. D., 1988. *Pflügers Arch.*, **411**, 76–79.
8. Chinet, A., Clausen, T. Girardier, L., 1977. *J. Physiol.*, **265**, 43–61.
9. Grynkiewicz, G., Poenie, M., Tsien, R. Y., 1985. *J. Biol. Chem.*, **260**, 3440–3450.
10. Hoffmann, E. K., 1987. *Curr. Topics Memb. Trans.*, **30**, 125–180.
11. Grassi de Gende, A. O., 1988. *J. Mol. Cell. Cardiol.*, **20**, 1087–1093.
12. Thevénod, F., Dehlinger-Kremer, M., Kemmer, T. P., Christian, A.-L., 1989. *J. Membrane Biol.*, **109**, 173–186.
13. Chinet, A., Giovannini, P., 1988. *Experientia*, **44**, A33.
14. Barany, E., 1987. *Acta Phys. Scand.*, **130**, 77–81.

32
Modelling of the Calcium-sensitivity Curves of Skinned Skeletal Muscle Fibres: the Non-linearity of the Hill Plot

E. DE BEER, R. GRUNDEMAN, T. MANUSSEN*, K. VAN BUUREN* AND P. SCHIERECK

*Departments of Medical Physiology and *Biomedical Pharmacology, University of Utrecht, Vondellaan 24 NL3521GG Utrecht, The Netherlands*

Introduction

It is generally assumed that the binding of Ca^{2+} to Troponin-C (Tn-C) initiates the contraction in skeletal muscle (1, 2). The binding process has been modulated by different physiological factors like [Mg^{2+}] (3, 4) or pH (3, 5). From biochemical studies (6) evidence has been supported that skeletal Tn-C has four binding sites for Ca^{2+}, two with a high affinity (K_h) and two with a low affinity (K_l). Mg^{2+} binds competitively to the high, but has neglible affinity for the low affinity sites.

The tension (T), the maximum tension (T_{max}) and the Calcium concentration [Ca^{2+}] can be related to each other with the Hill equation:
$T/T_{max} = [Ca^{2+}]^n/(K^n + [Ca^{2+}]^n)$. ($K$ is the calcium concentration at which $T/T_{max} = 0.5$.)
From this relation can be derived that the relation between $\log(T/(T_{max}-T)$ and the pCa (the negative logarithm of [Ca^{2+}]) is linear. However, it is already shown that this relation is systematically concave (5, 7). In the case of a Hill process the co-operativity between the constitual elements should be maximal. This condition will not be met in the initiation of the contraction because of the two different affinities for Ca^{2+}. To find the role of this deviation we model the

Muscle and Motility Vol. 2. Proceedings of XIXth European Conference in Brussels
© Intercept Ltd, PO Box 716, Andover, Hampshire, SP10 1YG, UK

initiation of the contraction by Ca^{2+} assuming that the binding to each binding site on Tn-C is independent.

Model and Methods

MODEL

The basic assumption of the model is presented by the equation:

$$X^{2+} + \text{Tn-C} \rightarrow X\text{Tn-C}$$

in which X represents Mg^{2+} or Ca^{2+}. Using the reaction constants for Mg^{2+} or Ca^{2+} (6), the occupation of Tn-C with Ca^{2+} was calculated for both the low- and the high-affinity sites.

We assumed that all sites have to be occupied by Ca^{2+} to induce force in the contractile apparatus. This implied that the developed tension is proportional to the product of the squares of the number of low and high affinity sites bound to Ca^{2+} (respectively Tn-C_{low} and Tn-C_{high}) as stated in the formula:

$$T_{rel} = [\text{CaTn-C}_{low}]^2 * [\text{CaTn-C}_{high}]^2$$

in which T_{rel} is the relative developed tension. From the calculated data the Hill plot was constructed.

METHODS

Single skinned fibres of the rabbit psoas muscle were obtained by means of freeze-drying (7, 8). Sarcomere length was measured by means of laser diffraction. The fibres were immersed in a solution containing the desired Ca^{2+} concentration. The developed force was measured with a strain gauge force transducer (AME 801, Sensonor, Norway). From the measured data the Hill plot was constructed.

Results

MODEL CALCULATION

The number of binding sites occupied with Ca^{2+} were calculated using the reaction constants as proposed (6) as a function of the Ca^{2+} concentration for two free Mg^{2+} concentration: 3 and 0 mM. The Hill plot, calculated from the number of occupied sites has been plotted in Fig. 1A (0 mM free $[Mg^{2+}]$) and Fig. 1B (3 mM free $[Mg^{2+}]$), together with the square of the number of sites (high and low) bound to Ca^{2+}. This figure shows that the Hill plot is not a straight line, but that the slope depends upon the Ca^{2+} concentration. The slope decreases from 4 at high pCa (low $[Ca^{2+}]$) to 1 at low pCa (high $[Ca^{2+}]$). The value of 4 at high pCa has been achieved with the explicit assumption that no kind of co-operative behaviour of the binding kinetics exists. Comparison of the two Hill plots shows that decreasing the free $[Mg^{2+}]$ results in a parallel leftward

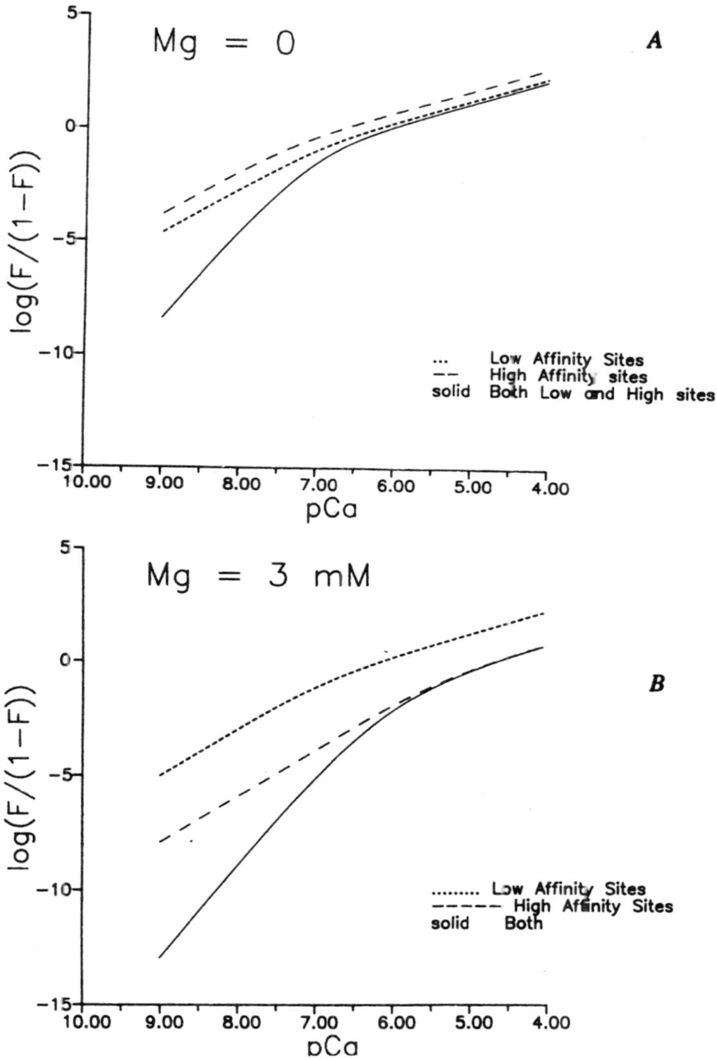

Fig. 1 Prediction of the Hill fit, the relation between log($F/(1-F)$) and pCa for two [Mg^{2+}]: 0 (A) and 3mM (B), together with the same transformation of the square of the number of occupied low- and high-affinity sites.

shift of the Hill plot. So the curvature of the Hill plot is not a result of the presence of Mg^{2+}. The leftward shift has been due to the competitive binding between Mg^{2+} and Ca^{2+} at the high-affinity sites. These plots show that at a physiological concentration of Mg^{2+} (1–3 mM) at low [Ca^{2+}] the low affinity sites regulates the contraction, while the majority of the high affinity sites are occupied by Mg^{2+}. At high [Ca^{2+}] the high-affinity sites regulate the contraction, because the low-affinity sites are saturated.

EXPERIMENTAL RESULTS

Fig. 2A shows typical calcium sensitivity curves of the same fibre for two different sarcomere lengths, 2.0 and 3.1 μm. This figure shows that in psoas muscle sarcomere length affects the binding of Ca^{2+} to the contractile apparatus mostly at high pCa values. This is further illustrated in Fig. 2B, in which the Hill plots as obtained from this experiment have been shown. It is clear from this figure that the Hill plot is not a straight line at a sarcomere length of 2.0 μm, but is concave. The curvature is in the same direction as in the model calculation. However, the slopes at both the high and the low pCa-value (6 and 2 respectively) are higher than those from the model calculation (4 and 1 respectively). At a sarcomere length of 3.1 μm the Hill plot is nearly linear. The threshold for activation has been shifted to higher pCa values. From the model

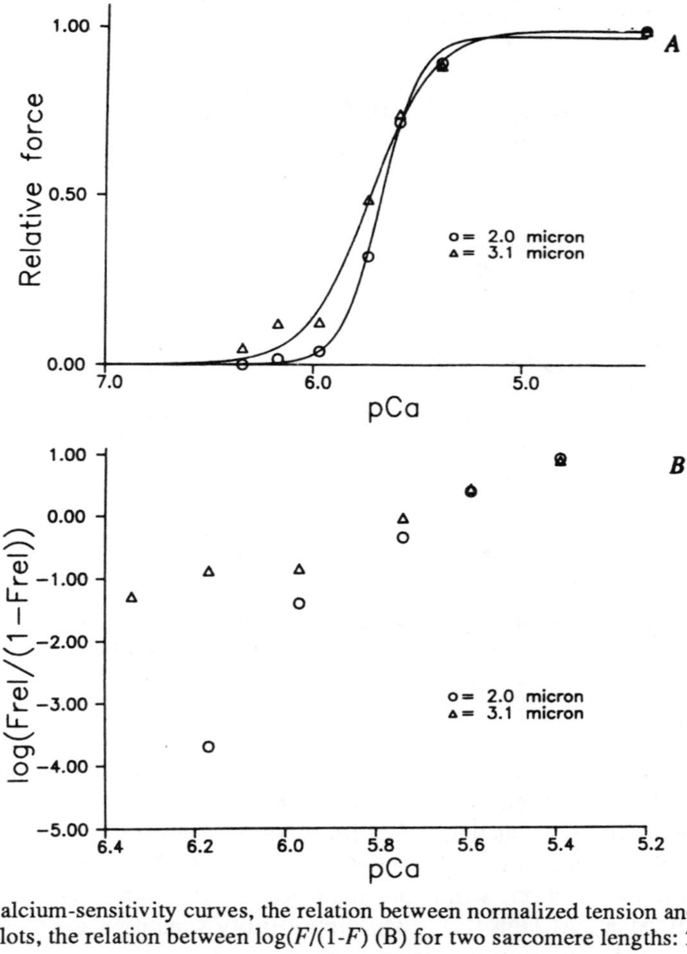

Fig. 2 Calcium-sensitivity curves, the relation between normalized tension and pCa (A) and Hill plots, the relation between log(*F*/(1-*F*) (B) for two sarcomere lengths: 2.0 and 3.1 μm.

calculation we have the indication that at high pCa values the low affinity sites are the regulatory sites. So, this suggests that at low Ca^{2+} concentration lengthening of the sarcomere induces a change in binding properties of the low-affinity sites.

A similar effect can be observed when the cardiotonic drug pimobendan has been added to the bathing solution. The Hill plots obtained at the same sarcomere length are shown in Fig. 3. The main effect of the drug can be observed at high pCa values. The threshold for contraction is shifted to lower $[Ca^{2+}]$, suggesting that the drug mainly affects the low affinity sites. The Hill plot is more linear when pimobendan has been added than without.

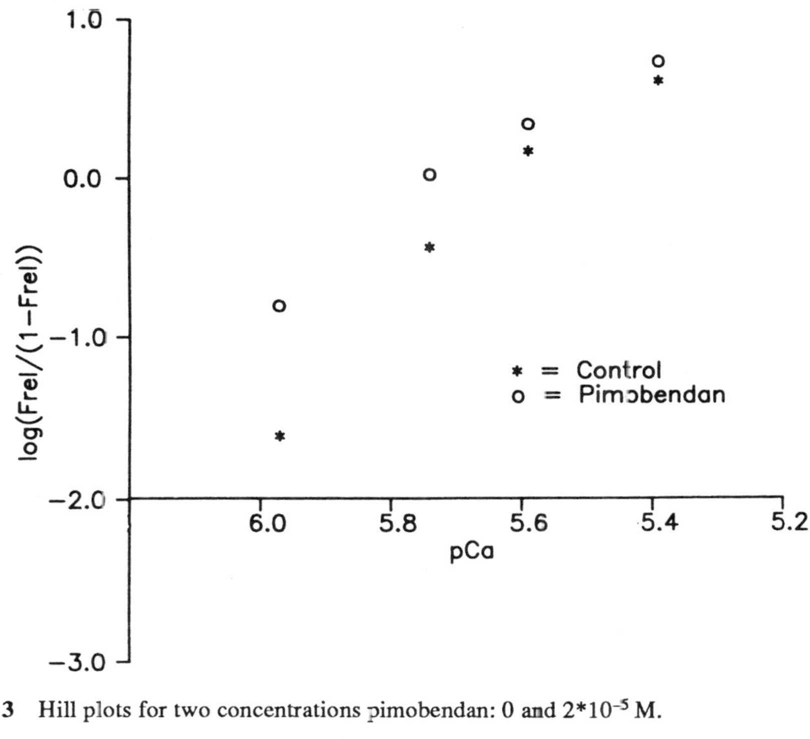

Fig. 3 Hill plots for two concentrations pimobendan: 0 and $2*10^{-5}$ M.

Discussion

The model calculation based on biochemical data (6) of the binding kinetics of Ca^{2+} and Mg^{2+} Tn-C shows that the curvature of the Hill plot can be explained by independent binding sites for Ca^{2+} on Tn-C. The used reactions constants are the only available from biochemical experiments. These reaction constants are dependent upon the biochemical environment, so in the intact muscle they can differ from the actual. However, they can give insight in the underlying mechanism, although the agreement between model and experiment is not perfect. At high pCa values a slope of the Hill curve of 4 has been predicted, while the slope for low pCa values equaled 1. The experimental values are 6 and 2 respectively. Such high values can only be understood if more than four binding sites have been involved in the regulation of the contraction. It has been suggested that the whole actin filament acts as co-operative sytem (8). From the model calculation it is evident that cooperativity itself is not an absolute condition, but that only the number of active sites, necessary for contraction, is important.

The calcium sensitivity curve has been often characterized with the fitted constants of the Hill equation. When this curve could be described by the Hill equation the Hill plot should be a straight line. This is not the case, especially at high pCa values the experimental points differ significantly from the fit. Therefore a scheme (3) in which Ca^{2+} was consecutively bound, has been proposed. However, fits with this model results in negative reaction constants (7, 3). From the model calculation it is evident that at high pCa values the majority of the high affinity sites has been bound to Mg^{2+}, while in the fitting procedure it is assumed that all sites are available for binding to Ca^{2+}. This explains the negative values of the fitted reaction constants.

The model calculations show that removal of Mg^{2+} results in a parallel shift of the Hill plot. This is also experimentally found (4). The shape of the observed Hill plots depends upon sarcomere length, and also upon the species used. In psoas muscle of vertebrates the main effect of sarcomere length on the calcium sensitivity curve and Hill plot has been observed at high pCa values (this study; Moss *et al.* (4)).

This suggests that sarcomere length mainly affects the low affinity sites, possibly by enhancing the binding of Ca^{2+} to the low affinity sites or decreasing the affinity for Mg^{2+}. The Hill plot seems to get linear, but this can be due to the shift of saturation of the low affinity sites to very low $[Ca^{2+}]$, while the high affinity sites are still saturated with Mg^{2+} thereby force development blocking.

The same observations hold for the interaction of the cardiotonic drug pimobendan with the contractile apparatus. The drug shows its main effect at low $[Ca^{2+}]$, thereby suggesting that it affects the binding kinetics to the low affinity sites.

References

1. Ebashi, S., 1984, *J. Moll. Cell. Cardio.* **16**, 129–126.
2. Ebashi, S., Endo, M, 1968, *Prog. Biopys. Mol. Biol.* **18**, 125–183.
3. Ashley, C. C., Moisecu D. G., 1977, *J. Physiol. (Lond.)*, **270**, 627–652.
4. Moss, R. L. Swinford, A. E., Greaser, M. L., 1983, *Biophys. J.* **43**, 115–119.
5. Martyn, D. A. Gordon, A. M., 1988, *J. Musc. Res. Cell Mot.*, **9**, 428–445.
6. Potter, J. D., Johnson, J. D. 1983. In: *Calcium, and Cell Function*, vol. II: p. 145.
7. De Beer, E. L., Gründeman, R. L. F., Wilhelm, A. J., Calijouw, C. J., Klepper, D., Schiereck, P. 1988, *Basic Res. Cardiol.*, **83**, 410–423.
8. Brandt, P. W., Diamond, M. S., Rutchik, J. S., Sachat, F. H., 1987, *J. Mol. Biol.*, **195**, 885–896.

33
Dependence of Myocardial Relaxation on Mechanical and Inotropic Factors: Experimental Data and Mathematical Model

V. YA. IZAKOV, V.S. MARKHASIN, L. B. KATSNELSON AND F. A. BLYAKHMAN

Department of Biophysics, Institute of Physiology, Ural's Branch of Academy of Sciences of the USSR, Sverdlovsk, USSR

Introduction

To date there has been a lot of data, proving the role of mechanical and inotropic factors in the process of myocardial relaxation (1, 2). The latest studies paid special attention to the contribution of mechanical heterogeneity of heart muscle and its lusitropic properties (3, 4). It is evident that clearing up the contribution of each of these factors in the process of this relaxation is closely connected with its molecular mechanism. Based on the existing data we have developed a mathematical model, which is able to connect the molecular and mechanical processes in myocardium (5).

The aim of this work is to show some experimental data indirectly proving the results, developed in the model, and also to use the results of numerical modelling for explaining possible mechanisms of mechanical heterogeneity contribution to the relaxation process.

For investigating the influence of myocardial mechanical heterogeneity on the process of contraction and relaxation a new approach, based on the behaviour investigation of a pair of connected papillary muscles (duplet) with different mechanical properties was developed. The process of contraction and relaxation

Muscle and Motility Vol. 2. Proceedings of XIXth European Conference in Brussels
© Intercept Ltd, PO Box 716, Andover, Hampshire, SP10 1YG, UK

is studied first in each separate muscle of the duplet and then at their interaction, when the load is applied to the duplet as a whole.

Methods

All the methods for investigation of mechanical activity of isolated heart muscles and for studying the mechanical heterogeneity of myocardium were stated previously (6). In short, the papillary muscles (length 3–5 mm, cross-section up to 0.5 mm^2), taken from right ventricular of warm-blooded animals were placed into parallel isolated chambers, perfused by physiological solution. Each muscle was attached by one end to the force transducer and by the other to the precision ergometer.

With the help of the computers, included in the feedback between the force transducer and ergometer, a load was applied to each isolated muscle as well as to the duplet as a whole. The muscle contracted in the physiological regime, imitating the characteristic phases of intact heart: isometrical contraction, shortening under alternating load, isometrical relaxation and rapid and slow diastolic tension. The stimulation was carried out through huge platinum electrodes, placed in each chamber, with current impulses 3–6 ms.

To estimate relaxation rate, t_{30} was used — the time of isometrical relaxation phase at the level of 70% from the contraction amplitude.

Results

In the experiment held on the isolated papillary muscles of a rabbit and a cat it was estimated that in the physiological loading regime there exists a linear relationship between t_{30} and end-systolic lengths (Fig. 1). The less L_{es}, the less t_{30}, i.e. the relative muscle relaxation is higher. Besides t_{30} at fixed L_{es} value depends on the character of loading (Fig. 1): the greater the muscle shortening rate up to the given L_{es}, the less t_{30}. There exists a strong correlation (r>0.95) between t_{30} and end-systolic force (P_{es}) at fixed L_{es} values.

For evaluation of lusitropic effect of inotropic factors t_{30}–L_{es} relationship was compared in control conditions and after the influence of the drugs. The doubling of extracellular concentration of Ca^{2+} increases the t_{30} values at all fixed L_{es} (Fig. 2).

The influence of stimulation frequency on the isometrical relaxation rate is defined by the animal species. In the myocardium of rabbits and cats the contraction force increases with the rise of stimulation rate. In those cardiac muscles, the increase of stimulation rate at fixed L_{es} values raises t_{30}, that is decelerates relaxation (Fig. 3). The opposite effect is seen in rat myocardium, where contraction force falls with the increase of stimulation frequency. In this type of muscle preparation isometrical relaxation is decreased with frequency stimulation increase. Thus, lusitropic effect under the influence of the stimulation frequency is completely determined by the amplitude variations.

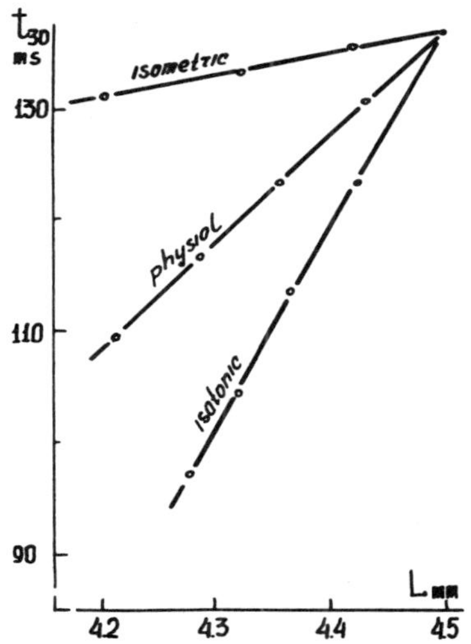

Fig. 1 Muscle length (*L*) relaxation time (t_{30}) relationship in three various types of contraction.

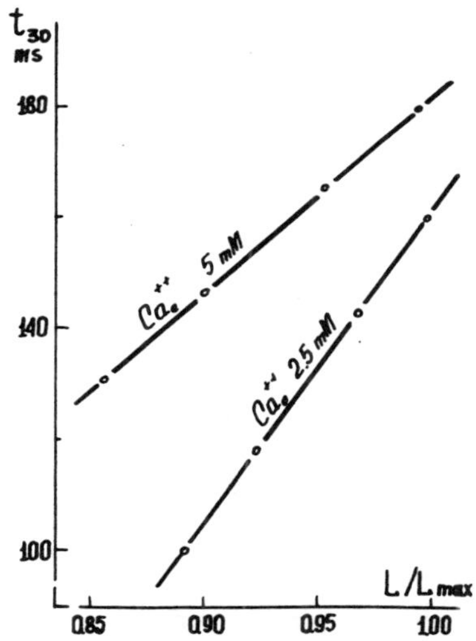

Fig. 2 Normalized muscle length (L/L_{max})–relaxation time (t_{30}) relationship for various extracellular Ca^{2+} concentrations.

The catecholamine influence on relaxation rate depends on their dosage. At low concentrations (10^{-9}–10^{-3} M) of epinephrine the contraction force increases, but at fixed L_{es} it decelerates relaxation (Fig. 4), i.e. the above effect correlates

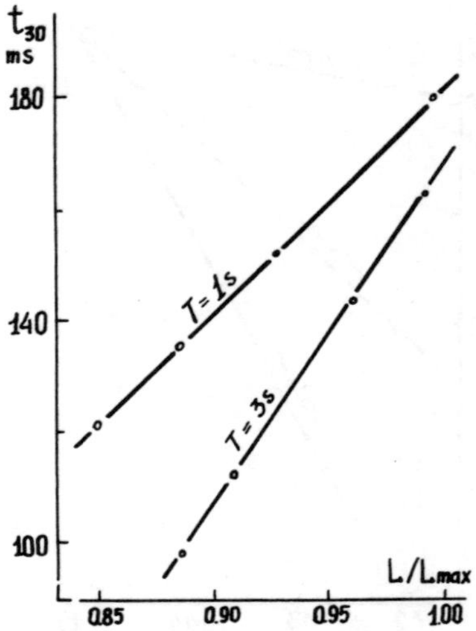

Fig. 3 Normalized muscle length (L/L_{max})–relaxation time (t_{30}) relationship at various intervals between stimuli.

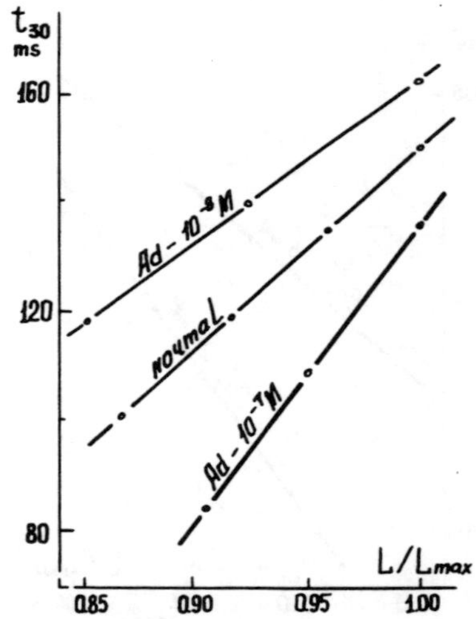

Fig. 4 Normalized muscle length (L/L_{max})–relaxation time (t_{30}) relationship of various concentrations of adrenalin.

with inotropic effect of catecholamines. At large concentrations (10^{-7}–10^{-6}) at L_{es}-const. epinephrine only slightly rises the contraction amplitude, but raises the relaxation rate to a much greater extent (Fig. 4).

As an example of mechanical heterogeneity the pairs of parallel-attached papillary muscles of one length and cross-section were examined. The muscle preparations in a duplet stimulated each other with the intervals, lasting 5–20% of the time of contraction–relaxation cycle (Fig. 5). The influence of asynchronic stimulation of duplet muscles on its relaxation is shown in Fig. 6. It is seen that, on lengthening the intervals, the characteristic time of duplet relaxation rises.

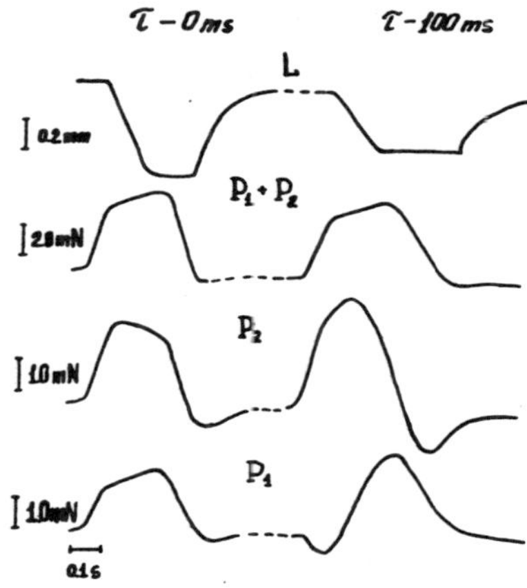

Fig. 5 Experimental records of a duplet activity for various delays between stimuli. From the top to the bottom: muscle shortening (L), total duplet force ($P_1 + P_2$), and force of the second (P_2) and first (P_1) papillary muscles in a duplet; τ is the delay between stimuli.

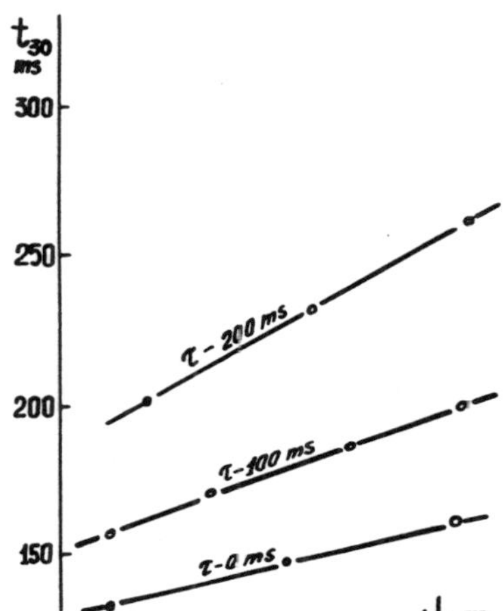

Fig. 6 Relaxation time (t_{30}) versus shortening of a duplet (ΔL) at various delays between stimuli.

Analysing the relaxation–contraction cycle of each separate muscle and of a duplet as a whole it was estimated that the time course of the cycle changes abruptly (see Fig. 5). It is connected with load redistribution in each muscle in the duplet, because of their asynchronic contraction. Besides, each of the duplet 'members' relaxes auxotonically under variable load, though the relaxation of the duplet is formally isometrical.

Another example of mechanical heterogeneity imitated the local catecholamine influence on myocardium and/or varying susceptibility of the myocardium sites to such influences. For this purpose epinephrine in concentration 10^{-7} was added into the perfusion solution of one of the muscles (Fig. 7).

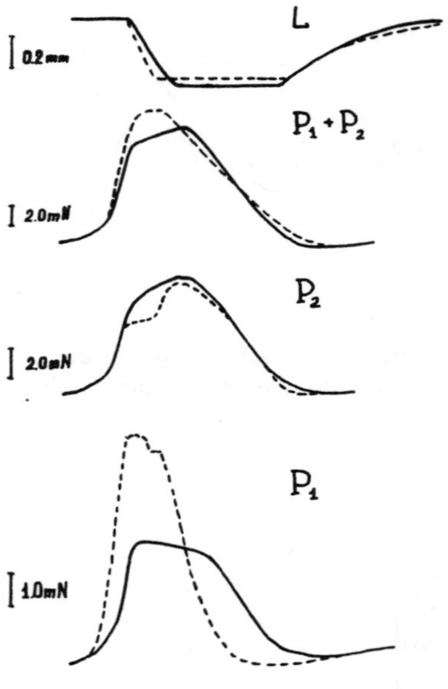

Fig. 7 Experimental records of a duplet activity: effect of adrenalin The Designations are the same as in Fig. 5. Dashed lines: effect of adrenalin (10^{-7}M) added to the solution bathing the first muscle. Continuous line: control experiment without adrenalin.

Fig. 8 shows that, in spite of relaxation acceleration of one of the muscles under the epinephrine influence, the relaxation of the duplet as a whole was retarded. Such a paradoxical result may be explained by the redistribution of loads on the muscles in the duplet in consequence of arising muscle contraction asynchronism in the duplet. The nature of this phenomenon is connected with catecholamine induced differences in the time course of the contraction-relaxation cycle.

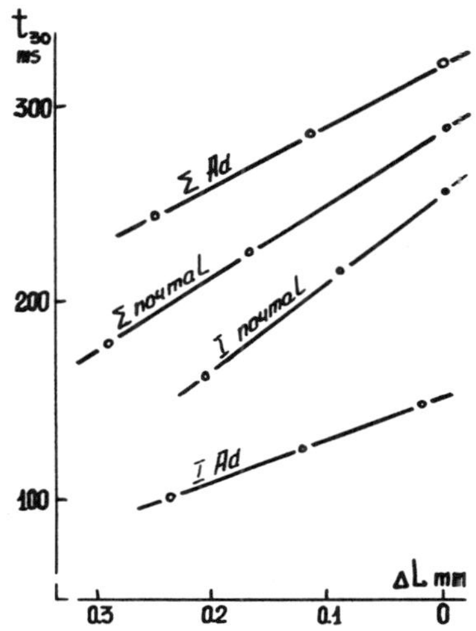

Fig. 8 Relaxation time (t30) versus shortening of a duplet (ΔL): effect of adrenalin. I, first muscle for adrenaline 10^7M (AD) and in control (normal); Σ, duplet.

Discussion

The results, described above, allow the following determinants of heart muscle relaxation to be distinguished: end-systolic length, shortening rate up to end-systolic length and inotropic state. Besides, an important factor, determining the time course of myocardium relaxation is its mechanical heterogeneity.

To understand the mechanism of the described determinant, a mathematical model of the contraction–relaxation cycle of the heart muscle was constructed. The results of numerical modelling were compared with various experimental data including those mentioned above. Good qualitative correlation of the experiment and the theory was found.

The central idea of the model is the co-operative character of calcium binding on troponin. Based on the data from the literature (7, 8), it is postulated that the rate constant of TnCa complexes dissociation depends on the number of myosin crossbridges in the zone of overlapping of thin and thick filaments and on the concentration of these complexes in the zone. The first type of cooperativity accounts for the momental load preceding relaxation; the second one for the influence of inotropic factors. The analysis of the model, shows that the lower the number of formed TnCa complexes, the greater the myocardium relaxation rate.

Thus, the greater the muscle shortening rate at the moment of its relaxation and the lower the level of contractile proteins activation by Ca^{2+} up to the given moment, the lower are the TnCa concentration and characteristic time of myocardium relaxation.

Thus the time for the muscle achieving its end-systolic length is an independent relaxation determinant.

Such time has to be considered as one of the relaxation determinants, because the time course of contractile protein activation and muscle shortening are relatively independent processes.

All the mentioned determinants are functioning together in case of mechanical heterogeneity of myocardium, determining its characteristic mechanical behaviour as well as the time course of relaxation (see Figs. 5, 7). The main mechanism here is the load redistribution between the muscles, changing both the rate of shortening of each of the muscles in the duplet and the time of reaching the end-systolic length. The analysis of the model, shows that mechanical heterogeneity also influences the time course activation of contractile proteins.

References

1. Blyakhman, F. A., Izakov, V. Ya., Markhasin, V. S., 1990, *Sechenov Physiol. J. of the USSR*, **76** (**1**), 41–49.
2. Brutsaert, D. L., Rademakers, F. E., Sijs, S. U., 1984, *Circulat.*, **69 (1)**, 190–196.
3. Brutsaert D. L., 1985, *Verh. Kon. Acad. yeneesk. Belg.*, **47** (**4**), 257–286.
4. Lew, W. Y. W., Rasmussen, Ch. M., 1989, *Am. J. Physiol.*, **256** (**1**), H222–H232.
5. Katsnelson L. B., Izakov V. Ya., Markhasin V. S. 1990. Gen. Physiol. Biophys. N 9, p. 219–244.
6. Blyakhman, F. A., Narfikov, Kh. M., Markhasin, V. S., 1988, *Sechenov Physiol. J. of the USSR*, **74** (**8**), 1191–1195.
7. Hoffman P. A., Fuchs F. 1987. J. Physiol., Vol. 253, p. 90–96.
8. Moss, R. L., Guilian, G., Greaser, M. L., 1985, *J. Gen. Physiol.*, **6** (**4**), 583–600.

34
Parvalbumin Saturation and Slowing of Relaxation

A. BERQUIN AND J. LEBACQ

Department of Physiology, UCL 5528, 1200 Brussels, Belgium

Introduction

Parvalbumin (PA) is believed to act as a soluble relaxing factor, enhancing relaxation rate by removing calcium from troponin C. In this view, a progressive saturation of PA calcium-binding sites during the first few seconds of contraction would explain the 'Abbott effect' (AE), a slowing of relaxation as a function of tetanus duration well described in amphibian fast muscles. But recent data show that factors other than PA may be involved in the slowing of relaxation. This suggests that the Abbott effect is due to a conjunction of several different processes and not solely to PA.

Abbott Effect and Parvalbumin: the Current Hypothesis

Relaxation of amphibian muscles is fastest after short contractions and slows down progressively with increasing tetanus duration (1). When relaxation rate is measured as the inverse of the time required for a 5% tension fall ($y = 1/t_{5\%}$), its time course can be described as follows:

$$y_t = y_0 e^{-\alpha t} + y_f \qquad (1)$$

where a is rate constant (s^{-1}) and t is tetanus duration (s) (2). The first term of this equation describes the 'Abbott Effect', an exponential decay of relaxation

Muscle and Motility Vol. 2. Proceedings of XIXth European Conference in Brussels
© Intercept Ltd, PO Box 716, Andover, Hampshire, SP10 1YG, UK

rate observed at the beginning of a tetanus. Its rate constant varies from 0.4 to 1.18 s^{-1} in frog muscle at 0°C (2, 3).

The AE is depressed by previous activity and reprimes progressively during the recovery period (2); it is relatively independent of muscle length, i.e. filament overlap (4).

Parvalbumin is a soluble cytoplasmic protein present in large amount in amphibian muscles. It presents two Ca-Mg-binding sites, with a higher (10^4 times) affinity for calcium than for Mg. In resting muscle (low Ca concentration), PA is probably mainly saturated with Mg (5, 6). During activation, Ca binding on PA is limited by a slow PA-Mg dissociation. It is commonly admitted that all the PA contained in a frog sartorius is Ca-saturated after a few seconds of stimulation (7). During the recovery period, pumping of Ca by SR allows a slow PA-Ca dissociation.

Because of these Ca-binding properties, PA is thought to act as a relaxing factor: after short tetani, it takes part in the free $[Ca^{++}]_i$ lowering, enhancing relaxation rate. After longer contractions, Ca-saturated PA is not able to play its relaxing role; this explains the slowing of relaxation observed and AE depression by previous activity. Recovery of muscle's ability to relax faster is then due to Ca-PA dissociation, induced by SR activity.

A number of experimental data argue in favour of this hypothesis (for a review, see ref. 7); the more recent ones are:

1. The ratio $y_0/(y_0 + y_f)$ (c.f. eqn 1) is higher at 0°C than at 10°C. This is understandable if the variable part of relaxation rate (y_o) is due to PA, while the constant part (y_f) is related to SR enzymatic activity, with a higher temperature coefficient (8).
2. AE is absent in mouse soleus (no PA) and present in mouse extensor digitorum longus (EDL; high PA contents) (9).
3. Rate constant of PA-Mg dissociation (0.99 s^{-1}) is close to that of the slowing of relaxation (1.18 s^{-1}). Moreover, rate constant of PACa dissociation (0.18 s^{-1}) is in good agreement with AE recovery (0.12 s^{-1}) (3).

However, during repeated fatiguing contractions, relaxation slows down even after a few seconds of stimulation (see, e.g. ref. 10). In mouse EDL at 20°C, relaxation rate decreases continuously during a 15 s tetanus (9), thus well after complete saturation of PA. Therefore, other factors must be involved in the late slowing of relaxation.

Other Possible Origins of AE

SLOWING OF FREE [CA] LOWERING

A slowing of the decrease in free Ca concentration has been described by several authors (11, 12). Most of it may be attributed to *saturation of PA*, as explained above, but a slowing of Ca pumping by SR may equally be involved.

During fatiguing activity, there is a strong correlation between the *relaxation rate* and the *affinity for ATP hydrolysis* (i.e. the maximum amount of work obtainable per mole of ATP hydrolysed). This could cause a slowing of Ca uptake by the Ca-ATPase of SR, and thus a decrease in relaxation rate (10).

Acidosis is also likely to slow down Ca pumping by SR (13).

MODIFICATION OF AM CYCLING RATE

The *decrease in affinity for ATP hydrolysis* may reduce the rate of crossbridge cycling, but this is less probable since no correlation exists between affinity and isometric force (10, 14).

The effect of *phosphate accumulation* is complex. In whole muscle it reduces the affinity for ATP hydrolysis, as discussed above. In skinned fibres, P_i reduces isometric force, stiffness and ATPase rate; it has no effect on velocity of shortening but increases relaxation rate. This can be explained by a product inhibition of the phosphate release step of the AM cycle. Relaxation would then occur by a reversal of two steps of the cycle, Actin binding to the complex myosin products, followed by phosphate release (15–22). Therefore, it is very difficult to predict which of these effects of P_i would dominate in living muscle, slowing of Ca uptake by SR or speeding up of AM detachment.

An *intracellular acidification* has been shown to reduce isometric tension, ATPase activity and velocity of shortening. Part of these effects is probably indirect, since acidosis modifies the equilibrium between mono and diprotonated forms of P_i in favour of the diprotonated form, probably responsible for P_i effects on AM cycle (16–23). But, unlike P_i, H^+ decreases velocity of shortening of skinned fibres. This suggests that protons by themselves have some direct effect on the AM cycle. It seems that acidosis reduces both the maximum force developed per cross-bridge (24, 25) and the number of active crossbridges (26). Moreover, Ca sensitivity of skinned fibres decreases with pH (27), perhaps because affinity of TNC for Ca is depressed (25). Acidosis may thus either enhance relaxation rate (increasing phosphate effect or reducing Ca affinity of TNC) or slow down relaxation (via some direct effect on crossbridges).

ADP accumulation has probably no effect on AM cycle in living muscle (17).

Conclusion

The role of parvalbumin in speeding up relaxation after short tetani is well established, and its saturation could well explain the slowing of relaxation observed after only a few seconds of contraction. But PA cannot account for the slowing of relaxation observed after long-lasting tetani, which may be due to complex effects of phosphate accumulation and acidification. However, the influence of phosphate and protons on crossbridge cycling in living muscle is unclear, and should be further explored.

References

1. Abbott, B. C., 1951, *J. Physiol.*, **112**, 438–445.
2. Peckham, M., Woledge, R. C., 1986, *J. Physiol.*, **374**, 123–135.
3. Hou, T. -t, Rall, J. A., Johnson, J. D., 1990, *XIXth EMC*, Brussels.
4. Curtin, N. A., 1976, *J. Physiol.*, **258**, 80–81.
5. Cannell, M. B., Allen, D. G., 1984, *Biophys. J.*, **45**, 913–925.
6. Gillis, J. M., Thomason, D., Lefèvre, J., Kretsinger, R. H., 1982, *J. Muscle Res. & Cell Mot.*, **3**, 377–398.
7. Gillis, J. M., 1985, *Biochem. et Biophys. Acta*, **811**, 97–145.
8. Hou, T., Rall, J. A., 1987, *Biophys. J.*, **51**, 475.
9. Berquin, A., Lebacq, J., 1990, *J. Muscle Res. Cell Mot.*, **11**, 76.
10. Dawson, M. J., Gadian, D. G., Wilkie, D. R., 1980, *J. Physiol.*, **299**, 465–484.
11. Blinks, J. R., Rüdel R., Taylor, S. R., 1978, *J. Physiol.* **277**, 291–323.
12. Cannell, M. B., 1986, *J. Physiol.*, **376**, 203–218.
13. MacLennan, D. H., 1970, *J. Biol. Chemistry*, **245**, 4508–4518.
14. Dawson, M. J., Gadian, D. G., Wilkie, D. R., 1978, *J. Physiol.*, **274**, 861–866.
15. Altringham, J. D., Johnston, I. A., 1985, *J. Physiol.*, **368**, 491–500.
16. Cooke, R., Franks, K., Lucianni, G. B., Pate, G., 1988, *J. Physiol.*, **395**, 77–97.
17. Cooke, R., Pate, G., 1985, *Biophys. J.*, **48**, 789–798.
18. Dawson, M. J., Smith, S., Wilkie, D. R., 1986, *Biophys. J.*, **49**, 268.
19. Hibberd, M. G., Dantzig, J. A., Trentham, D. R., Goldman, Y. E., 1985, *Science*, **228**, 1317–1319.
20. Kawai, M., 1986, *J. Muscle Res. Cell Motility*, **7**, 421–434.
21. Kawai, M., Güth, K., Winnikes, K., Haist, C., Rüegg, J. C., 1987, *Pflügers Arch.*, **408**, 1–9.
22. Lacktis, J. W., Homsher, E., 1986, *Biophys. J.*, **49**, 269.
23. Nosek, T. M., Fender, K. Y., Godt, R. E., 1987, *Science*, **236**, 191–193.
24. Curtin, N. A., Kometani, K., Woledge, R. C., 1988, *J. Physiol.*, **396**, 93–104.
25. Kentisch, J. C., Palmer, S., 1989, *J. Physiol.*, **410**, 67P.
26. Curtin, N. A., Rawlinson, S. R., 1984, *J. Physiol.*, **354**, 70P.
27. Fabiato, A., Fabiato, F., 1978, *J. Physiol.*, **276**, 233–255.

35
Parvalbumin, Labile Heat and Unexplained Enthalpy: Old Questions Revisited

A. BERQUIN AND J. LEBACQ

Department of Physiology, UCL 5528, 1200 Brussels, Belgium

Introduction

Ca-Mg interaction with parvalbumin (PA) was thought to explain several properties of fast striated muscles: labile heat (LH), unexplained enthalpy (UE) and Abbott effect (AE), which share several similar features and were thus likely to be due to a sole phenomenon. But some recent experimental data show that LH, UE and AE may be dissociated in some cases. In the view of ancient and new data, we propose that LH, UE and AE are due to a conjunction of several different origins, PA being their common, but not unique, factor. AE possible causes are discussed in the previous paper; in this paper we will focus mainly on LH and UE.

Labile Heat and Unexplained Enthalpy Properties

THE CLASSICAL HYPOTHESIS

The time course of heat rate during an isometric tetanus in frog sartorius can be described as consisting of two terms, a steady rate ('stable heat') and an exponentially decaying term, the 'labile heat' (1):

$$h_t = h_A e^{-\alpha t} + h_B \quad (1)$$

(where t is the time in s and α a rate constant in s^{-1}).

Muscle and Motility Vol. 2. Proceedings of XIXth European Conference in Brussels
© Intercept Ltd, PO Box 716, Andover, Hampshire, SP10 1YG, UK

In most energy balance-sheet experiments, the amount of enthalpy produced by amphibian muscles is higher than that expected from the measured consumption of energetic metabolites (mainly ATP and PCr) (2). The 'Unexplained Enthalpy' is the part of the enthalpy change which cannot be accounted for by known chemical reactions.

Like the Abbott effect, LH and UE follow very similar decaying kinetics with rate constants close to 1 s^{-1} at 0°C (3, 4). They are depressed by previous activity, reprime during recovery period (5, 6, 7) and are relatively independent of muscle length, i.e. filament overlap (1, 8). LH and UE amounts are respectively 20–40 and 10–40 $mJ.g^{-1}$ in frog sartorius (1, 3, 4, 6, 7, 9).

Main properties of parvalbumin are described in Chapter 34 and are in good agreement with properties of LH and UE. Ca-Mg exchange on PA is exothermic (33 $kJ.mol^{-1}$ sites of frog PA; (10)). Frog muscles containing $0.35\ 10^{-6}$ mol $PA.g^{-1}$, the enthalpy change of PA saturation with Ca should amount approximately to 23 $mJ.g^{-1}$.

NEW DATA AND THEIR CONSEQUENCES ON THE HYPOTHESIS

(a) LH production of isolated fibres from frog tibialis anterior is very different from fibre to fibre (11). This could be explained by different PA contents due to fibre type.

(b) The ratio of PA contents in muscles from Rana pipiens and Rana temporaria is similar to the ratios of LH and UE produced in these species (12).

(c) The relationship between amount of LH and PA contents in Xenopus fibres is linear, with a slope close to the enthalpy change of Ca binding to PA. But there is a significant intercept on the LH axis (13).

(d) Moreover, there is a linear relationship between amount of LH and enhancement of relaxation rate, with a positive intercept on the LH axis (14).

(e) LH is produced by mouse soleus, which contains only trace amounts of PA (15).

(f) LH repriming is biphasic (rate cts 0.1 and $2 * 10^{-3}$ s^{-1}, while UE recovery is monophasic (rate ct $\cong 0.17$ s^{-1}) (9).

(g) The rate constants of PA-Mg dissociation and LH production are similar (0.99 and 1 s^{-1}), as are the rate constants of PA-Ca dissociation, UE recovery and the fast term of LH recovery (0.18, 0.17 and 0.1 s^{-1}) (9, 16).

(h) Ca concentration in SR returns to pre-contractile level within 30 s after a 6 s tetanus (17). This is close to UE recovery duration (9).

(i) The energy balance-sheet is negative during 30 sec. after a 6 sec. tetanus. This could be due to a reversal of the process responsible for UE production (9).

The main properties of LH, UE, AE and PA are summarized in Table 1. Several conclusions may be drawn:

1. PA must take part in LH production (cf. a, b, c, g above), but part of LH must be due to some other process(es) (c, e, f).

Table 1 Main properties of labile heat (LH), unexplained enthalpy (UE), Abbott effect (AE, see chapter 34) and Ca-Parvalbumin interaction (PA). (Numbers refer to references.)

	LH	UE	AE	PA
Kinetics:	Expon. decay	Approx. expon. Decay(3, 4)	Expon. decay	PA-MG Dissociation:
Rate Constants (s^{-1}):	1(1)		1.18(16)	0.99(16)
Amount ($mJ.g^{-1}$):	20-40(7)	10-40(7)	–	23(10)
Reduction by previous activity	+(5)	+(6)	+(21)	PA-Ca saturation
Repriming	Biphasic	Monophasic	Monophasic	PA-CA dissociation:
Rate Constants (s^{-1})	0.002 & 0.1(9)	0.17(9)	0.12(16)	0.18(16)
Effect of Muscle Length	–(1)	–(8)	–(22)	–
Rana Pipiens/Temporaria[18]	0.5	0.5		0.5
Mouse Soleus	+(15)	?	–(23)	–(14)
Others	Linear rel. with PA & AE intercept +(13)	Neg. balance sheet: 30 s Post tetanus(9)	$y_0/(y_0+y_f)$: 0>10 °C(24)	

2. Only part of LH has the same origin as AE and UE (c, d, e, f).
3. UE could be relatively explained by Ca-PA interaction (b, g, h, i) but quantitative correlation is poor.

Therefore, Ca-PA interaction must explain part of LH and UE: it is exothermic, occurs at the beginning of tetani only, follows a decaying kinetics, is reduced by previous activity, reprimes progressively, and rate constants are in good agreement (cf. Table 1). All muscles containing PA must then produce LH and UE. However, PA-Ca interaction cannot account for the whole LH and perhaps UE. Therefore, PA is not the unique origin of these phenomena.

OTHER POSSIBLE ORIGINS OF LH AND UE

Calcium interaction with troponin

P sites These Ca-binding sites have the same properties as Ca-Mg sites of PA (Table 2). Their saturation could produce 5 $mJ.g^{-1}$ (9), with the same rate constants.

T sites The behaviour of these regulatory sites is very different: they must be nearly saturated before the peak of tension, and have lost their Ca before the end of relaxation (thus they cannot take part in the depression/repriming mechanism). The enthalpy change of Ca binding to T sites of TnC is not well known, probably modified by association of TnC to other components of Tn and by an important temperature sensitivity. Homsher *et al.* (9) consider that this reaction could produce 5 $mJ.g^{-1}$ muscle, but if the enthalpy change is 78 (and not 32) $kJ.mol^{-1}$ sites (18) and assuming 0.14 10^{-6} mol $sites.g^{-1}$, total enthalpy change could reach 10.9 $mJ.g^{-1}$. An accurate estimation of the enthalpy pro-

Table 2 Main properties of labile heat, unexplained enthalpy and abbott effect. Properties of their possible underlying mechanisms, Ca-parvalbumin interaction, Ca interaction with P- and T-sites of troponin, exponential hydrolysis of PCr and incomplete acto-myosin cycle. (Numbers refer to references.)

	During one Tetanus			Recovery		
	Enthalpy produced ($mJ.g^{-1}$)	Total duration (s)	Rate constant (s^{-1})	Total duration (min)	Rate constant 1 (s^{-1})	Rate constant 2 (s^{-1})
LH	20-40[7]	6–8	1[7]	15–30	0.1[9]	0.002[9]
UE	10-40[7]	6–8	±0.7	0.5[9]	0.17[9]	—
AE		8–10	1.18[16]		0.12[16]	—
PA	20[10]		0.99[16]		0.18[16]	—
TNC P	5[9]		0.99[16]			
T	5[9]	<tension rise	5[9]	<relax. duration		
PCr	10-15[9]		1[9]	30-40[9]		2 10^{-3}[9]
AM	?	<tension rise		<relax. duration		

duced by Ca binding to T sites of TnC is thus quite difficult to obtain, but it is probable that at least 5 $mJ.g^{-1}$ is produced.

Exponential PCr hydrolysis

From results of Kushmerick and Paul (19), Curtin and Woledge (3) and Homsher *et al.* (4), it has been suggested (9) that there exists an exponential term of PCr hydrolysis during the first few seconds of a tetanus, producing 15 $mJ.g^{-1}$ with a rate constant of 1 s^{-1}. Muscle's ability to split PCr probably recovers slowly, in 30–40 min (time constraint of approximately 500 s), which could account for the slow phase of LH recovery.

This exponential terms is quite difficult to observe on raw data from the authors cited above, but it appears from a mean of their results that PCr hydrolysis may be faster during the first 1 or 2 seconds of contraction. Excess heat produced during the first second only could amount approximately to 10 $mJ.g^{-1}$. This analysis shows that a slight increase in PCr hydrolysis rate at the beginning of a tetanus is sufficient to account for a significant part of LH (but not UE, by definition of UE).

Incomplete AM cycle

Some steps of the AM cycle are strongly exothermic and thus the enthalpy change of part of the cycle may be very large. When muscle switches from rest to activity, a redistribution of myosin states (M.ADP → AM.ADP.P) constitutes an incomplete AM cycle, which could take part in LH and UE production (20).

If frog sartorius contains 0.3 10^{-3} mol myosin ATPase sites and if 80% of myosin switches from M.ADP.P to AM.ADP.P states, enthalpy change of the incomplete cycle should be only 42 $kJ.mol^{-1}$ in order to produce 10 $mJ.g^{-1}$ muscle. Therefore, part of LH and UE produced at the very beginning of a tetanus could be due to a change in myosin repartition.

Others

LH and UE are produced in presence of iodoacetate or when O_2 is replaced by N_2. Glycolysis and oxidative phosphorylations are thus probably not involved in these phenomena.

Errors in estimates of ΔH values are ever possible, mainly because of uncertainties concerning real *in vivo* conditions (ionic strength, pH, ...), particularly during long lasting tetani. For a review of this problem, see (20).

Thermoelastic, shortening and stimulus heats are negligible (20).

Conclusion

The role of parvalbumin in labile heat and unexplained enthalpy production is well established. However, PA is not sufficient to account for these phenomena (Table 2). Production of LH and UE may be partly due to Ca binding to TNC, partly to a redistribution of myosin states, and an exponential term of PCr hydrolysis contributes perhaps to LH production.

However several points remain unclear. The understanding of a possible incomplete AM cycle requires a more accurate knowledge of the different steps of this cycle with their individual enthalpy changes, and detailed measurements of the rate of PCr hydrolysis at the beginning of tetani would provide more informations about the possibility of an exponentially decaying term.

References

1. Aubert, X., 1956, *Le Couplage Energétique de la Contraction Musculaire.* Arscia Ed., Brussels.
2. Gilbert, C., Kretzschmar, K. M., Wilkie, D. R., Woledge, R. C., 1971, *J. Physiol.*, **218**, 163–193.
3. Curtin, N. A., Woledge, R. C., 1979, *J. Physiol.*, **288**, 353–366.
4. Homsher, E., Kean, C. J., Wallner, A., Garibian-Sarian, V., 1979, *J. Gen. Physiol.*, **73**, 553–567.
5. Aubert, X., Maréchal, G., 1963, *Arch. Int. Physiol. Bioch.*, **71**, 282–283.
6. Curtin, N. A., Woledge, R. C., 1977, *J. Physiol.*, **270**, 455–471.
7. Woledge, R. C., Curtin, N. A., Homsher, E., 1985, *Monographs of the Physiological Society.*, **41**, 357p.
8. Curtin, N. A., Woledge, R. C., 1981, *J. Physiol.*, **316**, 453–468.
9. Homsher, E., Lacktis, J., Yamada, T. Zohman, G., 1987, *J. Physiol.*, **393**, 157–170.

10. Smith, S. J., Woledge, R. C., 1985, *J. Muscle Res. Cell Mot.*, **6**, 757–768.
11. Curtin, N. A., Howarth, J. V., Rall, J. A., Wilson, M. G. A., Woledge R. C., 1986, *J. Muscle Res. Cell Mot.*, **7**, 327–332.
12. Hou, T., D'Anniballe, L. J., Rall, J. A., 1988, *Biophys. J.*, **53**, 569.
13. Elzinga, G., Lännergren, J., Simonides, W. S., 1989, *J. Physiol.*, **414**, 48P.
14. Elzinga, G., Lännergren, J., 1988, *J. Physiol.*, **407**, 72P.
15. Berquin, A., Lebacq, J., 1989, *J. Physiol.*, **415**, 117P.
16. Hou, T.-t, Rall, J. A., Johnson, J. D., 1990, XIXth European Conference on Muscle and Motility, Brussels.
17. Tormey, J., Homsher, E., 1986, *Biophys. J.*, **49**, 423.
18. Gillis, J. M., 1985, *Biochem. et Biophys. Acta*, **811**, 97–145.
19. Kushmerick, M. J., Paul, R. J., 1976, *J. Physiol.*, **254**, 711–727.
20. Curtin, N. A., Woledge, R. C., 1978, *Physiol, Reviews*, **58**, 690–761.
21. Peckham, M. Woledge, R. C., 1986, *J. Physiol.*, **374**, 123–135.
22. Curtin, N. A., 1976, *J. Physiol.*, **258**, 80–81P.
23. Berquin, A., Lebacq, J., 1990, *J. Muscle Res. Cell Mot.*, **11**, 76.
24. Hou, T., Rall, J. A., 1987, *Biophys. J.*, **51**, 475.

36
Intracellular pH and Vascular Tone

N. C. SPURWAY

Department of Physiology, University of Glasgow, Glasgow G12 8QQ, UK

Introduction

Blood vessels dilate when external pH falls—thus acidotic tissues get more blood. Both the property and its adaptive benefit were noted more than 100 years ago (1). However, in smooth muscle, though not striated, intracellular pH (pH_i) works the other way: cytoplasmic acidity increases force. Here I shall briefly collect the main evidence for this assertion, including experiments described at the present Conference.

In 1983, Ighoroje and I began to apply ammonium chloride (NH_4Cl) pulses to vascular tissue. Such pulses had been shown (2, 3) to cause transient alterations of pH_i in large cells impaled by pH-microelectrodes, effects attributable to the rapid diffusion of covalent ammonia (NH_3) in advance of NH_4^+ ions. We found (4) that NH_4Cl application, which would raise pH_i, relaxed vascular smooth muscle; ammonium washout (lowering pH_i) was vasoconstrictive (Fig. 1).

Methods

The experiments illustrated were on isolated lengths of rat tail arteries, or the entire vascular beds of rabbit ears. Preparations were peristaltically perfused, at constant rates, with normal or modified Ringer's fluid, and the perfusion pressure was monitored as an indicator of vascular tone. Noradrenaline (NA) was normally introduced continuously to give background tone 2–4 times that of the unstimulated preparations. Main experiments were done at room temperature, since these appendage vessels function normally in this range. However, trials at 37°C showed no material differences except that the responses were faster.

Muscle and Motility Vol. 2. Proceedings of XIXth European Conference in Brussels
© Intercept Ltd, PO Box 716, Andover, Hampshire, SP10 1YG, UK

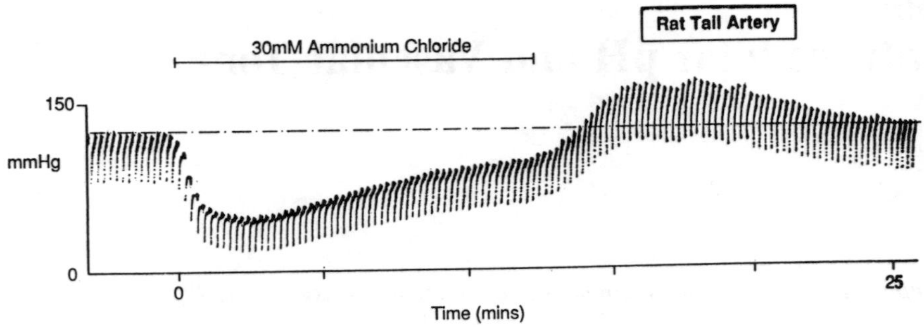

Fig. 1 Response of perfused rat tail artery, activated by 5.10^{-6} M NA at 20°C, to a 12.5 min pulse of 30 mM NH_4Cl.

Modifications to the Ringer's were made by substitution of the neutral salt (NH_4Cl), sodium salt of organic acid, hydrochloride of organic base), 5–40 mM, for equivalent NaCl. The pH values of normal and modified Ringer's were always identical : 7.2 or 7.4 in illustrations here. Periods of perfusion ('pulses') of modified Ringer's were 2–15 min; recovery in normal Ringer's was for at least twice the duration of the preceding pulse.

Results

The dilator response to NH_4Cl, and overshooting constriction on its withdrawal, were unfailingly observed, provided background tone lay somewhere on the steep part of the NA dose-response curve. At the extremes of the curve, however, responses were essentially unidirectional. Thus in unactivated vessels, or any others whose tone was exceptionally low, only the constriction occurred. In those at the plateau of activation the dilatation phase predominated.

The phenomena were interpreted as pH_i effects within the muscle cells for the following reasons. First, they were unaffected by chemical sympathectomy or methylene blue; thus neither nerves nor endothelium need be considered (5). Second, even the shapes and time-courses of pH_i change induced by NH_4Cl pulses in other cells were reminiscent of our tone responses. Finally, carbon dioxide (CO_2), which is of course highly permeant but acidic, had been shown 10 years earlier (6, 7) to be vasonconstrictive when applied at neutral extracellular pH: the dilator property usually associated with CO_2 is a consequence of extracellular pH fall. It had previously seemed possible that the effect of intracellular CO_2 was a specific molecular action. However, the only imaginable feature common to CO_2 entry and NH_3 washout was that both lowered pH_i.

Nevertheless, a check seemed desirable that pH_i responses to NH_4Cl in the population of small cells making up a vascular media were the same as in large, impalable cells. NMR experiments (8), confirmed in another laboratory by fluorescent indication (9), showed that the responses were indeed the same.

Our account of the tone responses of skin vessels to NH_4Cl was supported by independent observations on NA-activated ring and strip preparations of rat mesenteric arterioles (10) and aortae (11, 12) and of porcine coronary vessels under carbachol (13); these reports also include confirmation that only the constrictor effects can be obtained from unactivated vessels. We found that the basic response pattern applied—particularly clearly at 37°—to a variety of deeper vessels from rabbits, rats and cats. It also applied to K-depolarized as well as a range (5-HT, vasopressin, thromboxane derivative) of other pharmacologically activated preparations (5, 14). In addition we recently discovered a report (15) of the dilator effect of ammonia upon the cerebral circulation which preceded all the above work; though in that study, washout constrictions were not noted and the relevance of pH_i was not suspected.

We know of only two findings in any conflict with those above. Guinea pig ductus arteriosus, contracted with 100% O_2, *relaxed* when pH_i was reduced by NH_4Cl washout and other procedures (16). We have not reproduced this finding with rabbit ductus, but it is the case that in this tissue the constriction elecited by NH_4Cl washout was followed, especially at 100% O_2, by an undershoot of tone below control level. An interpretation of these results will be proposed in the discussion. A second conflicting finding is that K-activated rat mesenteric artery rings usually *constricted* to NH_4Cl application, though under NA activation they always dilated (10). Among the factors probably contributing to this latter divergence are the cationic properties of NH_4.

The NH_4^+ ion is sufficiently small and K^+-like to pass through the same channels; where figures are available its permeability has been found to be ~1/10th P_K. Thus when placed outside a cell at a concentration of 10–40 mM it must have a powerful depolarizing influence, while at washout, after accumulating intracellularly, it must have a hyperpolarizing one. These membrane potential changes cannot fail to complicate the tone responses to pH_i. Furthermore, the fairly rapid transmembrane passage of NH_4^+ ions appears to be a major factor making the pH_i changes themselves transient (17) (though normal pH-regulating mechanisms are present in these cells (9, 18) and contribute also). Again there are several complications, including membrane potential and buffering effects, in the interpretation of CO_2 experiments. However, certain other weak electrolytes, molecularly too large to be likely to pass through conducting channels, have been shown in large cells to cause simple, monophasic displacements of pH_i over experimental time-scales (3). Our recent work has been with these.

Neutral salts of weak, organic acids (notably including lactic) elicited marked and virtually monophasic constrictions (Fig. 2) from both superficial and deep preparations (14, 19). Washout produced brisk return of tone to near the original level, but rarely a significant swing beneath this. These results powerfully support the generalization that the effect of intracellular acidification in vascular muscle is constrictor.

The complementary experiments are those with weak bases (20). All seven bases tested produced rapid and marked vasodilatations (Fig. 3), those elicited by the larger molecules showing little or no sign of recovery during applications lasting 15 min. We conclude that the effects of intracellular alkalinization, at

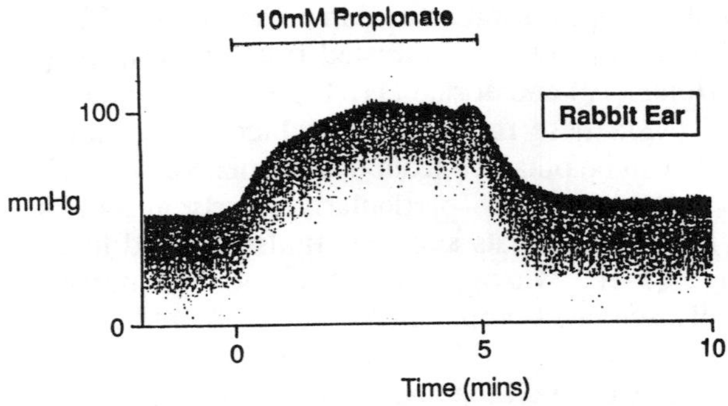

Fig. 2 Response of perfused rabbit ear vessels, activated by 10^{-6} M-NA at 20°C, to a 5 min pulse of 10 mM Na propionate.

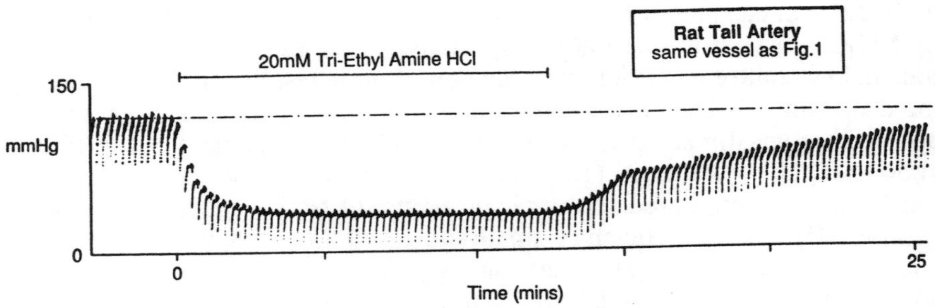

Fig. 3 As Fig. 1, except that pulse is of 20 mM triethylamine HCl.

least in NA-activated ear and tail vessels, are always dilator: any complications with NH_4^+ may now be ascribed with virtual certainty to its membrane actions, not to pH effects. It is tempting to think that this result might have clinical potential.

Meanwhile, there is a final set of experiments to cite. Bamosa and I recently (21) found it possible to load the muscle cells of intact arteries with fura-2. NH_4Cl application reduced the indicated $[Ca^{2+}]_i$, NH_4Cl washout or propionate/lactate application raised it. In the light of these results we can propose a simple mechanism by which pH_i changes, however caused, might modify tone.

Discussion

Assume that H^+ and Ca^{2+} ions compete at all binding sites. Within the smooth muscle cell one such site is calmodulin; at constant $[Ca^{2+}]_i$, raised $[H^+]_i$ would therfore reduce tone. The other sites, however, are Ca stores: sarcoplasmic reticulum, mitochondria and so on. When $[H^+]_i$ rises, these will tend to release more Ca^{2+}: as has just been noted, $[Ca^{2+}]_i$ is indeed seen to rise (21). Our postulate is that in the majority of vascular preparations the $[Ca^{2+}]_i$ elevation more than compensates for calmodulin's reduced Ca affinity, so tone is raised when $(H^+)_i$ increases. In an unusual vessel or metabolic state, however, it is presumably possible for the converse to happen (10), analogous to the situation which is the norm for striated muscles.

And what is the advantage of the vascular pattern? Perhaps (22) it permits the maintenance of profound vasoconstriction, as in a cold limb, when the vessels have closed off their own blood supply.

References

1. Gaskell, W. H., 1880, *J. Physiol.*, **3**, 48–75.
2. Boron, W. F., de Weer, P., 1976, *J. Gen. Physiol.*, **81**, 53–91.
3. Thomas, R. C., 1984, *J. Physiol.*, **354**, 3P–22P.
4. Ighoroje, A. D., Spurway, N. C., 1984, *J. Physiol.*, **357**, 105P.
5. Ighoroje, A. D., 1987, PhD thesis, University of Glasgow.
6. MacLellan, D. G., Pickard, J. D., Spurway, N. C., 1974, *J. Physiol.*, **242**, 97P.
7. Pickard, J. D., Simeone, F. A., Vinall, P., 1976. In: Betz, E. (ed.) *Ionic Actions on Smooth Muscles*, pp. 101–4. Springer, Berlin.
8. Spurway, N. C., Wray, S., 1987, J. Physiol., 393, 57–71.
9. Aalkjaer, C., Cragoe, E. J., 1988, *J. Physiol.*, **402**, 391–410.
10. Aalkjaer, C., Mulvaney, M. J., 1988, *Prog. Biochem. Pharmacol.*, **23**, 150–8.
11. Furtado, M. R., 1987, *Life Sci.*, **41**, 95–102.
12. Danthuluri, N. R., Deth, R. C., 1989, *Am. J. Physiol.*, **256**, 867–75.
13. Nguyen-Duong, H., 1988, *Pflügers Arch.*, **411**, R203.
14. Bamosa, A. O., 1989, PhD Thesis, University of Glasgow.
15. Andersson, K. E., Brandt, L., Hindfelt, B., Ljunggren, B., 1981, *Acta Physiol. Scand.*, **113**, 349–53.
16. Garnier, D., Roulet, M. J., 1986, *J. Physiol.*, 377, 121P.
17. Bamosa, A. O., Ighoroje, A. D., Spurway, N. C., 1987, *J. Physiol.*, **392**, 47P.
18. Ighoroje, A. D., Spurway, N. C., 1985, *J. Physiol.*, **367**, 46P.
19. Bamosa, A. O., Spurway, N. C., 1988, *Med. Sci. Res.*, **16**, 469–70.
20. Spurway, N. C., Godfrey, R. J., Cham, C-K., 1991, *J. Muscle Res. Cell Motil.*, abstracts of the XIXth European Conference on Muscle and Motility, Brussels, in press.
21. Bamosa, A. O., Spurway, N. C., 1990, *J. Physiol.*, **423**, 64P.
22. Jennett, S. M., pers. comm.

37
Conduction Velocity of Isolated Rat Muscles Depending on Stimulation, Modified Potassium Concentration and Catecholamines

F. KÖSSLER, F. LANGE, G. LINSEL AND G. KÜCHLER

Central Institute of Occupational Medicine, Department of Work Physiology, Berlin 1134, Germany

Introduction

During muscular activity ionic shifts occur between intra- and extracellular spaces resulting in an increase of Na^+ inside the cell and a loss of K^+ (1). Following intensive stimulation K^+ is accumulated within the interstitium of the muscle up to 10 nM/kg or higher concentrations (2). Such high extracellular K^+ concentrations are able to change membrane properties and the excitability of muscle cells. Possible relations between K^+ accumulation and muscle fatigue have been discussed in a few papers, only (3–6). One of the processes involved in muscle fatigue seems to be an impaired signal propagation along the muscle fibre. It was of interest to investigate the effect of solutions with increased K^+ concentration on the conduction velocity (c.v.) of resting muscles *in vitro* and to compare the changes with that obtained after fatiguing stimulations. Fast- and slow-twitch muscles of the rat with different sensitivity to fatigue were used in the present study. Further, epinephrine was applied as a stimulus of the Na^+K^+-ATPase to test the response of the two muscle types as well as to investigate whether the K^+ induced changes of the c.v. may be modified by increased ATPase activity.

Muscle and Motility Vol. 2. Proceedings of XIXth European Conference in Brussels
© Intercept Ltd, PO Box 716, Andover, Hampshire, SP10 1YG, UK

Methods

Fibre bundles were prepared from m. soleus (SOL) and m. extensor digitorium longus (EDL) isolated from rats (Wistar, 250–450 g) and stored at resting length in oxygenated Tyrode solution (mM: NaCl 137; KCl 5.0; $MgCl_2$ 1.0; $CaCl_2$ 2.0; Na_2HPO_4 1.0; $NaHCO_3$ 12.0; glucose 11.0) bubbled with carbogen gas (95% O_2, 5% CO_2); pH 7.4, 2°C. The c.v. was measured in a temperature-controlled perspex chamber (3–4 ml) at 22 ± 1°C. The fibre bundle was fixed on one end but connected with the lever of a force transducer on the other one. The preparation was adjusted to give maximum twitch tension by means of a micromanipulator and a stimulation unit (supramaximum biphasic pulse of 0.2 ms with 0.02 Hz). For fatigue experiments the muscles were stimulated by trains of 50–100 Hz for 20 s. The c.v. was estimated as the ratio between the distance of two annular platinum electrodes and the time delay of the peaks of the myoelectric signals (extracellular action potentials) measured by these electrodes and recorded on UV-sensitive paper by an oscilloscope. For variation of extracellular potassium concentration K^+ was omitted (0 K^+) or added to Tyrode solution to give 10 or 15 mM K^+. Epinephrine was prepared as a stock solution (1 mM) and diluted in Tyrode to a final concentration of 10 μM. The data were analysed by standard statistical methods.

Results

The c.v. of fast and slow twitch muscles differs more or less with the tendency to somewhat higher values in EDL muscles. In the present study (Table 1) the c.v. of SOL muscles was 1.02 ms^{-1} but that of EDL muscles 1.39 ms^{-1}. Immediately after tetanic stimulation the c.v. declined to 82% (SOL), and 60% (EDL), respectively. In contrast to an overcompensated restoration in SOL muscles, EDL did not recover completely within 30 min. A similar decrease of the conduction velocity was induced in resting muscles by increasing the K^+ concentration of the Tyrode solution (Table 2). K^+ 10 mM caused a reduction to about 85% in SOL but to 75% in EDL muscles. The recovery after replacement of normal Tyrode (5 mM K^+) was not completed within 30 min. With higher K^+ concentrations (15 mM) the drop was accelerated and the changes became

Table 1 Conduction velocity ($m.s^{-1}$) of EDL and SOL muscles before (control), immediately after three tetanic stimulations (50 Hz, 20 s each), and during 20 and 30 min of recovery; 22 ± 1°C; *n* in parenthesis.

Muscle	Control	After stimulation	Recovery 20 min	30 min
SOL (7)	1.02 ± 0.24	0.82 ± 0.17	1.20 ± 0.20	1.16 ± 0.20
EDL (7)	1.39 ± 0.26	0.83 ± 0.22	1.15 ± 0.17	1.18 ± 0.22

Table 2 Changes of the conduction velocity ($m.s^{-1}$) of EDL and SOL muscles during 20 min of exposure to 10 mM K^+ and 30 min recovery in Tyrode; 22 ± 1°C, *n* in parenthesis.

Exposure (min)		EDL(8) ($m.s^{-1}$)	%	SOL(10) ($m.s^{-1}$)	%
Tyrode		1.14 ± 0.27	100	0.97 ± 0.18	100
100mM K^+	2	1.22 ± 0.26		0.98 ± 0.17	
	4	1.14 ± 0.18		0.94 ± 0.12	
	8	1.07 ± 0.17		0.89 ± 0.10	
	12	0.96 ± 0.18		0.87 ± 0.10	
	16	0.92 ± 0.18		0.84 ± 0.12	
	20	0.86 ± 0.18	75.5	0.82 ± 0.11	85
Tyrode	15	0.85 ± 0.18		0.81 ± 0.09	
	30	0.93 ± 0.25	81.5	0.88 ± 0.13	91

irreversible. In parallel with the decreased c.v. the amplitudes of extracellular action potentials were flattened and changed their typical shape (Fig. 1, left). In additional experiments, 10 mM K^+ reduced the amplitude of intracellular action potentials, the rate of rise, and the repolarisation phase.

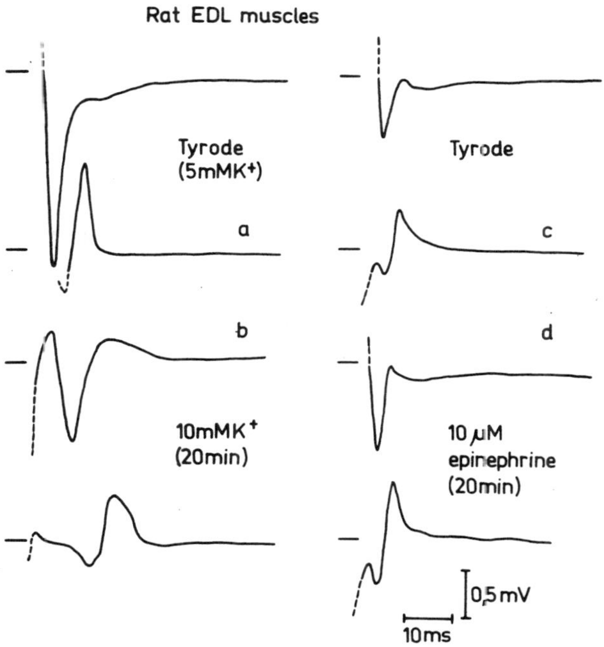

Fig. 1 Typical records of extracellular action potentials of rat EDL muscles in Tyrode solution (a, c) and after 20 min of exposure to 10mM K^+ (b) or to 10 µM epinephrine (d); 22± 1°C. Note the opposite effect of K^+ and adrenaline on the potential peaks.

The addition of epinephrine (10 μM) to the Tyrode solution increased the peaks of the extracellular potentials and accelerated the c.v. (Fig. 1, right). Application of epinephrine together with 10 mM K^+ resulted in a decrease of the c.v. like that in 10 mMK^+ without epinephrine. After return to normal Tyrode solution the recovery was facilitated in epinephrine-exposed muscles.

Discussion

In the present study it was shown that similar decreases of the conduction velocity of the action potentials in muscle fibres could be produced by electrical stimulation as well as by increased extra-cellular K^+ in Tyrode if the concentration of K^+ corresponded to that measured in fatigued muscles (2). Both after electrical stimulation and in high K^+ solution there was a stronger depression of the c.v. in EDL muscles.

The mechanism responsible for K^+ accumulation within the interstitium may be an insufficient capacity of the Na^+K^+-ATPase. The basal activity of this enzyme was found to be not very different in EDL and SOL. But its activation by electrical stimulation or catecholamines was much higher in SOL muscles (7). The experiment with epinephrine (10 μM) demonstrates an increase of the c.v. in both electrical stimulation and high K^+ solution. The simultaneous application of epinephrine and 10 mM K^+ was not able to avoid the decrease of the c.v. but the recovery was facilitated. In conclusion it may be suggested that the capacity of the Na^+K^+-ATPase and its adrenergic stimulation during exercise are important factors in avoiding K^+ accumulation during fatigue.

References

1. Juel, C., 1986, *Pflügers Arch.* **406**, 458–463.
2. Hnik, P., Vyskocil, F., Ujec, E., Vejsada, R., Rehfeldt, H., 1986, In: Saltin, E. (ed.) *Biochemistry of Exercise*, VI, pp. 345–363. *Int. Ser. Sport Sci.*, **16**, Hum. Kin. Publ. Champaign, III.
3. Juel, C., 1988, *Muscle & Nerve*, **11**, 714–719.
4. Kössler, F., Caffier, G., Lange, F., 1990, *Z. ges. Hyg.* **36** 354–356.
5. Sjøgaard, G., 1987, *Medicine Sport Sci.* **26**, 98–109.
6. Vollestad, N. K. Sejersted, O. M., 1988, *Eur. J. Appl. Physiol.*, **57**, 336–347.
7. Everts, M. E., Retterstøl, K., Clausen, T., 1988, *Acta Physiol. Scand.*, 134, 189–198.

38
The Action of Cholinesterase Blocking Agents on Mechanical Properties of Slow-tonic and Fast-twitch Fibres of Rabbit Extraocular Muscles

G. ASMUSSEN AND U. GAUNITZ

Carl-Ludwig-Institute of Physiology, University of Leipzig, Germany

Introduction

Mammalian extraocular muscles (EOM) contain two motor systems. One is composed of a variety of single innervated twitch-muscle fibres which are among the fastest fibres in mammals, the other contains two types of multiple innervated slow fibres which possess characteristics of avian or amphibian slow-tonic muscle fibres (see (1) for review). Thus EOM have a high sensitivity to drugs like acetylcholine (ACh) and an application of such substances evokes sustained contractures (see (2) for review). In this study the action of cholinesterase (ChE) blocking agents on the mechanical properties of both systems were investigated *in vitro* using the isolated rabbit inferior oblique muscle (3). The slow-fibre system was activated by an application of ACh in the bathing solution, for the activation of the fast-fibre system the direct massive stimulation of the muscles with short-lasting impulses was used. The data obtained confirm and extend results presented earlier (4, 5).

Muscle and Motility Vol. 2. Proceedings of XIXth European Conference in Brussels
© Intercept Ltd, PO Box 716, Andover, Hampshire, SP10 1YG, UK

Material and Methods

The inferior oblique muscle (IO)of nine adult rabbits (4 female, 5 male; body, weight: 2.2–3.8 kg) were used. The preparation of the IO and the arrangement of the muscle *in vitro*, the composition of the physiologic saline, the stimulation and the tension recording of twitches, tetani and contractures are described in detail elsewhere [4, 5]. ACh chloride in concentrations of 55 nM to 5.5 mM and physostigmine or prostigmine-salicylate in concentrations of 1–5 μM were added to the physiologic saline.

Results

ACTION OF CHE-BLOCKERS ON THE SLOW-FIBRE SYSTEM

Physostigmine or prostigmine in concentrations of 2–5 μM added to the bathing solution blocked the ChE almost completely, on the average 2.5 μM were sufficient and the application time necessary was 10–15 min. The ACh-sensitivity of the slow-tonic fibres increased. The threshold dose to evoke a sustained contracture decreased from 3–9 μM ACh in normal Ringer solution to 80–300 nM ACh in a Ringer solution containing 2.5 μM physostigmine or prostigmine. ChE-blocking agents shifted the dose-response curve of ACh to the left (Fig. 1). The maximum contracture tension was observed at concentrations between 5 and 50 μM ACh and it was almost the same as in physiologic saline without ChE-blockers at concentrations of 2–5 mM ACh. The shape of the contractures was nearly unchanged, only the relaxation period was somewhat prolonged. After a rinse with normal physiologic saline of 20–30 min the effects of the ChE-blocking agents were completely reversible.

EFFECTS OF CHE-BLOCKERS ON THE FAST-FIBRE SYSTEM

The result of a typical experiment (Fig. 2) shows the action of physostigmine, in a concentration blocking the ChE completely, on twitches and tetanic contractions of the fast fibres. There were no remarkable differences in the actions of physostigmine or prostigmine. ChE-blocking agents increased the amplitude of single twitches 2–3 times. An application time of 8–15 min was necessary to obtain the maximum twitch potentiation response; longer application was without any further effect. The contraction and half-relaxation times of untreated muscles of 6.4 ± 0.5 ms and 7.0 ± 0.5 ms increased under the influence of ChE-blockers to 13.1 ± 0.8 ms and 14.0 ± 1.2 ms ($n = 8$) respectively, and consequently the fusion frequency dropped from 320 ± 30 Hz to 170 ± 20 Hz. The maximum tetanic tension output was unchanged, however, an unusual biphasic relaxation time course was observed (Fig. 2). After a fused tetanus the muscle preparation suspended in Ringer solution containing ChE-blocking agents exhibited a regular quick drop of the tension but only to about 20% of the maximum tetanic tension followed by a very slow final relaxation. Also these

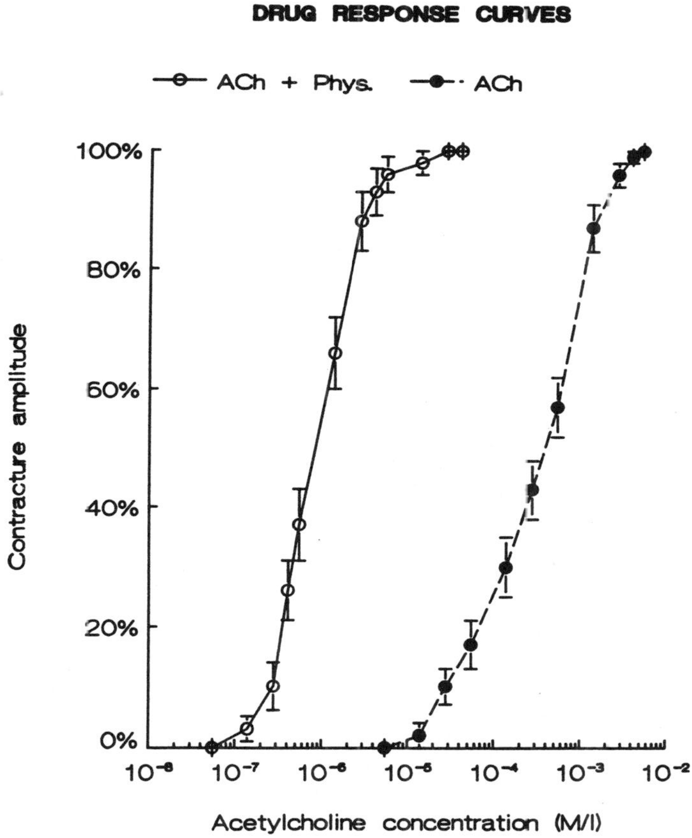

Fig. 1 Influence of physostigmine (2.5 μM) on the relation between contracture tension (in % of the maximum contracture) and ACh-concentration of isolated rabbit IO *in vitro* at 35°C. Data are presented as mean ± SD ($n = 7$).

effects of ChE-blockers were completely reversible after a rinse of 20–30 minutes duration in normal physiologic saline.

Discussion

It was demonstrated that mammalian EOM contain a multiple innervated slow fibre system (1, 2, 3) quite similar to that of amphibians and birds. Iontophoretic applications of ACh to slow fibres of amphibians (6) and birds (7) have shown that these fibres are sensitive to ACh along their whole surface membrane and that they react to an application of ACh in the bathing solution with sustained contractures (8, 9). Furthermore, not fully developed mammalian twitch-fibres exhibit besides a high junctional ACh-sensitivity a detectable extrajunctional one (10) and produce transient ACh-contractures (11), whereas adult common

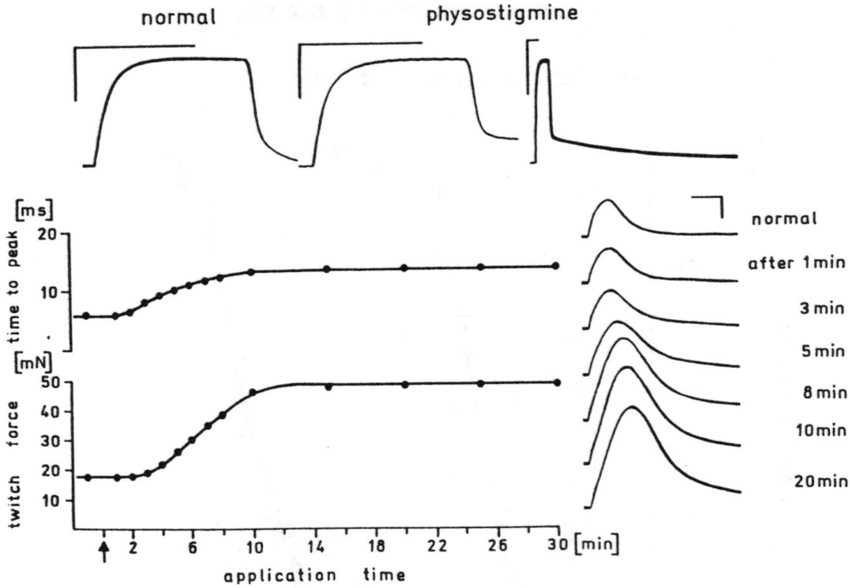

Fig. 2 Action of physostigmine (2.5 μM) on single twitches and tetanic contractions of an isolated IO *in vitro* at 35°C evoked by direct massive stimulation (rabbit, male, 3.3 kg body mass).
Below left: Changes of the twitch force and the contraction time in dependence of the application time of the drug. The arrow marks the start of the application.
Below right: Typical twitches (calibration: 10 ms, 10mN). *Above*: Characteristic fused tetanic contractions (stimulation frequency 300 Hz). Note the late retarded relaxation after administration of physostigmine (calibration: 200 ms. 100 mN).

mammalian twitch fibres are sensitive to ACh only below the sole-plate of the neuromuscular junction and do not exhibit any reaction to an application of ACh in the bathing solution (10, 11).

Recently it was shown that some fast fibres of EOM of full grown animals express also neonatal and/or embryonic myosin during their whole life span (12, and own unpublished observations), and one could speculate that these fibres, if they have also some other properties of not fully developed fibres, could contribute to the ACh-contractures of the EOM. The fact that the maximum tension output of the ACh-contractures is about 15–20% of the maximum tetanic tension is in good accordance with the amount of the muscle fibres belonging to the slow-fibre system (1, 4).

Physostigmine or prostigmine exhibit, at concentrations completely blocking the ChE, a considerable and reversible potentiation of the single twitch. It was shown in frog muscles that ChE-blockers belong to those drugs which prolong the duration of the action potential and therefore increase the duration of the 'active state' (13). The prolongation of the 'active state' explains the twitch-potentiation connected with increased contraction and relaxation times.

The biphasic relaxation after tetanic stimulation could be explained by the following considerations. The massive direct stimulation used excites not only the fast-twitch muscle fibres but also the intramuscular nerves. This leads to a liberation of ACh at the motor nerve endings and, because of the blockage of the ChE, to an accumulation of ACh in the muscle evoking a contracture of the slow-tonic fibres. Therefore after the tetanic stimulation the tension developed by the fast fibres drops quickly to the level held by the slow fibres (about 20%) followed by a very slow relaxation because of the decrease of the intramuscular ACh-concentration by the steady replacement of the physiologic saline.

Acknowledgements

We should like to thank Miss R. Rätze for excellent assistance during the experiments.

References

1. Spencer, R. F., Porter, J. D., 1988. In: Büttner-Ennever (ed.) *Neuroanatomy of the Oculomotor System*, pp 33–79.. Elsevier, Amsterdam.
2. Asmussen, 6., 1974, *Wiss. Z. Karl-Marx-Univ. Leipzig, Math.Naturwiss. R.*, **23**,125–140.
3. Mayr, R., Gottschall, J., Gruber,H., Neuhuber, W., 1975, *Anat. Embryol.*, **148**, 25–34.
4. Asmussen, G., Gaunitz, U., 1981, *Pflügers Arch.*, **392**, 183–190 .
5. Asmussen, G., Gaunitz, U., 1981, *Pflügers Arch.*, **392**, 191–197.
6. Kiessling, A., 1964, *Pflügers Arch.*, **280**, 189–192.
7. Fedde, M. R., 1969, *J. Gen. Physiol.*, **53**, 624–637.
8. Ginsborg, B. L., 1960, *J. Physiol.*, **154**, 581–598.
9. Kuffler, S. W., Vaughan Williams, E. M., 1953, *J. Physiol.*, **121**, 318–340.
10. Diamond, J., Miledi, R., 1962, *J. Physiol.*, **162**, 393–408.
11. Gutmann, E., Hanzlikova, V., *Physiol. Bohemoslov.*, **15**, 508–524.
12. Sartore, S., Mascarello, F., Rowlerson, A., Gorza, L., Ausoni, S., Vianello, M., Schiaffino, S., 1987, *J. Muscle Res. Cell Motil.*, **8**, 161–172.
13. Pagala, M. K. D., Sandow, H., 1976, *J. Pharm. exp. Ther.*, **197**, 439–451.

SECTION SIX

Mechanics and Crossbridges

39
Calcium Requirement for Aluminofluoride Inhibition of Force in Skinned Muscle Fibres

P. B. CHASE* AND M. J. KUSHMERICK**

*Departments of *Radiology and **Physiology and Biophysics, University of Washington, Seattle, Washington 98195, USA*

Introduction

Metallofluoride complexes, such as aluminofluoride, can form stable complexes with F-ADP-actin and with G-proteins and GDP; the metallofluoride complexes are thought to be high-affinity analogues of inorganic phosphate (1, 2, 3). We find that aluminofluoride is also a potent inhibitor of force generation by skinned muscle fibres. An apparent requirement for calcium was demonstrated for both the onset of and recovery from inhibition of force by aluminofluoride. The kinetics of aluminofluoride binding to and dissociation from its site of inhibition were greatly enhanced in the presence of Ca^{2+}, i.e. during crossbridge cycling, in a manner that is qualitatively similar to orthovanadate (4). Unloaded shortening velocity was also inhibited by aluminofluoride, but to a lesser extent than force. Aluminofluoride may be a useful reagent in the study of intermediates in the actomyosin interaction in fibres as well as *in vitro*.

Methods

Single glycerinated fibre segments from rabbit psoas muscle were prepared and mounted for mechanical measurements as described (5). During the experiment, the fibre segment was shortened periodically (0.2 s^{-1}) at a rate approxi-

Muscle and Motility Vol. 2. Proceedings of XIXth European Conference in Brussels
© Intercept Ltd, PO Box 716, Andover, Hampshire, SP10 1YG, UK

mately equal to V_{max}, reducing force to the baseline, followed by rapid restretch to the original length (6); force was measured during the steady-state period between shortenings (7). The control conditions during activation were (mM): 5 MgATP, 15 PCr, 1 Pi, 4 EGTA, 250 MOPS, 200 ionic strength adjusted with acetate and Tris, 50 Na^+, 50 K^+, 3 Mg^{2+}, pCa 4.5 (pCa = $-\log_{10}[Ca^{2+}]$), pH 7.1, 250 units/ml of creatine kinase (CK), 12°C. $Al(NO_3)_3$ and NaF were added to solutions from concentrated stock solutions. The final concentration of NaF added was 10 or 16 mM with varying amounts of $Al(NO_3)_3$. Solution of the multiple binding equilibrium between F^- and EGTA with Al^{3+}, Ca^{2+}, and Mg^{2+} (using the binding constants in refs. 8, 9, 10) indicated that the major fraction of Al^{3+} was complexed with F^- at the total concentrations of Al^{3+} and F^- used in these experiments. This method of adding Al^{3+} and F^- increased the ionic strength by 10 to 12 mM above control for solutions with 10 mM total fluoride. Orthovanadate (Vi) stock solutions were prepared and assayed as described (11, 12). Force in the presence of inhibitor was normalized to the average of bracketing control contractions before and after. Unloaded shortening velocity (V_{us}) was measured using the slack method (13) as previously described (5).

Results

Maximum Ca-activated force of skinned fibres was inhibited by aluminofluoride (Fig. 1A). In the record shown, the steady-state force at pCa 4.5 was inhibited by >90% in the presence of aluminofluoride. The inhibition was fully reversible. Fig. 1A also demostrates that both the onset of and recovery from inhibition occurred much more slowly in relaxing conditions (pCa 8) than at full activation (pCa 4.5). The end of the record shows that recovery occurred over several minutes at pCa 4.5 ($t_{1/2} \sim 3.5$ min), but that little or no recovery occurred during 8 min in relaxing solution which did not contain aluminofluoride. If recovery had occurred at pCa 8, then the force upon activation should have rapidly risen to ~50% of control. At the beginning of the record in Fig. 1A, pre-incubation of the relaxed fibre with aluminofluoride did not result in any significant inhibition of the subsequent activation (without aluminofluoride). The dependence on active crossbridge cycling of both the onset of and recovery from inhibition are qualitatively similar to that of Vi (Fig. 1B; ref. 4). The records in Fig. 1 suggest that recovery from inhibition may be slower with aluminofluoride than with Vi, although this point has not been examined in detail.

The inhibition of force by varying the total amount of $Al(NO_3)_3$ added (0.01-3 mM) while maintaining total fluoride constant (10 mM) is shown in Fig. 2. Addition of $Al(NO_3)_3$ alone did not inhibit force, even at concentrations which caused a precipitate to form. These data implicate a complex of F with Al, rather than with Ca or other ions, as the inhibitor because an increase in the concentration of Al caused greater inhibition; if, for example, CaF^+ was responsible, the extent of inhibition would have decreased with increasing Al due to an increase in the fraction of F complexed with Al. Note that the relation between the total concentration of Al and the free solution concentration of the inhibitory complex (which is likely to be the AlF_4^- complex (1), although this

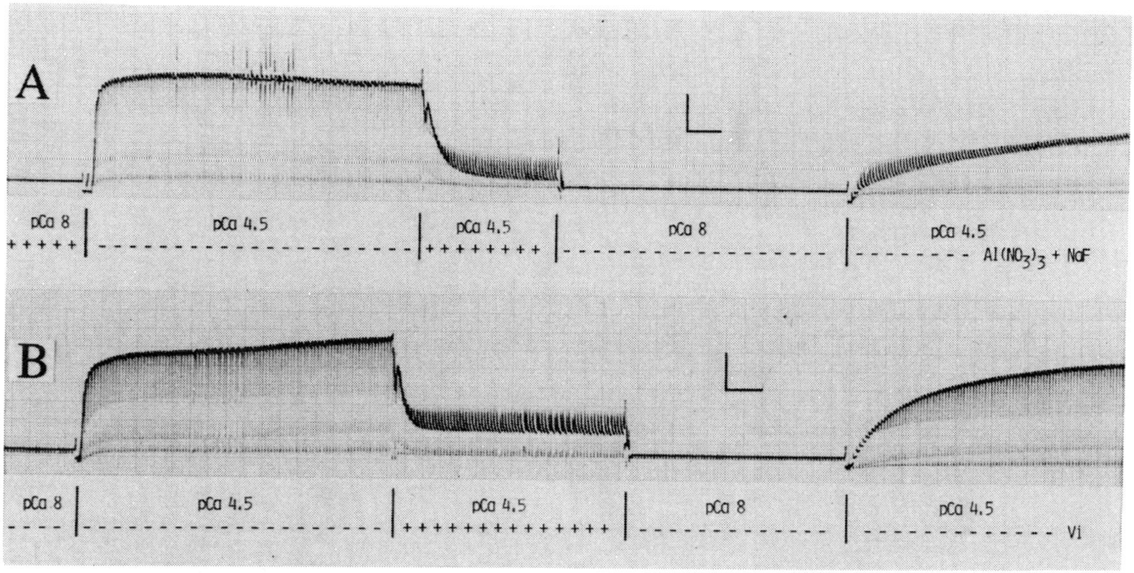

Fig. 1 Calcium-requirement for inhibition of maximum Ca-activated force of skinned fibres from rabbit psoas. Original force records demonstrating the Ca-requirement for the onset of and recovery from inhibition of skinned fibre force by aluminofluoride (A) or Vi (B). Vertical lines below force records indicate solution changes; the pCa and the presence (+) or absence (-) of aluminofluoride or Vi are given for each solution.

(A) Protocol demonstrating that inhibition by aluminofluoride (4 mM $Al(NO_3)_3$ + 16 mM NaF) and recovery from inhibition occur much more slowly in relaxing conditions (pCa 8) than during maximum Ca-activation (pCa 4.5). Calibrations: force, 20 mg; time, 1 min.

(B) Protocol similar to that used in (A) demonstrating that recovery from inhibition by Vi (100 μM) occurs much more slowly in relaxing conditions than during maximum Ca-activation. Calibrations: force, 20 mg; time, 1 min. Effects of Vi in the absence of Ca^{2+} were not examined in this fibre.

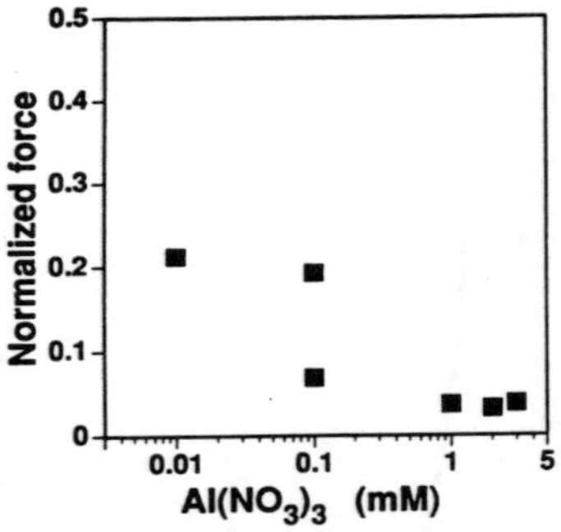

Fig. 2 Inhibition of skinned fibre force by aluminofluoride. Force in the presence of aluminofluoride was normalized to the average of bracketing control contractions. The total concentration of fluoride was constant (10 mM) and the concentration of the inhibitory complex was varied by altering the total amount of Al (0.01-3 mM).

has not yet been demonstrated for actomyosin) is probably influenced significantly by aluminofluoride binding to CK.

V_{us} was also inhibited by aluminofluoride. V_{us} was 40% of control at 0.1 mM $Al(NO_3)_3$ and 10 mM NaF; it was less inhibited than force under these conditions (Fig 2). In contrast, Pi has been reported to have little or no effect on V_{us} (14–19). However, since the inhibition of force by Pi in those studies was substantially less (force was inhibited by 30 to 50% of control) than by aluminofluoride (force inhibited by >80%), further work will be necessary to determine whether aluminofluoride and Pi have similar effects on V_{us} at comparable levels of force inhibition.

Discussion

The results are consistent with aluminofluoride being a Vi-like analogue of Pi which inhibits the force-generating interaction of actomyosin in skinned fibres. The long-lived inhibition of Ca-activated force (at pCa 8, subsequent to incubation with aluminofluoride at pCa 4.5) has the biochemical correlate of S1 with vanadate trapped in the active site of myosin (11, 12, 20). We interpret the Ca-requirement for inhibition by aluminofluoride as being due to the acceleration of Pi release during active crossbridge cycling (21), thus increasing the flux of myosin heads through states with only ADP bound; it is the activation of crossbridge cycling which is mediated by Ca^{2+}. The enhancement of the rate of release of aluminofluoride or Vi by active crossbridge cycling would therefore correspond to the actin-mediated increase in rate of Vi release from S1 and HMM (12, 20).

These results also have practical significance. Fluoride and aluminum contaminate some batches of peptides synthesized by the *t*-Boc strategy (data not shown); some fluoride is carried over from the HF cleavage step, while Al (as well as Si) most likely derives from glass containers which are etched by the residual fluoride. Contaminant aluminofluoride was an inhibitory agent in some previous studies involving synthetic peptide effects on skinned fibre contraction (15, 22); fluoride and aluminum have now been identified by ^{19}F-NMR and electron probe X-ray microanalysis, respectively, as the previously unknown contaminant inhibitor proposed by us (24).

Acknowledgements

We thank Dr T.W. Beck for assistance in all aspects of the project, Drs R.W. Wiseman and C.D. Hardin for assistance with ^{19}F-NMR measurements, Dr. M.E. Cantino for electron probe X-ray microanalysis, and Ms E. Egan for typing and editorial assistance.

Supported by the US National Institutes of Health.

References

1. Chabre, M., 1990, *Trends Biochem. Sci.*, **15**, 6–10.
2. Combeau, C., Carlier, M.-F., 1989, *J. Biol. Chem.*, **264**, 19017–19021.
3. Combeau, C., Carlier, M.-F., 1988, *J. Biol. Chem.*, **263**, 17429–17436.
4. Dantzig, J. A., Goldman, Y. E., 1985, *J. Gen. Physiol.*, **86**, 305–327.
5. Chase, P. B., Kushmerick, M. J., 1988, *Biophys. J.*, **53**, 935–946.
6. Brenner, B., 1983, *Biophys. J.*, **41**, 99–102.
7. Sweeney, H. L., Corteselli, S. A., Kushmerick, M. J., 1987, *Am. J. Physiol.*, **252**, C575–C580.
8. Martell, A. E., Smith, R. M., 1974, *Critical Stability Constants*, vol. 1. Plenum Press, New York.
9. Smith, R. M., Martell, A. E., 1989, *Critical Stability Constants*, vol. 6. Plenum Press, New York.
10. Smith, R. M., Martell, A. E., 1976, *Critical Stability Constants*, vol. 4. Plenum Press, New York.
11. Goodno, C., 1979, *Proc. Natl. Acad. Sci. USA*, **76**, 2620–2624.
12. Goodno, C. C., 1982, Methods Enzymol., **85**, 116–123.
13. Edman, K. A., 1979, *J. Physiol.*, **291**, 143–159.
14. Altringham, J. D., Johnston, I. A., 1985, *J. Physiol.*, **368**, 491–500.
15. Chase, P. B., Kushmerick, M. J., 1986, *Biophys. J.*, **49**, 10a.
16. Cooke, R., Franks, K., Luciani, G. B., Pate, E., 1988, *J. Physiol.*, **395**, 77–97.
17. Cooke, R., Pate, E., 1985, *Biophys. J.*, **48**, 789–798.
18. Luney, D. J., Godt, R. E., 1987, *Biophys. J.*, **51**, 468a.
19. Luney, D. J., Godt, R. E., 1986, *Biophys. J.*, **49**, 85a.
20. Goodno, C. C., Taylor, E. W., 1982, *Proc. Natl. Acad. Sci. USA*, **79**, 21–25.
21. Hibberd, M. G., Webb, M. R., Goldman, Y. E., Trentham, D. R., 1985, *J. Biol. Chem.*, **260**, 3496–3500.
22. Chase, P. B., Kushmerick, M. J., 1988, *J. Cell Biol.*, **107**, 258a.
23. Chase, P. B., Kushmerick, M. J., 1989, *Biophys. J.*, **55**, 406a.
24. Chase, P. B., Bursell, J., Kushmerick, M. J., 1990, *Biophys. J.*, **57**, 538a.

40
Influence of Phosphate on the Force Response During Controlled Lengthening of Skinned Psoas and Soleus Muscle Fibres

P. G. A. VERSTEEG, G. J. M. STIENEN, Z. PAPP* AND G. ELZINGA

Laboratory for Physiology, Free University, van der Boechorststraat 7, 1081 BT Amsterdam, The Netherlands
**Department of Physiology, University Medical School of Debrecen, PO Box 22, Debrecen, H4012 Hungary*

Introduction

There exists considerable variation in fatigue properties between fibre types of skeletal muscles. Fatigue resistance derived from the decline in isometric force during repeated contractions, for instance, is high in slow-twitch fibres and low in fast-twitch fibres.The decline in force results, in part, from incomplete activation of the contractile system. There exists, however, also considerable evidence that accumulation of metabolites directly affects crossbridge function and thereby contributes to the effects of fatigue. The concentration of inorganic phosphate (P_i) rises to up to 30 mM as a result of sustained ATP hydrolysis. It has been shown that P_i has an inhibiting effect on isometric force but not on the mechanical properties during shortening (1). In our experiments we studied the effect of phosphate not only during isometric contraction but also during controlled lengthening at constant velocity. Our results are not only relevant to the understanding of the crossbridge action in different fibre types but also in relation to the functioning of fatigued muscles during eccentric contraction.

Muscle and Motility Vol. 2. Proceedings of XIXth European Conference in Brussels
© Intercept Ltd, PO Box 716, Andover, Hampshire, SP10 1YG, UK

Methods

Psoas and soleus muscles were obtained from adult rabbits weighing about 3 kg. Fibre bundles of about 2 mm in diameter and 2–3 cm long were tied to glass rods and chemically skinned according to the procedure described by Goldman, Hibberd and Trentham (2). The fibres were stored for up to two months in a storage solution at 18°C. Experiments were performed on single-fibre segments (2–5 mm in length) separated from the skinned bundles under storage solution.

The segment was mounted by means of aluminium clips between a thin glass rod extending an Akers AM801 force transducer element and a displacement servosystem. The fibres could be incubated in different solutions by changing troughs. Temperature was kept at 15°C and initial sarcomere length, measured by laser diffraction, was adjusted to 2.4 μm. Force and length signals were recorded with a digital oscilloscope (Gould 1425) and after A to D conversion (Labmaster) at a sampling rate of 1 kHz stored in a computer (Olivetti).

The solution used during the experiments contained 1 mM free Mg^{2+}, 5 mM MgATP, 20 mM creatine phosphate and 1 mg/ml creatine kinase. The relaxing solution, the preactivating solution and the activating solution contained 20 mM EGTA, 0.5 mM EGTA and 19.5 mM HDTA, and 20 mM CaEGTA, respectively. Preactivating and activating solutions were also made with 15 mM phosphate by adding KH_2PO_4. The ionic strength was adjusted to 200 mM and the pH was adjusted to 7.0 by KOH. The composition of the solutions was calculated with a computer program, using published equilibrium constants.

We examined force responses to a change in length. During isometric contraction in the activating solution the fibre was stretched by 10% of its initial length at velocities of 0.1, 0.25, 0.5 and 1.0 L_o/s. During the measurements, the segments were incubated in relaxing solution for 4 min, in pre-activating solution for 4 minutes, in activating solution until a steady force level was attained and from there placed back again in relaxing solution. After that the same sequence was followed with the solutions containing phosphate. In general, the eight activation–relaxation cycles required could be repeated before the isometric force during control activation was less than 80% of the first activation. The experiments were terminated when control force became less.

Results

The average isometric force of fully activated fibres was about 150 kN/m² with no significant difference between psoas and soleus fibres. In the presence of 15 mM phosphate the force was less for both muscles, but for psoas more than for soleus (respectively 58% and 78% of the force in the absence of phosphate).

In Fig. 1, the force responses to a controlled lengthening are shown of a psoas and soleus fibre in the absence and presence of 15 mM phosphate in the activating solution. The force response during lengthening, especially at the faster rates, e.g. 0.5 L_o/s (Fig. 1), showed a distinct initial rapid increase followed by a slow rise in force. The amplitude of stretch at this breakpoint was determined at the point of intersection of the initial and final tangents of the

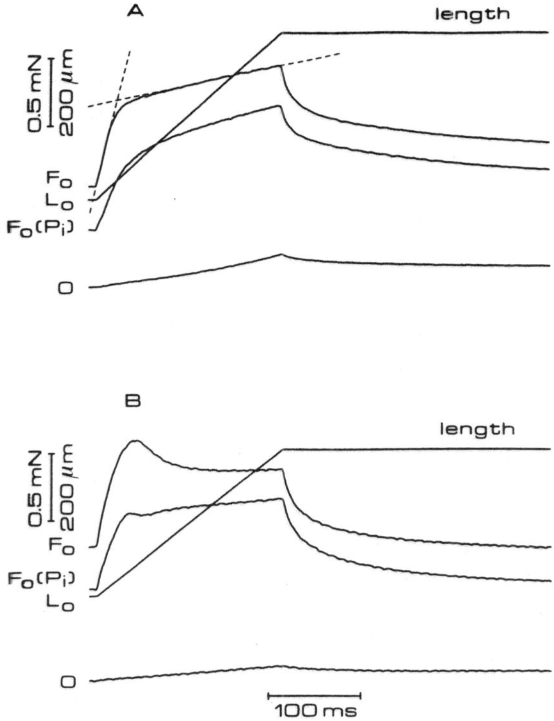

Fig. 1 Length clamp measurements in skinned psoas (A) and soleus (B) muscle fibres in the absence and presence of 15 mM inorganic phosphate (P_i). In each panel the force response (F) is shown to length changes of 10% of initial length (=L_o) at a velocity of 0.5 L_o/s. The trace marked with $F_o(P_i)$ shows the results in the presence of 15 mM phosphate and the lower trace shows the passive force response obtained in relaxing solution.

force responses during lengthening (cf. Fig. 1A). As can be seen from the responses obtained in relaxing solution, part of the slow rise in force was caused by the increase in passive force during lengthening. Correction for the changes in passive force was carried out by subtracting these relatively small responses from the response in the activating solution.

The breakpoint in psoas fibres during lengthenings at 0.5 L_o/s (n = 5), was found at a relative amplitude of stretch (±SEM) of 0.0126 ± 0.0006 of L_o. With 15 mM phosphate, it was found at 0.024 ± 0.002 of L_o. These values differ significantly (P<0.05). In soleus fibres (n = 4) the breakpoint occurred at a relative amplitude of 0.0134 ± 0.0002 of L_o with, in the presence of 15 mM phosphate, only a small increase to 0.016 ± 0.002 of L_o. This latter difference was not significant at P<0.05.

Another difference between the results of psoas and the majority of the soleus fibres is the presence in soleus fibres of an overshoot in the force response. This overshoot diminished when phosphate was added to the activating solution.

Discussion

It is well established that there are differences in phosphate sensitivity of the isometric force between fast and slow fibres. The results presented here show that compared with psoas fibres, soleus fibres are less sensitive to phosphate. The decrease of isometric force at saturating Ca^{2+} concentration in the presence of 15 mM phosphate is 42% for psoas and 22% for soleus.

Moreover, we found a quantitative difference in the influence of phosphate on the mechanical properties during stretch. The force response during steady lengthening showed at least two phases defined by the breakpoint where the slope of the force increment is changed. This breakpoint marks a large reduction in the slope stiffness of the muscle, probably by reaching a degree of extension required to induce yielding of the sarcomeres. This may represent the maximum range of sliding movement between the myosin and actin filaments that a crossbridge can accommodate before detachment. In the presence of phosphate, the breakpoint in the force responses occurs at a larger amplitude of stretch. This is particulary clear for psoas fibres; the breakpoint is shifted from a lengthening of 0.013 of L_o to 0.024 of L_o. In soleus fibres, this breakpoint was also found at a relative amplitude of 0.013 of L_o while with 15 mM phosphate it was found at 0.016 of L_o. In psoas as well as in soleus fibres, these values in the absence of phosphate correspond to a relative sliding movement within each half sarcomere (1.2 μm) of 15 nm. With phosphate this value increase to 29 nm in psoas but is much less in soleus, namely 19 nm.

Assuming that phosphate induces a change in the distribution of attached crossbridges towards a non- or low-force producing state (3), we conclude that the degree of extension of non- or low-force producing crossbridges before the yield point is reached, is larger. Our results, concerning both isometric force development and mechanical properties, indicate that the phosphate sensitivity (or the affinity for P_i binding (4)) in soleus fibres is less than in psoas fibres. The high fatigue resistance of the slow soleus fibres therefore might in part be caused by a smaller affinity for P_i binding in slow than in fast fibres.

References

1. Cooke, R., Pate, E., 1985, *Biophys. J.*, **48**, 789–798.
2. Goldman, Y. E., Hibberd, M. G., Trentham, D. R., 1984, *J. Physiol.*, **345**, 577–604.
3. Hibberd, M. G., Dantzig, J. A., Trentham, D. R., Goldman, Y. E., 1985, *Science*, **228**, 1317–1319.
4. Bowater, R., Sleep, J., 1988, *Biochemistry*, **27**, 5314–5323.

41
ATPase Activity of Skinned Cardiac Trabeculae of the Rat during Repetitive Changes in Length

G. J. M. STIENEN AND G. ELZINGA

Laboratory for Physiology, Free University, van der Boechorststraat 7, 1081 BT Amsterdam, The Netherlands

Introduction

It is well known that energy turnover in a muscle and work performance are coupled (1). For skeletal muscle near 0°C, it has been found that the energy liberation in excess of what is produced in an isometric contraction is approximately proportional to the work done. For cardiac muscle, however, energy liberation during isotonic shortening is less than during an isometric contraction. Several factors may be involved in this discrepancy such as temperature, degree of activation, muscle length, and type of muscle studied (cf. 2). We studied the effect of length changes in maximally activated skinned cardiac trabeculae of the rat at 20°C and a sarcomere length of 2.1 μm to investigate the reason for the discrepancy. We avoided large isotonic shortenings by changing the length of the activated preparations repetitively. Energy turnover was obtained by measuring the ATPase activity. Total ATPase activity was corrected for the basal metabolic activity to relate the mechanical activity specifically to the actomyosin ATPase activity during an isometric contraction.

Muscle and Motility Vol. 2. Proceedings of XIXth European Conference in Brussels
© Intercept Ltd, PO Box 716, Andover, Hampshire, SP10 1YG, UK

Methods

Trabeculae isolated from rat right ventricle were skinned in relaxing solution supplemented with 1% (vol/vol) Triton X-100 for 30 min or by treatment with saponin (50 μg/ml, for 60 min). The rate of ATP hydrolysis was measured photometrically by coupling the regeneration of ATP enzymatically to the oxidation of NADH. The breakdown of NADH was measured via the absorption of UV light at 340 nm, that passed through the bath beneath the preparation as described previously (3). Calibration of the signal was carried out before and after each measurement by injecting a known amount of ADP in the bathing solution. The experimental solutions contained 100 mM BES (pH = 7.1, adjusted with KOH), 3 mM free Mg^{2+}, 5 mM MgATP, 10 mM PEP, 0.9 mM NADH, 4 mg/ml pyruvate kinase, 0.24 mg/ml lactate dehydrogenase, and 0.2 mM Ap_5A. The relaxing solution contained 20 mM EGTA (pCa = 9.0), the preactivating solution contained 0.5 mM EGTA and 19.5 mM HDTA, and the activating solution contained 20 mM CaEGTA (pCa = 4.53). The ionic strength was adjusted to 200 mM by K-propionate. Sarcomere length was adjusted to 2.1 μm. The temperature was kept at 20 ± 1 °C.

Results

The basal ATPase activity of Triton X-100 skinned trabeculae, measured in relaxing solution, was larger than that of saponin treated preparations. The basal ATPase rate after saponin treatment corresponded to 0.52 ± 0.05 μmol. s^{-1}.gram dry weight^{-1} (mean ± SEM., n = 3). When these preparations were further treated with Triton X-100, the basal ATPase activity increased 2.1 times. The basal ATPase activity after Triton treatment could be inhibited by oligomycin B (10 μM) or by prolonged incubation in Triton (24 h).

The contraction-related ATPase activity, obtained by subtracting the basal ATPase activity from the ATPase rate in activating solution, was 2.15 ± 0.22 μmol. s^{-1}.gram dry weight^{-1} (n = 6). By repetitively changing the length, the ATPase activity increased above the isometric value, as illustrated in Fig. 1. This figure shows that force increased rapidly to a steady isometric level when the trabecula was transferred from the pre-activating solution to the activating solution, present in the measuring chamber, in which the NADH absorbance was measured. After the ATPase measurement the preparation was incubated in relaxing solution, which resulted in force relaxation. The continuous decrease in absorbance before immersion and after removal is caused by a contaminating ATPase in the enzymes used and NADH bleaching. From the slope of the absorbance trace relative to this background change, the rate of ATP hydrolysis can be obtained. It can be seen that the ATPase rate increased, when the length of the preparation was repetitively changed (between the interrupted lines in Fig. 1). Thereafter the ATPase rate returned to its isometric level. In this example, the amplitude of the length changes was 5% of the initial length of the trabecula and the repetition frequency was 24 Hz. Under these conditions the contraction related ATPase activity in saponin treated preparations increased by

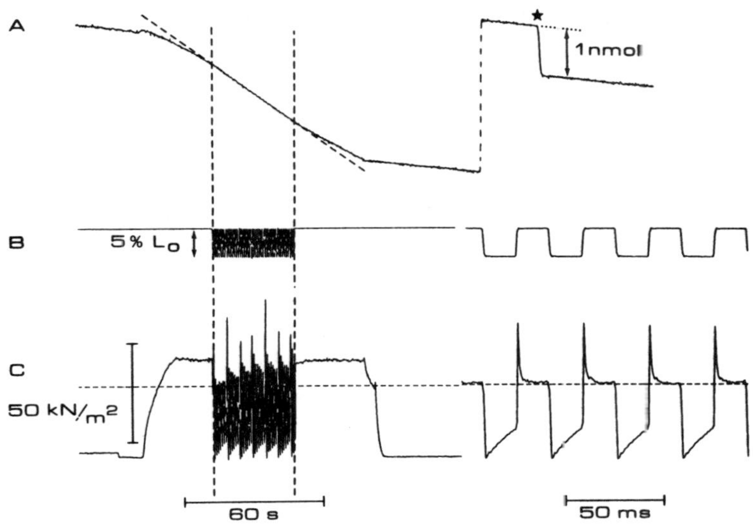

Fig. 1 *Left:* overview of simultaneously measured ATPase activity and force development. (A) NADH absorbance, (B) length, (C) force. *Right:* calibration of the NADH absorbance signal. At (*) 1 nmol ADP was injected into the measuring chamber. *Below:* length changes and force responses at higher time resolution.

32 ± 4% (n = 3). The increase in the rate of ATP hydrolysis was both amplitude and frequency dependent: it was maximal at amplitudes between 5 and 10% and at about 24 Hz.

Discussion

The basal ATPase activity found in Triton X-100 skinned preparations is rather high. It constitutes a considerable fraction (~ 45%) of the total ATPase activity at saturating Ca^{2+} concentrations (4). Miller *et al.* (5) showed that Triton X-100 caused damage to mitochondrial membranes while saponin, in the concentration used, left the function of the mitochondria and the sarcoplasmic reticulum (SR) intact (6, 7). Our experiments with oligomycin, prolonged treatment with Triton, and with saponin treated preparations indicate that Triton X-100 enhances the mitochondrial ATPase activity. This activity diminishes by prolonged treatment due to mitochondrial damage or washout of membrane fragments.

We found that repetitive length changes resulted in an enhanced rate of ATP hydrolysis. This indicates that active movement causes an increase in the contraction related rate of ATP hydrolysis above the isometric value and suggests that the Fenn effect is also present in maximally activated cardiac trabeculae of the rat. The force-related ATPase activity might include some contribution of SR-activity. This contribution could be rather small because under our experimental conditions the SR-ATPase rate would be equal to the rate of calcium induced calcium release. If present, however, the enhancement of the force related ATPase activity would even be larger.

The basic mechanism of the enhancement of ATPase activity is not clear. As the repetitive length changes contained equal periods of shortening and restretch, ATPase activity could be enhanced during shortening, restretch or both. Further experiments with length changes of different shapes are therefore needed.

Acknowledgements

Financial support of the Netherlands Heart Foundation is gratefully acknowledged. We thank F. Dijkema, W. Engel, K. v.d. Lubbe and P. Sneekes for expert technical assistance.

References

1. Fenn, W. O., 1923, *J. Physiol.*, **58**, 175–203.
2. Rall, J. A., 1982, *Am. J. Physiol.*, **242**, H1–H6.
3. Stienen, G. J. M., Roosemalen, M. C. M., Wilson, M. G. A., Elzinga, G., 1990, *Am. J. Physiol.*, **259**, C349–357.
4. Van den Berg, C. J., Elzinga, G., Stienen, G. J. M., 1989, *J. Physiol.*, **415**, 114P.
5. Miller, D. J., Elder, H. Y., Smith, G. L., 1985, *J. Muscle Res. Mot.*, **6**, 525–540.
6. Endo, M., Iino, M., 1980, *J. Muscle Res. Cell Mot.*, **1**, 89–100.
7. Harrison, S. M., Miller, D. J., 1984, *J. Physiol.*, **353**, 55P.

42
The Role of Weakly Attached Crossbridges and Filament Compliance in the Contractile Mechanism of Skeletal Muscle Fibres

D. W. G. JUNG AND T. BLANGÉ

Department of Physiology, University of Amsterdam, Meibergdreef 15, 1105 AZ Amsterdam, The Netherlands

Introduction

In his 1957 paper, A. F. Huxley (1) proposed a model of muscle contraction, based on sliding of myosin and actin filaments and the interaction between these filaments by means of temporary actomyosin complexes or crossbridges. These crossbridges undergo a cyclic process of attachment to and detachment from actin. Later it was shown (2) that elasticity of the muscle preparation depends on the history of force development, thus favouring the Huxley-model above the, until then current, Hill-model with its passive elastic component(s). Therefore, it became necessary to study elasticity at ever faster time scales in order to gain insight in fast changes of state of crossbridges. It appeared (3) that time relationships had to be measured well under 1 ms. Ford, Huxley and Simmons (4) were the first to study the elasticity of intact muscle fibres in this time range. They (5) concluded that the results of their experiments were in agreement with the concept of rigid myofilaments. More recently, results, in particular of elasticity measurements of Brenner *et al.* (6) led to the concept of weakly bound crossbridges. This concept had its origin in the biochemical

Muscle and Motility Vol. 2. Proceedings of XIXth European Conference in Brussels
© Intercept Ltd, PO Box 716, Andover, Hampshire, SP10 1YG, UK

studies of, e.g. Eisenberg and Greene (7). The introduction of weakly attached states in the crossbridge cycle gave rise to questions about the number of crossbridges in activated and rigor fibres. Recently, Jung *et al.* (8) showed that the elastic properties of rigor and activated fibres did not match. It was suggested that filament compliance could account for part of the observed difference. The question whether a mix of weakly attached and rigor crossbridges in the force generating state could account for the observed difference in the elasticity of rigor and activated fibres remained unanswered.

We studied the elasticity of frog muscle fibres at a time scale close to 1 μs. In terms of frequency, we measured elasticity in a frequency range of 500 Hz to 100 kHz. The aim of these measurements was to study in more detail the high frequency characteristics of the elasticity of muscle fibres under specific conditions, taking into account the phase- or time-relationship between fibre force and fibre length, in order to discuss whether the myofilaments were indeed rigid and whether it is possible to distinguish mechanically weakly-attached crossbridges from rigor crossbridges. Goldman and Simmons (9) showed that stiffness of muscle fibres also depends on the interfilament distance of the myofilaments. We therefore measured the changes in the ultrastructure of the muscle fibres as a function of the applied conditions. Interfilament distance and intensity ratio were measured by means of recording the equatorial X-ray diffraction pattern of muscle fibres.

Methods

As already mentioned in the introduction, elasticity measurements are an important tool for exploring the mechanism of muscle contraction. Furthermore, equatorial X-ray diffraction-patterns give additional information about the mechanism of contraction. In both types of measurements we used skinned fibres because of the possibility to induce different states (rigor [defined as the state which develops if the fibre is kept isometric and is deprived of ATP in the absence of Ca^{2+}], activated state [defined as the state which develops if the fibre is kept isometric in the presence of ATP and Ca^{2+}] and relaxed state [defined as the state in which the fibre is kept isometric in the presence ATP and the absence of Ca^{2+}]) by changes in the composition of the incubation solution. The fibres obtained from the iliofibularis muscle of the frog were skinned by means of freeze-drying (10). The elastic properties were measured by recording tension responses to quick changes in fibre length. Therefore, the muscle fibres (on average 5 mm length and 125 μm diameter) were glued between a force transducer (resonant frequency 50–70 kHz, sensitivity 4.1 mV/mN) and a displacement system (length changes of the fibre of 10–100 μm completed within 30–40 μs) (see Fig. 1). The resolution in the time domain was approximately 1 μs. The interpretation of elasticity measurements is based on the simulation of the tension transients by estimation of the value of the parameters of the elastic impedance of the half-sarcomere, which is supposed to consist of three or four damped elastic elements. So, the model consisted of 5 or 7 model parameters ($E_1, E_2, E_3, (E_4), \mu_2, \mu_3, (\mu_4)$) (See Fig. 1). This simulation procedure

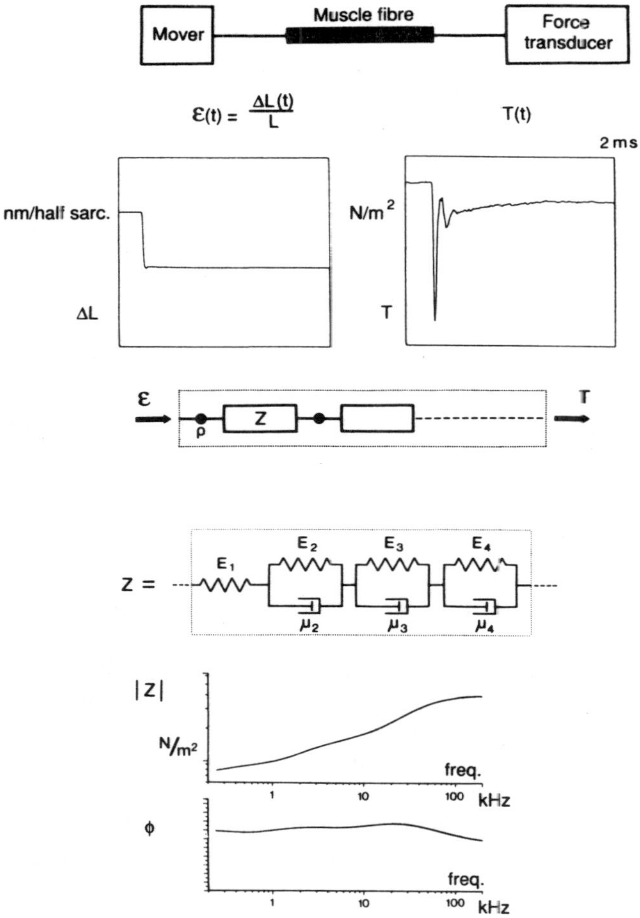

Fig. 1 Top: Schematic diagram of the experimental device used to perform small and rapid length changes of muscle fibres. *Upper middle:* Example of the change of fibre length [$\varepsilon(t)$] and the tension transient [$T(t)$] of an activated fibre that follows this length change. *Middle:* Model used to describe the visco-elastic properties of muscle fibres in terms of impedance [Z] and density [r] of the half sarcomere. *Lower middle:* Model in which the impedance [Z] is thought to consist of an undamped [E_1] and three damped [(E_2, μ_2), (E_3, μ_3), (E_4, μ_4)] elastic elements in series. *Bottom:* Impedance of the half sarcomere of an activated fibre plotted as a function of frequency.

is valid if it is supposed that the unregistered part of the force-response (that part of the response after the end of registration, in this case after 2 ms) can be neglected. Expressed in frequencies, the procedure yields the elastic impedance value in the region 500 Hz to 100 kHz (see Fig. 1). Equatorial X-ray diffraction-patterns of muscle fibres were measured using the small-angle X-ray diffraction facility of the Synchrotron Radiation Source at Daresbury (UK).

Mass-shift and lattice spacing were measured by fitting the small-angle diffraction patterns (8).

Results

GENERAL FEATURES OF THE TENSION TRANSIENTS

The tension transients of fully activated (see e.g. Fig. 1), relaxed and rigor fibre segments have similar features. In all three cases it appeared that the delay between onset of length change and onset of force change is shorter than half the period of resonance due to propagation and the evoked standing wave of the mechanical perturbance in the fibre segment. This illustrates why for the very first perturbance, or at high frequencies, a higher Young's modulus (E_1) is needed in the simulation than at lower frequencies. This is realized by introduction of the damped elastic element (E_2, μ_2) in the model. The corresponding relaxation time is a few microseconds. Simulation, over the first two milliseconds, of tension responses of relaxed and rigor fibres is possible if another relaxation time (about 60 μs) is introduced (E_3, μ_3). In case of activated fibres a third relaxation time (about 500 μs) is needed for a satisfying simulation (E_4, μ_4).

The values of the model parameters E_2 and E_3 of relaxed fibres are small compared to the values of the corresponding parameters in activated and rigor fibres, whereas E_1 is about one-fifth of the corresponding value in rigor and during activation. Since, at physiological conditions, strong binding crossbridges are supposed not to be present in the relaxed state, we have to look for another source of this stiffness in the relaxed state.

Comparison between the parameter values of rigor and activated fibres showed that E_1 and E_2 are about equal, whereas E_3 is about twice as high in rigor as during activation. This is not compatible with the supposition, that in rigor there are just some more of the same crossbridges as during activation.

FILAMENT COMPLIANCE

The force of the isometrically contracting fibre can be varied through alteration of the Ca^{2+} concentration of the incubation solution, thus giving rise to a partial activation. It appeared, from the simulation of the tension responses by means of the fitting procedure (averaged results of nine experiments), that E_2, E_3 and E_4 were proportional to isometric force (see Fig. 2). This means that these parameters could be proportional to the number of attached crossbridges. E_1 deviated from proportionality. The value of this parameter is not close to zero in the relaxed state and the increase with tension levels off at tension values higher than 75% of the fully activated fibre. This suggests that at least partially the source of compliance related to E_1 has to be sought outside the crossbridges.

It appeared that, as a function of frequency, the impedance of fully activated fibres is alternatingly dominated by elasticity and damping (see Fig. 1). When the fibre is partially activated, i.e. at lower Ca^{2+}-concentrations of the incubation solution and thus lower isometric force, the distinction between elastic and

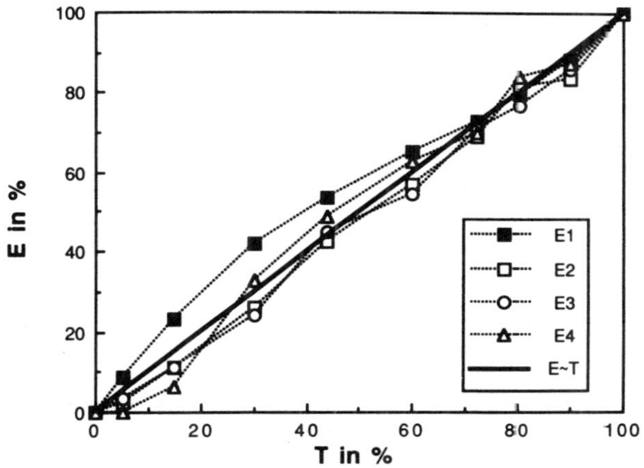

Fig. 2 Normalized values of the four elastic model parameters of Ca^{2+}-activated fibres as a function of isometric tension. The data points at zero tension represent the values of relaxed fibres. Sem is approximately the size of the symbols. Average of nine fibres. Maximum tension: 93.5 kN/m^2. Maximum: $E_1 = 27$ MN/m^2, $E_2 = 58$ MN/m^2, $E_3 = 22$ MN/m^2, $E_4 = 10$ MN/m^2. Data corrected for passive stiffness. (Conditions: 4°C, 2.15 μm sarc. length, 160 mM ionic strength, pH 7.0)

frictional (damped) behaviour faded out with decreasing force and disappeared almost completely in the relaxed fibre. The elastic impedance of relaxed fibres increased rather smoothly with the square root of the frequency (see Fig. 3). The phase angle remained almost constant at about 45°. This specific dependence on the frequency is in between elastic behaviour (impedance independent

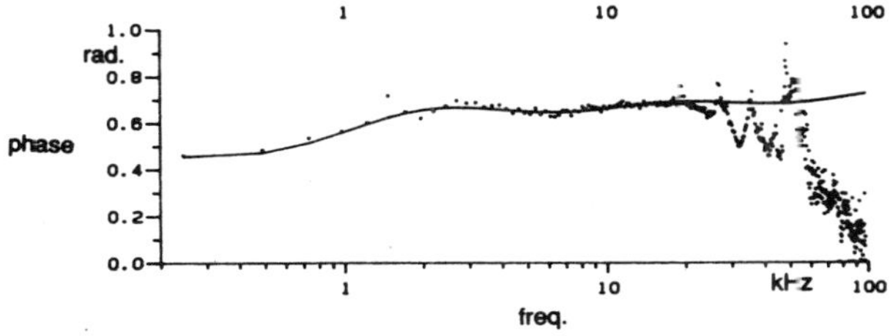

Fig. 3 The elastic impedance of a relaxed fibre as a function of frequency. The dotted curves represent the measured impedance and phase shift whereas the solid curves represent the simulated impedance and phase shift. Conditions as in Fig. 2.

of frequency and a constant phase angle of 0°) and frictional behaviour (impedance linearly dependent on frequency and a phase angle of 90°). Apparently, in the relaxed fibre at frequencies larger than 1 kHz the sarcomere still has an appreciable mechanical impedance left, which cannot simply be explained in a sliding-filament model in which filaments are rigid. A few (weakly attached) crossbridges, acting in parallel and with relaxation times as displayed in the response of the activated fibre, could perhaps explain the deviations of the square root relation between impedance and frequency of relaxed fibres, but not the relation itself.

WEAKLY ATTACHED CROSSBRIDGES

As reported by Brenner *et al.* (6), stiffness of the relaxed fibre increased after incubation in solutions of low ionic strength, whereas interfilament distance decreased. In order to eliminate effects due to a changing interfilament distance, we also measured stiffness and lattice spacing of fibres compressed osmotically by means of the high molecular weight polymer Dextran. As a result of adding Dextran to the relaxing solution, interfilament distance decreased, stiffness increased and the intensity ratio of the 11 and 10 reflection remained constant. Lowering the ionic strength of the relaxing solution resulted in a decreasing interfilament distance, an increasing stiffness and an increasing intensity ratio (see Fig. 4). It is obvious that almost half of the changes that occur in E_1 as a function of ionic strength can be explained by changes in interfilament distance, whereas only a small part of the changes in E_2 and E_3

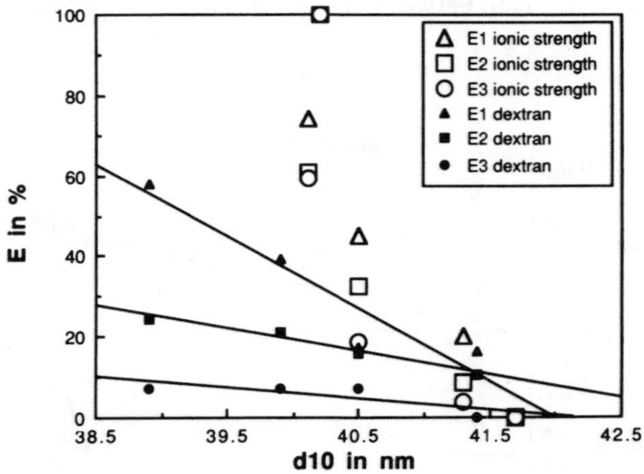

Fig. 4 Normalized values of the elastic model parameters of relaxed fibres, incubated in solutions with varying ionic strength -30 to 220 mM (no dextran added) or in solutions with 1–3% dextran added (160 mM ionic strength), as a function of lattice spacing. The solid lines are fitted to the data points of relaxed fibres with dextran added to the solution. Sem is approximately the size of the symbols. Average of six fibres. Maximum: $E_1 = 20$ MN/m^2, $E_2 = 14.7$ MN/m^2, $E_3 = 3.7$ MN/m^2. Minimum: $E_1 = 9.5$ MN/m^2; $E_2 = 3.3$ MN/m^2, $E_3 = 1.0$ MN/m^2. Conditions as in Fig. 2.

can be explained by changes in interfilament distance. In addition, just as in the response of the activated fibre, a third relaxation showed up at extreme low ionic strength conditions. This suggests that weakly bound crossbridges (if equivalent to crossbridges at low ionic strength) have their main contribution at low frequencies (<20 kHz). Changes in interfilament distance induced by osmotic compression might be correlated with 'high-frequency' stiffness. The number of weakly attached crossbridges is not correlated with the 'high-frequency' stiffness (11).

CROSSBRIDGES IN Ca^{2+}-ACTIVATED FIBRES

The values of the model parameters E_1, E_2 and E_3 increased as rigor force ('low rigor') was raised either by stretching the fibre or by evoking rigor in the presence of Ca^{2+} ('high rigor' (12), see Fig. 5). Model parameter E_1 is not pro-

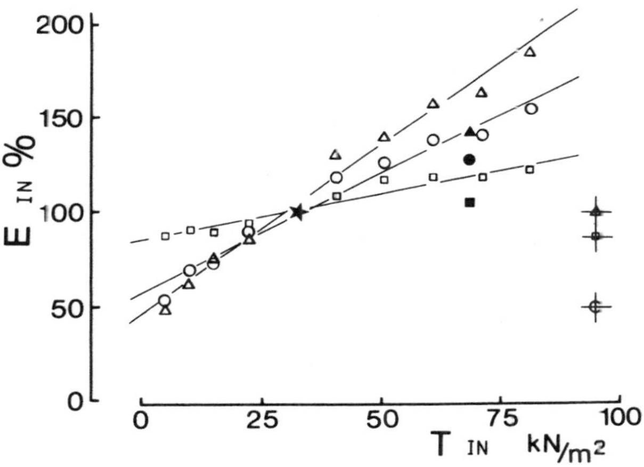

Fig. 5 Normalized values of the three elastic model parameters of rigor fibres (★) and strained rigor fibres as a function of tension (E_1 squares, E_2 triangles, E_3 circles). The solid lines are fitted to the data points of rigor fibres. All values are given as a percentage of the values of the rigor control measurements. Sem is approximately the size of the symbols. Average of four fibres. Open symbols: 'low tension' rigor. Filled symbols: 'high tension' rigor. Open symbols with cross hair: fully activated fibres. Conditions as in Fig. 2.

portional to the number of crossbridges, as stated above, and is therefore not of interest in the comparison of stiffness of rigor and activated fibres. However, the discrepancy between E_2 and E_3 became larger as rigor tension reached the isometric tension level of activated fibres. This means that the difference between activated and rigor fibres cannot be reduced to a difference in the number of crossbridges in these two states. It is also of no use to assume that the stiffness (E_2 and E_3) of activated fibres is the result of a mix of weakly attached crossbridges (as found in relaxed fibres at low ionic strength) and rigor crossbridges because stiffness (E_2 and E_3) of weakly-attached crossbridges is low compared to rigor crossbridges (11). So, a combination of weakly attached and rigor crossbridges cannot account for both the stiffness and the intensity ratio data of activated fibres. Remains the possibility that the elastic properties of rigor crossbridges differ from the properties of force-generating crossbridges. In that case, stiffness as well as the intensity ratio of activated fibres could be the result of a combination of weakly attached, rigor and force-generating crossbridges.

Conclusions

Study of the tension reponses of skinned fibre segments to small ramp-shaped changes in fibre length led to a method of analysis in terms of an elastic model of the half-sarcomere consisting of an undamped and two or three damped elastic elements in series. The values of the parameters of the model were estimated by fitting these tension responses. Evaluation of the elastic parameters of activated fibres as a function of isometric force (by means of variations of the Ca^{2+}-concentration) showed that the elastic moduli increased with isometric force. This is compatible with the idea that the number of crossbridges increased with increasing isometric force. However, analysis of the undamped elastic element E_1 revealed that this parameter is not proportional to force. This suggests that at least part of the source of compliance related to E_1 has to be sought outside the crossbridges. The parameter E_1 of relaxed fibres is far more sensitive to changes in the interfilament distance, and less sensitive to changes in the number of weakly attached crossbridges, than E_2 or E_3. Furthermore, detailed analysis of the elastic impedance of activated fibres at very low tension values and of relaxed fibres, suggests that a distribution between specific elastic parameters might be partly artificial. An explanation would be that in this case filament elasticity and friction with a viscous character determines the tension response of the sarcomere at fast changes of length.

The stiffness of relaxed fibres is increased at low ionic strength, which suggests the presence of weakly bound crossbridges. If we suppose that these weakly bound crossbridges have the same elastic characteristics as crossbridges in the rigor state, then the quantitative analysis shows that these crossbridges cannot contribute much to the elasticity in the activated fibre.

A comparison between the values of the elasticity of fully activated fibres and fibres in rigor (whether or not stretched to an isometric tension level comparable to the tension level of fully activated fibres) shows that the model parameters in these two states are not related by a single scaling factor. If elasticity of the half-sarcomere is due to elasticity of the crossbridges acting in parallel, while the filaments are rigid, one has to conclude that the elastic characteristics of the crossbridges are different in rigor and during activation. Taking into account the difference in the stiffness of rigor and activated fibres and the stiffness of weakly attached crossbridges, whether or not they have similar characteristics as rigor crossbridges, it is likely that the activated state is not a simple combination of a number of rigor crossbridges and a number of weakly attached crossbridges. It is suggested that in the activated state under isometric conditions a mix of weakly attached crossbridges, rigor crossbridges and force-generating crossbridges occurs. The elastic properties of the latter remain to be measured.

Acknowledgements

This study was supported by the Netherlands Organization for the Advancement of Research (NWO), also in connection with the agreement between the Science and Engineering Research Council and NWO concerning the Synchrotron Radiation Source at Daresbury (UK).

References

1. Huxley, A. F., 1957. *Prog. Biophys.*, **7**, 255–313.
2. Blangé, T., Karemaker, J. M., Kramer, A. E. J. L., 1972. *Eur. J. Physiol.*, **336**, 277–288.
3. Blangé, T., Stienen, G. J. M., 1985. *Eur. J. Physiol.*, **405**, 12–18.
4. Ford, L. E., Huxley, A. F., Simmons, R. M., 1977. *J. Physiol. (London)*, **269**, 441–515.
5. Ford, L. E., Huxley, A. F., Simmons, R. M., 1981. *J. Physiol. (London)*, **311**, 219–249.
6. Brenner, B., Schoenberg, M., Chalovich, J. M., Greene, L. E., Eisenberg, E., 1982. *Proc. Natl. Acad. Sci. USA*, **79**, 7288–7291.
7. Eisenberg, E., Greene, L. E., 1980. *Ann. Rev. Physiol.*, **42**, 293–309.
8. Jung, D. W. G., Blangé, T., Graaf, H. de, Treijtel, B. W., 1988. *Biophysical J.*, **54**, 897–908.
9. Goldman, Y. E., Simmons, R. M., 1986. *J. Physiol. (London)*, **378**, 175–194.
10. Stienen, G. J. M., Güth, K., Rüegg, J. C., 1983. *Eur. J. Physiol.*, **397**, 272–276.
11. Jung, D. W. G., Blangé, T., Graaf, H. de, Treijtel, B. W., 1989. *Biophysical J.*, **55**, 605–619.
12. Kawai, M., Brandt, P. W., 1976. *J. Gen. Physiol.*, **68**, 267–280.

43
Temperature Sensitivity of Isometric Tension in Glycerinated Mammalian Muscle Fibres

K. W. RANATUNGA

Department of Physiology, Medical School, Bristol BS8 1TD, UK

Introduction

Following the observations of Hadju (1), it is well recognized that the tetanic tension of skeletal muscle is depressed at low temperatures. Data from several studies show that the cooling depression of tetanic tension in mammalian muscles is particularly marked at temperatures below about 20°C so that the tension at 10°C is about 50% of that recorded at 37°C (see 2 and references therein). The maximal active tension of mammalian skinned muscle fibres shows a very similar behaviour (3).

From heat measurement experiments (4), it has been shown that elasticity in resting intact muscle is rubber-like (i.e. passive tension increases with temperature (5)), whereas active and rigor muscle exhibit normal elasticity (6, 7 and refs therein). Indeed, in glycerinated muscle fibres, the rigor tension is depressed, and the active tension is transiently depressed, following a rapid jump in temperature (8, 9, 10) indicating normal elasticity.

The observations summarized above have been made on different preparations, using different techniques and typically within limited temperature ranges. The aim of the present study was to reexamine the temperature sensitivity of isometric tension in single glycerinated mammalian muscle fibres over a temperature range of 0–40°C, by continuous monitoring of muscle fibre tension when the temperature was changed at an average rate of 0.5–2°C/s. Preliminary findings describing the temperature dependence of tension in relaxed (passive tension), rigor and maximally active conditions are presented.

Muscle and Motility Vol. 2. Proceedings of XIXth European Conference in Brussels
© Intercept Ltd, PO Box 716, Andover, Hampshire, SP10 1YG, UK

Methods

The data presented in this paper are from rabbit Psoas fibres; they were chemically skinned and glycerinated as described previously (11). Similar data were obtained from rat fast and slow muscle fibres.

The basic design of the trough system used in the experiments was similar to that used in T-jump experiments (10), but it had a especially designed end-trough which was lined with aluminium foil and surrounded by a number of small Peltier devices. The temperature of the end-trough (50 µl) could be varied and clamped; a small (diameter 0.4 mm) thermistor in the trough provided feedback to the control circuit.

The solutions used in the experiments were similar to those used in T-jump experiments and contained 15 mM of glycerol-2-phosphate as the pH buffer (pH 7.1 at 20°C), creatine phosphate and phosphokinase. A fibre segment of 2–4 mm was mounted between two steel hooks, one attached to a tension transducer (Akers AE 801) and the other attached to a Piezo-electric element. The average sarcomere length was 2.3–2.6 µm in rigor tension and active tension experiments; the fibre was stretched 50–80% further for passive tension recording. The maximal active tension recorded in the present experiments was (mean ±SE) 148 ±17 kN/m^2 ($n = 5$) at 15–20°C. In an experiment, the fibre was transferred to the end-trough (already filled with the appropriate solution and temperature clamped at about 15°C), the temperature was lowered to 0–5°C and then increased in 5°C steps to 35–40°C. A Tandon computer with a CED 1401 peripheral (Cambridge Electronic Design Ltd) was used to digitize and store the tension transducer output and the thermistor and/or thermocouple outputs.

Results

The tension (upper) and temperature (lower) traces recorded from a Psoas fibre are shown in Fig. 1 and they illustrate the basic observations. Thus, the passive tension (A) was relatively insensitive to temperature, but showed a reversible increase with temperature (rubber-like). Both the rigor tension (B) and the active tension (C) changed markedly as the temperature was raised; the rigor tension decreased (normal elasticity), whereas the active tension increased with temperature.

For comparison with other studies on passive tension, temperature coefficient of tension (β) was calculated as $(1/P)(dP/dT)$, where P is passive tension and T temperature. The coefficient (β) was 5.7 (±0.75) × 10^{-3} per °C ($n = 7$). Using the force/extension ratio determined at constant temperature, the thermal strain (contraction), α, in relaxed muscle fibre was estimated to be –2.3 (±0.3) × 10^{-3} per °C.

Fig. 2A shows rigor tension versus temperature plots from a muscle fibre; the temperature sensitivity of tension (dP/dT) was estimated from the slope of the straight line fitted to each set of data. Force-extension relations were also determined, and the slope (dP/dL, where L is muscle fibre length) was found to be similar at different temperatures. The coefficient of thermal expansion (as

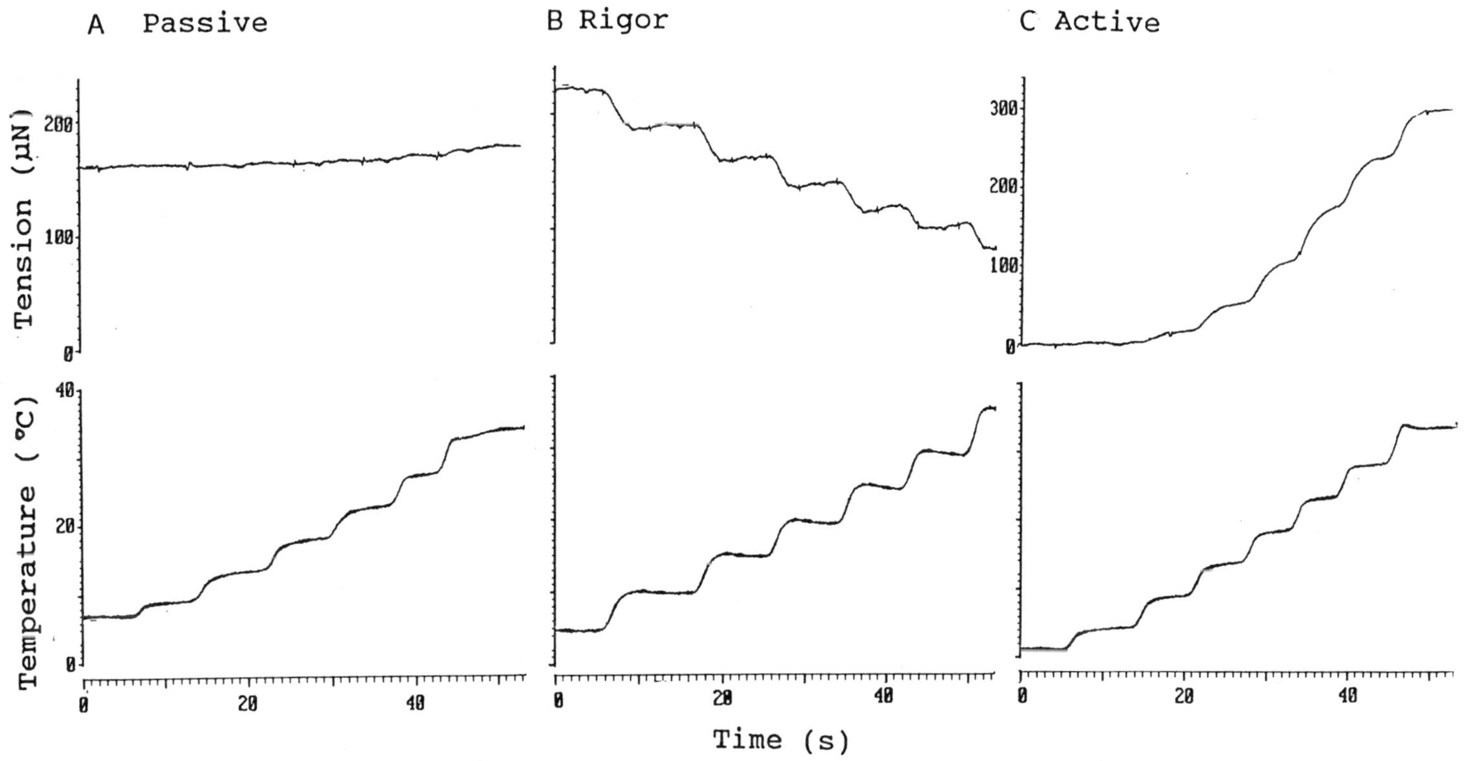

Fig. 1 Tension (upper) and temperature (lower) records from a Psoas muscle fibre in relaxed condition (A-passive tension), in rigor condition (B) and in maximally Ca-activated condition (C). The labelling in the left vertical axes apply to other unlabelled vertical axes.

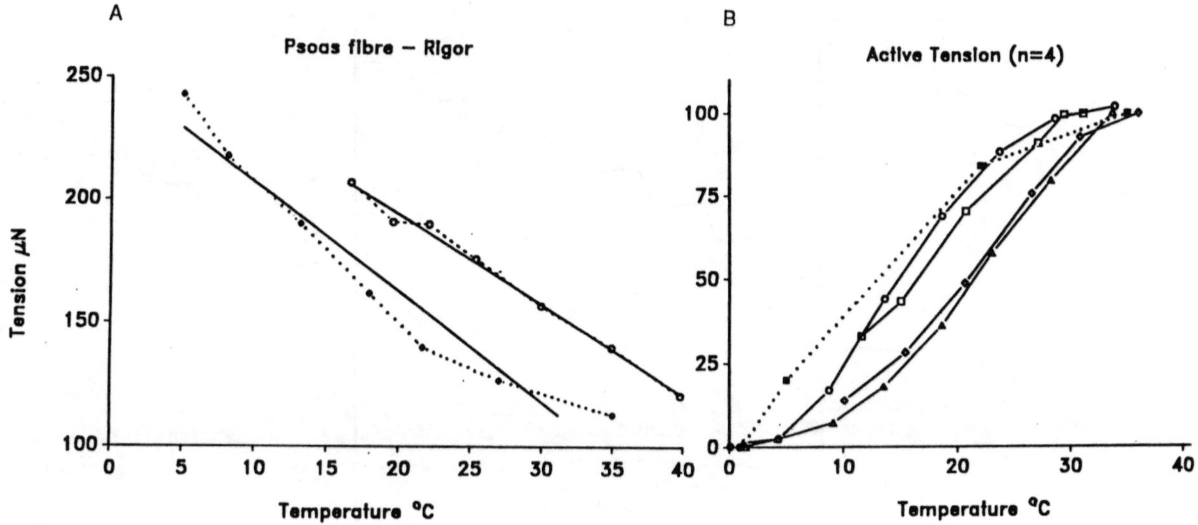

Fig. 2 (A) Rigor tension versus temperature plots from a muscle fibre; tension data were collected during warming (filled circles) and cooling (open circles). Straight lines are the calculated linear regressions. (B) Percentage active tension versus temperature plots from several fibres (solid curves). Dotted curve indicate data of Stephenson and Williams (3) from steady-state experiments.

$\alpha = (dP/dT)/(dP/dL))$ estimated from these measurements was $5.1(\pm 0.7) \times 10^{-5}$ per °C ($n = 7$) after correcting for expansion in the steel hooks attached to the transducers (as described in 8). The value for thermal expansion (α) is slightly higher than that obtained by Bershitsky and Tsaturyan (8) from temperature-jump experiments. The thermoelastic coefficient (R, calculated as αT, where T = 293 K, see 7) was 0.015 ±0.002; this is within the range of values obtained in heat measurement experiments on active and rigor muscle (range 0.008–0.018; see 4, 12).

Fig. 2B shows plots of active tension versus temperature obtained in the present experiments (solid curves) during warming, and, for comparison, mean steady state data reported by Stephenson and Williams (3) from contractions initiated at different temperatures are also shown (dotted curve). Assuming correspondence at normalized higher temperatures, the data from present experiments appear to deviate from the steady state curve at the lower temperatures.

Discussion

Temperature dependence of passive tension indicated rubber-like elasticity in glycerinated rabbit Psoas muscle fibres in the relaxed condition, as had been reported for resting tension of intact muscle (see 5 and refs therein). Interestingly, rubber-like behaviour was also obtained in the passive tension response to increased hydrostatic pressure (13). There is evidence now that the passive tension in a skinned muscle fibre may be borne largely or entirely by the Titin (=Connectin) containing gap filaments found in sarcomeres (see 14 and refs therein). Indeed, experiments of Maruyama *et al.* (15) demonstrated that a length of isolated Connectin showed rubber-like elasticity; the temperature coefficient of tension (β) obtained in this study for rabbit muscle fibres is well within the range reported for isolated Connectin (15) and for frog fibres (16).

Present results showed that rigor tension decreased approximately linearly over the entire temperature range of 0–40°C and the decrease could be accounted for by thermal expansion (normal elastic behaviour). Temperature-dependent changes in length (17) and local melting (18) of the subfragment-2 region of myosin molecule have been reported; indications of such changes occurring in rigor crossbridges are not clearly apparent in the present data.

Data for temperature dependence of active tension presented here should be considered with caution, since they were collected from relatively long (1 min) contractions. The *S*-shaped relation of active tension versus temperature and the discrepancy with the steady state data require further investigation.

References

1. Hadju, S., 1951. *Enzymologia*, **14**, 187–193.
2. Ranatunga, K. W., Wylie, S. R., 1989. *Biomed. Biochim. Acta*, **48**, S530–S535.
3. Stephenson, D. G., Williams, D. A., 1985. *J. Physiol.*, **360**, 1–12.
4. Hill, A. V., 1953. *Proc. Roy. Soc. B*, **141**, 161–178.

5. Hill, D. K., 1970. *J. Physiol.*, **208**, 161–172.
6. Woledge, R. C., 1961. *J. Physiol.*, **155**, 187–208.
7. Woledge, R. C., Curtin, N. A., Holmsher, E., 1985. *Energetic Aspects of Muscle Contraction.* Monograph of Physiol. Soc., 41, Academic Press, London.
8. Bershitsky, S. Y., Tsaturyan, A. K., 1989. *Biophys. J.*, **56**, 809–816.
9. Davis, J. S., Harrington, W. F., 1987. *Proc. Natl. Acad. Sci.*, **84**, 975–980.
10. Goldman, Y. E., McCray, J. A., Ranatunga, K. W., 1987. *J. Physiol.*,**392**, 71–95.
11. Fortune, N. S., Geeves, M. A., Ranatunga, K. W., 1989. *J. Musc. Res. & Cell Motil.*, **10**, 113–123.
12. Gilbert, S. H., Ford, L. E., 1986. *J. Musc. Res. & Cell Motil.*, **7**, 35–46.
13. Ranatunga, K. W., Fortune, N. S., Geeves, M. A., 1990. *Biophys. J.*, in press.
14. Horowits, R., Podolsky, R. J., 1987. *J. Cell Biol.*, **105**, 2217–2223
15. Maruyama, K., Matsubara, S., Natori, R., Nonomura, Y., Kimura, S., Ohashi, K., Murakami, F., Handa, S., Eguchi, G., 1977. *J. Biochem.*, **82**, 317–337.
16. Matsubara, S., 1975. *Jap. J. Physiol.*, **25**, 227–240.
17. Walker, M., Trinick, J., 1986. *J. Mol. Biol.*, **192**, 661–667.
18. Ueno, H., Harrington, W. F., 1986. *J. Mol. Biol.*, **190**, 59–68.

44
Tension Transients Initiated by the Joule Temperature Jump during Steady Shortening of Skinned Muscle Fibres

S. Y. BERSHITSKY* AND A. K. TSATURYAN**

*Department of Biophysics, Institute of Physiology, Urals Branch of Academy of Sciences of the USSR, Popova 30 Sverdlovsk 620014, USSR
** Institute of Mechanics, M. V. Lomonosov Moscow State University, Leninsky Gory, Moscow 117234, USSR

Introduction

Recently it was shown that the temperature jump (T-jump) induces a bi-exponential tension rise in activated skinned skeletal muscle fibres (1–5). The first (fast) component of the rise is sensitive to the length change applied a few milliseconds before the T-jump (1, 6) and is therefore due to some force-generating process of the crossbridges attached to the thin filaments. The rate constant and the final tension achieved at the end of the second (slow) component of the rise are insensitive to the preceding length change (6, 7) and, hence, this component is accompanied by the crossbridge detachment and re-attachment.

For further examination of the nature of the crossbridge processes involved in tension transients initiated by the T-jump, a series of experiments was performed where the T-jump was applied during ramp shortening of the rabbit psoas muscle fibre.

Methods

Fibre preparation, mechanical measurements and the T-jump method were described earlier (3). The T-jump was induced by the high-voltage (up to 3 kV),

Muscle and Motility Vol. 2. Proceedings of XIXth European Conference in Brussels
© Intercept Ltd, PO Box 716, Andover, Hampshire, SP10 1YG, UK

high-frequency (30 kHz) current pulse of 0.2 ms duration which was passed through the fibre suspended in air between two electrodes attached to the force transducer and the length change generator. The amplitude of the T-jump was calculated using an analogue circuit determining Joule heat liberated in the fibre during the pulse.

The fibre was activated with the solution containing (in mM): temperature insensitive cacodylate pH-buffer [100], pH 7.0 at 25°C, CaEGTA [10], MgATP [4], Mg^{2+} [1], creatine phosphate [15], creatine phosphokinase [100 U/ml], NaN_3 [3], the ionic strength [220]. In the relaxing solution CaEGTA was substituted by 10 mM of EGTA.

Results

The fibre was activated at 0–2°C and sarcomere length ~2.6 µm. When tension became steady the activation solution was removed from the trough and both temperature and tension increased slightly to 5–9°C and 50–72 kN/m² respectively. In 5 s after the solution removal, the fibre was forced to ramp shortening with velocity of 0.15–0.6 µm/s per a half-sarcomere (hs). The scheme of the experimental protocol is shown in Fig. 1. A step length release by 5–10 nm/hs was applied just before the ramp shortening to accelerate the steady tension achievement (8). At the moment of the T-jump, when the total (step plus ramp) shortening amounted 30–40 nm/hs tension achieved a plateau at the sarcomere length 2.5 µm. Tension transient initiated by the T-jump was monitored for 50 ms during the shortening. Then the shortening was ended and an isometric tension developed at high temperature.

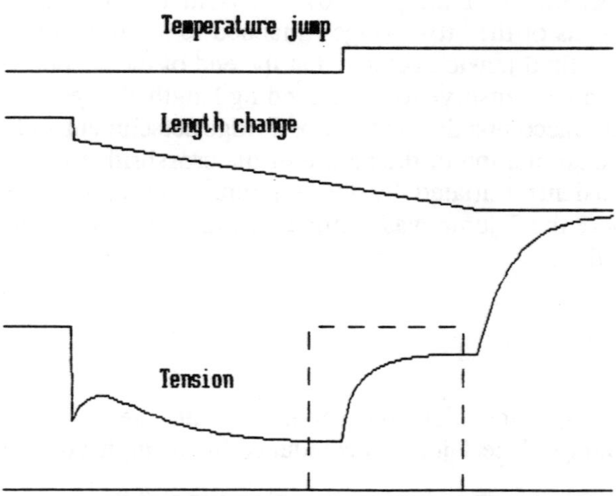

Fig. 1 Scheme of experimental protocol.

The T-jump to 28–33°C induced a biexponential tension rise similar to that during an isometric contraction. Squares in Fig. 2 are the experimental points from a fibre, and continuous lines are the best bi-exponential approximations obtained with the computer program (9).

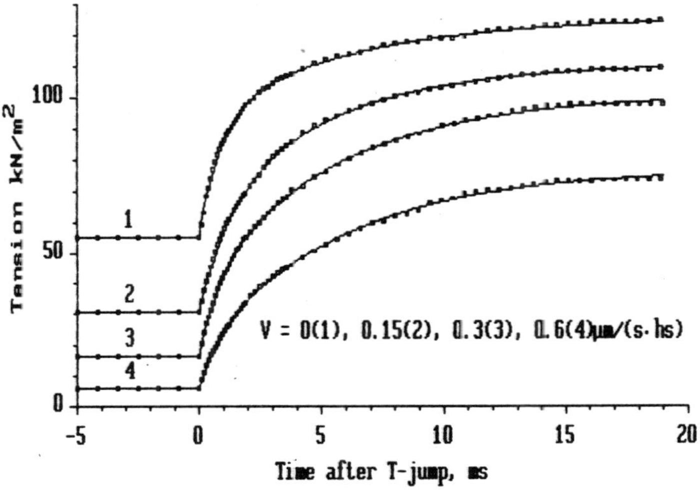

Fig. 2 Tension responses to T-jump at different velocities.

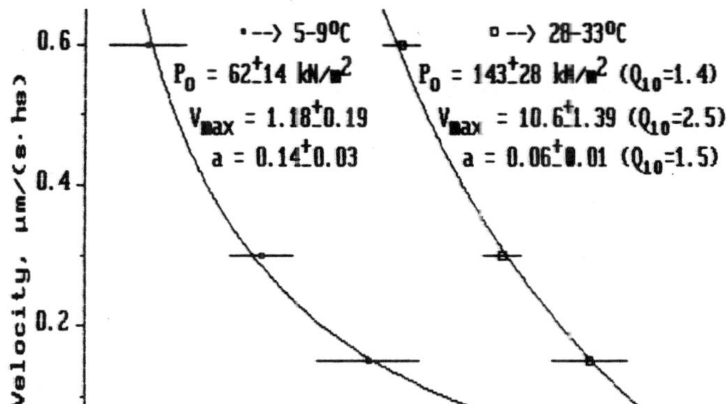

Fig. 3 Force–velocity relation at starting and final temperature.

The higher the velocity, the lower steady tension at both the starting and final temperature. The averaged force–velocity relations before and after the T-jump for five fibres are shown in Fig.3. The bars show standard deviations. The parameters of the Hill's equation V_{max} and a estimated using the least squares method as well as the Q_{10} values for these parameters and isometric tension P_o are also shown in Fig. 3.

The dependence of the rate constants of the first and second components of the tension rise initiated by the T-jump on the shortening velocity is shown in Fig. 4. It is seen that the rate constant k_1 of the first (fast) component increases with the velocity, whereas the rate constant k_2 of the second component is practically independent of the velocity.

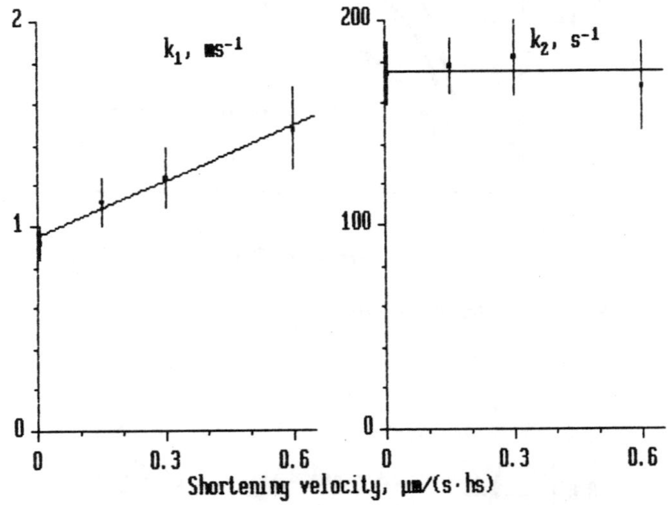

Fig. 4 Dependence of rate constants k_1, k_2 on shortening velocity.

Discussion

The parameters of the Hill's equation at the starting temperature in the experiments presented here (Fig. 2) were close to those in the steady temperature experiments (10) at 10°C. The Q_{10} values for P_o and V_{max} were also similar to those for the fast fibres of the warm-blooded animals (11). So, our data correspond with other studies of the temperature dependence of the force–velocity relationship.

The acceleration of the fast component of the tension rise initiated by the T-jump during the ramp shortening (Fig. 4) agrees with the data of our previous experiments with the step shortening of the fibre before the T-jump (1, 6). This acceleration also agrees with the results of the study of the fast mechanical transients during the ramp shortening of the frog fibres (8). Thus, the data presented here confirm the assumption that the fast components of the tension transients initiated by both length and temperature perturbations are length sensitive and, hence, result from fast force-generating process in the crossbridges attached to the thin filaments.

The absence of dependence of the rate constant of the slow component of the tension rise on the shortening velocity is in accordance with our previous data showing insensitivity of this component to the preceding length step (7). It offers additional evidence for the assumption that this component is accompanied by the crossbridge detachment and re-attachment to new sites in the thin filaments (1, 6, 7). On the other hand, the slow component of the mechanical transients accelerates significantly during the fibre shortening at 0–2°C (8). The apparent difference in these data and ours can be explained if one assumes that there is some rate-limiting step between the cross bridge attachment and force generation (12) and, if this step is strongly sensitive to both length change and temperature and, at high temperature, becomes too fast to limit the force generation.

References

1. Bershitsky, S. Y., Tsaturyan, A. K., 1986, In: *Trends in Biomechanics in Medicine and Surgery*, vol. 1, pp. 74–79. Riga,
2. Bershitsky, S. Y., Tsaturyan, A. K., 1988, *Biophysics*, 32, 147–149.
3. Bershitsky, S. Y., Tsaturyan, A. K., 1989a, *Biophys. J.*, **56**, 808–816.
4. Davis, J. S., Harrington, W. F., 1987, *Proc. Natl. Acad. Sci. USA*, **84**, 975–979.
5. Goldman, Y. E., McCray, C., Ranatunga, K., 1987, *J. Physiol.*, **393**, 71–95.
6. Tsaturyan, A. K., Bershitsky, S. Y., 1988, *Biophysics*, 32, 505–507.
7. Bershitsky, S. Y., Tsaturyan, A. K., 1989b. In: *Abstr. XVIIIth Europ. Conf. on Muscle & Motil.*, Lunteren, p. N28.
8. Ford, L. E., Huxley, A. F., Simmons, R. M., 1985, *J. Physiol.*, **361**, 131–150.
9. Provencher, S. W., 1976, *Biophys. J.*, **16**, 27–50.
10. Cooke, R., Pate, E., 1985, *Biophys. J.*, **48**, 789–798.
11. Ranatunga, K., 1982, *J. Physiol.*, **329**, 465–483.
12. Huxley, A. F., 1973, *Proc. Roy. Soc.*, **B183**, 83–86.

45
Storage and Release of Mechanical Energy during Contraction

G. A. CAVAGNA, N. C. HEGLUND, J. D. HARRY* AND L. PASSERINI

Istituto di Fisiologia Umana I, Università degli Studi di Milano, Via Mangiagalli 32, 20133 Milano, Italy

Introduction

In only a few instances does muscle behave purely as a motor, converting energy from chemical reactions into mechanical work and heat. In most cases, active muscles must also serve as a brake. When the braking action of muscle immediately precedes a subsequent shortening contraction, the possibility exists that part of the mechanical energy absorbed during the stretching is stored and successively released during the shortening. What mechanisms may perform this task?

Active muscle resists stretching with a force greater than the isometric force; since the force is transmitted to the external load through the tendons and the elastic elements within the crossbridges, stretching necessarily implies a greater storage of elastic energy within these structures (undamped elastic elements). In addition, if the length of the muscle as a whole, or of some sarcomeres, exceeds slack length, energy may be stored within damped parallel elastic elements. Several previous studies presented evidence suggesting an 'enhancement' of muscle mechanical performance due to previous stretching (1–7). In particular, experiments have been done on isolated fibres at slack length where isometric tension was at a maximum (P_o), passive tension was negligible and the possibility of extreme inhomogeneity of sarcomeres was slim. The active

**Present address: Division of Engineering, Box D, Brown University, Providence, RI 02912 USA*

Muscle and Motility Vol. 2. Proceedings of XIXth European Conference in Brussels
© Intercept Ltd, PO Box 716, Andover, Hampshire, SP10 1YG, UK

fibre, after a ramp stretch, was able to shorten against P_o, suggesting storage and release of mechanical energy by the contractile machinery itself (6). In this paper we present some preliminary data obtained by repeating the experiment described above with an improved technique and the additional measurement of the length changes of the sarcomeres.

Methods

The fibre, dissected from the caput laterale of the tibialis anterior muscle of the frog (usually *Rana temporaria*), was mounted in a solid quartz chamber with a slit in each of the walls of its smaller dimension to allow entry of glass hooks, one attached to a force transducer (8) and the other (on the opposite side) to a motor used to change fibre length. The fibre was suspended between these hooks using a clip attached to each tendon (9). The chamber was filled with Ringer solution (5). Two platinum-stimulating electrodes were submerged in the bath on either side of the fibre. Stimulation consisted of alternating polarity square pulses with an amplitude about 1.5 threshold and the frequency necessary to have a fused tetanus. A thermocouple monitored the Ringer temperature which was controlled by circulating chilled fluid through an aluminum block that jacketed the chamber. A slit in the block allowed illumination of the fibre from below. The motor was constructed from a Ling 101 magnet and a hand-made aluminum wire coil. A carbon tripod was mounted on the coil to support the glass hook for the fibre and a flag shadowing two photodiodes from an infrared emitting diode to monitor fibre length. Current was fed through the coil by a power amplifier. Step length changes were completed in about 100 ms. The average length change of sarcomeres in a tendon-free segment of fibre was measured using a striation-follower (10). The output of this apparatus (ΔV) is given by the difference between two voltages which are proportional to the number of sarcomeres moving passed two laser spots delimiting the segment of fibre. The signal corresponding to the whole length of one sarcomere is 0.03922 V. The initial number of sarcomeres between spots is $N_i = l_s/l_o$ where l_s is the segment length and l_o the initial sarcomere length. The final number of sarcomeres between spots is $N_f = N_i-(\Delta V/0.03922)$ since lengthening of the sarcomeres results in a positive signal ΔV. The length change of the sarcomeres (final-initial) is $(l_s/N_f)-(l_s/N_i)$ and the

$$\text{Strain (\%)} = (\Delta V.l_o.100)/(0.03922.l_s - \Delta V.l_o)$$

Sometimes the signal of the striation follower showed obvious jumps due to deterioration of the image of the striation pattern; these tracings were discarded. Two digital oscilloscopes (Nicolet 4094), labelled END and SEG, were used to record force and length changes of the fibre and the sarcomeres respectively. Before the experiment, the sarcomere length was set at about 2.2 µm, the cross-sectional area and length of the fibre were measured (5), and tuning of the striation follower was tested during strains of the unstimulated fibre and isometric contractions (Fig. 1). The procedure of the experiment was the same as that described in detail by Cavagna *et al.* (5). In short the fibre was first shortened,

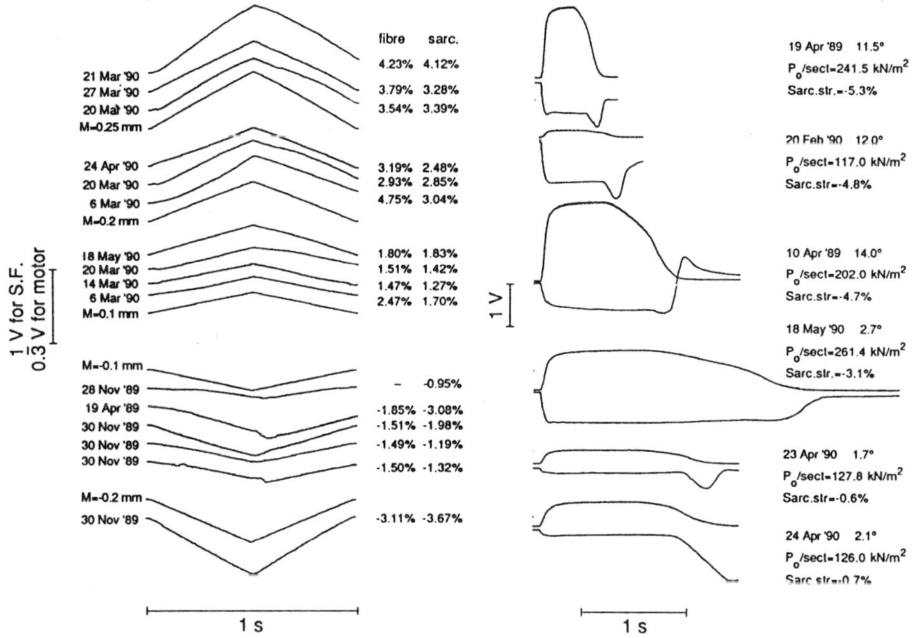

Fig. 1 Pre-experiment striation follower and fibre tests. Left: length changes imposed by motor (*M*) to unstimulated fibre and simultaneous output of striation follower (*S.F.*). Numbers near each trace give maximum percentage strain of fibre and of sarcomeres. The maximum strain of the fibre and of the sarcomeres occur at about the same time except in some shortenings, possibly as a consequence of the fibre becoming slack. Right: tension development (top trace in each pair) and sarcomere strain during isometric tetani: numbers give maximum stress and corresponding strain. Sarcomere strains were calculated from text equation.1 (l_o = 2.09–2.19 mm; l_s = 800–1360 mm). Fibres from ***Rana temporaria*** (lengths, l_f = 5.0–6.2 mm, cross section, S = 12 400–27 500 mm^2), and ***Rana esculenta*** (20 February 1990 and 6 March 1990, l_f = 3.75 mm, S = 3700 and 8900 mm^2).

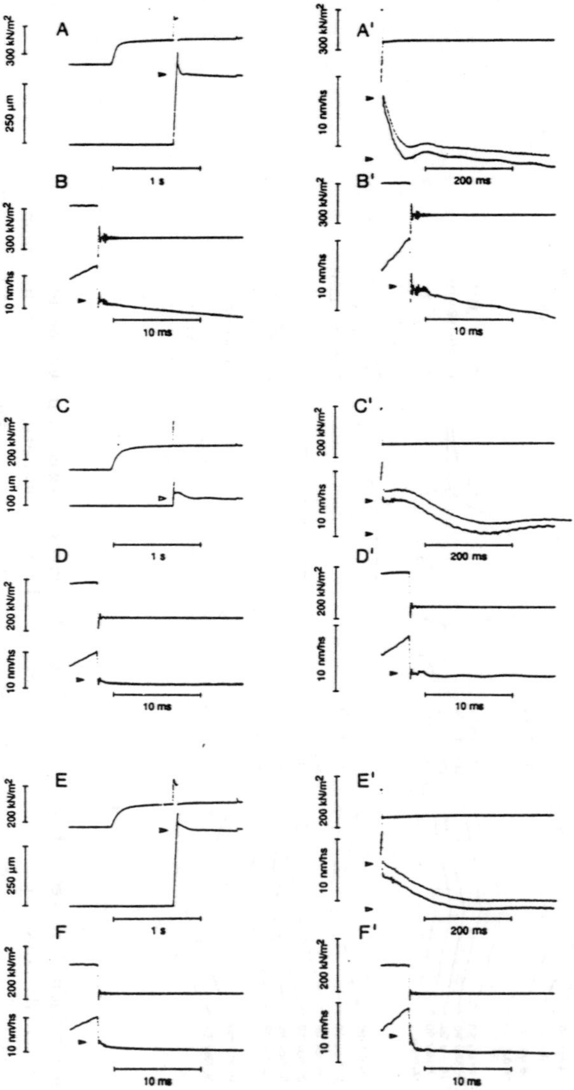

Fig. 2 Velocity transients taking place when force is suddenly reduced, immediately after stretching, to the isometric value developed before stretching (P_o). Tracings on the left (A–F) give force and length changes of fibre, those on the right (A′–F′) force and length changes of sarcomeres. Tracings A, C and E show the whole experiment on a slow time base, whereas the corresponding tracings A′, C′ and E′ are five times faster and begin 3 ms before release. The other tracings (100 times faster than A, C and E) were used to detect the beginning of phase 2 (closed arrows). End of phase 2 is indicated by the open arrows. Dim traces above A′, C′ and E′ are expanded insets of simultaneous length changes in A, C and E. Note the similarity of shape and duration of phase 2 shortening measured at the end of the fibre (A, C, E) and on the sarcomeres (A′, C′, E′). Tracings A, B, A′, B′: experiment of 18 May 1990, l_f = 5.25 mm, S = 17990 mm², l_o = 2.09 mm (at the end of 0.4 mm stretch), l_s = 1.2 mm, 2.7°C. Other tracings: 24 April 1990, l_f = 5.8 mm, S = 23340 mm², l_o = 2.16–2.24 mm (end of 0.1–0.4 mm stretch), l_s = 1.2 mm, 2.0°C.

unstimulated, by the amount of subsequent stretching. At the end of shortening, the fibre was twitched to take up its slack, the slow trace (500 ms/point) of the END scope was triggered and the fibre was stimulated to tetanus in isometric condition with the motor in fibre length feedback mode. When the isometric force reached a plateau, P_o, the active fibre was stretched (1.7–7.8% at 1.1–1.4 fibre lengths/s). Three ms before the end of the ramp stretch, the fast traces (5 ms/point) of both the END and the SEG scopes and the slow trace of the SEG scope (100 ms/point) were triggered and the striation follower was activated. At the end of the ramp the motor was switched to force feedback mode with the command to maintain a force equal to P_o. This produced a fast recoil of the fibre (phase 1) followed by a well defined rapid shortening (phase 2) against an average force, P_{iso}, equal to or slightly greater than P_o: only experiments where $1 < P_{iso}/P_o < 1.04$ were accepted. Phase 2 shortening was calculated by transferring oscilloscope cursor positions, specified by hand, directly to a microcomputer: the beginning of phase 2 was specified on the fast traces, the end on the slow traces (Fig. 2).

Results

The preliminary experiments reported here show that the mechanical transients initiated by a ramp stretch and release to P_o in frog muscle fibres (5) are due to properties of the sarcomeres and not to stress-relaxation of tendons or other end compliance. In fact, the amplitude, duration and shape of phase 2 shortening were consistently similar when measured at the fibre end or in the sarcomeres of a fibre segment, by means of two completely independent methods (Fig. 2).

Also in the present experiments, as in those described by Cavagna *et al.* (5), the amount of phase 2 shortening increased with the amplitude of the stretch.

The smaller stretches, around 2% of fibre length, occurred almost entirely during the initial rise of the force, ending near the peak of force, before the 'give' of the fibre (Fig. 2C). As a consequence, the total stretch, given to the fibre end, was shared between tendons and sarcomeres; in the two experiments illustrated in Fig. 2, sarcomere stretch was about 1.3% (24 April 1990) and 1.2% (18 May 1990). In this condition, phase 2 may begin with a plateau or even with a negative velocity (i.e. with a lengthening) before the shortening, and the fibre may exhibit an oscillatory response (N, 12, 12). This tends to be damped by increasing the amplitude of the stretch.

After large stretches the transient response consists of a well-defined sarcomere shortening against P_o: 8.3 nm per half sarcomere in the experiment of 18 May 1990 (7.4% sarcomere stretch), and 7.5 nm per half sarcomere in the experiment of 24 April 1990 (6.7% sarcomere stretch) (Fig.2).

Discussion

Phase 2 shortening after large stretches suggests the discharge of a damped system within the sarcomeres characterized by a limited travel. According to

Morgan (14), stretching at high speeds makes some sarcomeres lengthen beyond filament overlap, so that energy can be stored in their parallel elastic elements. It is difficult to imagine how the discharge of a number of viscoelastic elements may simulate the features characterizing phase 2: (a) a sharp reduction, often to zero, of the velocity of shortening at its end (a reversal, exceptionally, in the 18 May 1990 experiment of Fig. 2); (b) shortening taking place, in some conditions, in two distinct steps (5). In addition, the velocity of shortening against P_o after a stretch was found, in the whole muscle, to increase with temperature more than expected for a passive viscoelastic system (3). Furthermore, length changes were kept within or just beyond the plateau of the isometric force-length diagram (5), a region where inhomogeneity of sarcomeres may indeed occur, but possibly not to the extent required by Morgan's theory.

Cavagna *et al.* (5) considered phase 2 shortening after large stretches as direct evidence of the discharge against P_o of the damped element within the bridges described by Huxley and Simmons (15): this would take place from levels of potential energy unattainable during an isometric contraction and reached only because of the ramp stretch. The mechanical work done is given by the product of phase 2 shortening times the average load applied after stretching (approximately P_o). In the experiment of 18 May 1990, the average isometric stress was 251 kN m^{-2} and the maximum work was 2.07 J Kg^{-1} (2.17 mJ m^{-2} per half sarcomere). In a mirror experiment made by Cavagna *et al.* (5), 15 October 1984, the isometric stress was 285 kN m^{-2} and the maximum work was 2.91 J Kg^{-1} (2.96 mJ m^{-2} per half sarcomere). In an isometric contraction the number of attached bridges m^{-2} per half sarcomere can be approximated to be 4.10^{16} (11). Assuming the same number at the end of stretching, the work done by each crossbridge during phase 2 would be, for the two experiments, 54.10^{-21} J and 74.10^{-21} J respectively; these figures are of the same order of magnitude of those predicted for the maximum work done in a crossbridge cycle without previous stretching (16, p. 294). In the 24 April 1990 experiment, the maximum work was only 0.89 J Kg^{-1} (1.0 mJ m^{-2} per half sarcomere), mainly due to an abnormally low stress (about 130 kN m^{-2}); in this fibre the duration of phase 2 was also unusually large (Fig. 2).

In conclusion, the described experimental evidence suggests that the contractile component itself, not a passive structure, can store and release mechanical energy during contraction. This may be related to the negative rate of the energy balance (heat produced minus work done on muscle) during fast stretches (16, p. 216). Our working hypothesis is that the stressed structure responsible for it is the damped structure within the crossbridges (9, 15), and that its subsequent discharge is primarily responsible for phase 2 shortening against P_o.

Acknowledgements

The authors whish to thank Professors A. F. Huxley and V. Lombardi for kindly providing the original plans and the parts list for the construction of the striation follower, Dr A. Vignati for his precious contribution in the assembling and understanding of the striation follower, and Miss R. Garofolo for her enthusiasm and skill in isolating the fibres.

References

1. Bergel, D. H., Brown, M. C., Butler, R. G., and Zacks, R. M., 1972, *J. Physiol.*, **225**, 21–22.
2. Cavagna, G. A., Citterio, G., 1974, *J. Physiol.*, **239**, 1–14.
3. Cavagna, G. A., Citterio, G., Jacini, P., 1975, *J. Physiol.*, **251**, 65–66.
4. Cavagna, G. A., Dusman, B., Margaria, R., 1968, *J. Appl. Physiol.*, **24**, 21–32.
5. Cavagna, G. A., Mazzanti, M., Heglund, N. C., Citterio, G., 1986, *Am. J. Physiol.*, **251**, C571–C579.
6. Edman, K. A. P., Elzinga, G., Noble, M. I. M., 1978, *J. Physiol.*, **281**, 139–155.
7. Sugi, H., Tsuchiya, T., 1981, *J. Physiol.*, **319**, 239–252.
8. Huxley, A. F., Lombardi, V., 1980, *J. Physiol.*, **305**, 15–16.
9. Ford, L. E., Huxley, A. F., Simmons, R. M., 1977, *J. Physiol.*, **269**, 441–515.
10. Huxley, A. F., Lombardi, V., Peachey, L. D., 1981, *J. Physiol.*, **317**, 12–13.
11. Armstrong, C. F., Huxley, A. F., Julian, F. J., 1966, *J. Physiol.*, **186**, 26–27P.
12. Pringle, J. W. S., 1949, *J. Physiol.*, **108**, 226–232.
13. Ruegg, J. C., Steiger, G. J., Schadler, M., 1970, *Pflügers Arch.*, **319**, 139–145.
14. Morgan, D. L., 1990, *Biophys. J.*, **57**, 209–221.
15. Huxley, A. F., Simmons, R. M., 1971, *Nature*, **233**, 533–538.
16. Woledge, R. C., Curtin, N. A., Homsher, E., 1985, *Energetic Aspects of Muscle Contraction*. Academic Press, London.

46
Force–Velocity Relationship of Isomyosins SM_2 and IM of Mouse Soleus (Strain NMRI)

G. MARÉCHAL AND G. BECKERS-BLEUKX

Department of Physiology, Catholic University of Louvain 554C, 1200, Brussels, Belgium

Introduction

Mouse soleus is a slow-twitch muscle although its fibres are mixed, approximately 70% being of type I and 30% of type IIA (1). It also contains two isoforms of myosin: slow-twitch isomyosin SM_2, which is localized in type I fibres, and slow-twitch isomyosin IM which is present in type IIA fibres (2). Isomyosin SM_2 of mouse is similar in its subunit composition to that of guinea-pig soleus, a muscle containing only type I fibres: the myosin heavy chain is of the slow type, MHC_1, combined with slow type myosin alkali light chains, LC_{1s}; isomyosin IM which is absent from guinea-pig soleus has a myosin heavy chain of the fast type, MHC_{2A}, hybrid in its light chains, since it has both slow and fast alkali light chains LC_{1s} and LC_{1f} (3, 4). The unloaded velocity of contraction V_u (5) is higher for type IIA (IM) fibres (6.1 fibre length/s) than for type I fibres (1.7 fibre length/s) according to Asmussen & Maréchal (2). We show in this paper that in spite of the fact that V_u of type IIA fibres is fast, its force–velocity relationship is intermediate between the slow variety of type I fibres and the fast variety of type II fibres of fast-twitch muscle, such as mouse EDL.

Muscle and Motility Vol. 2. Proceedings of XIXth European Conference in Brussels
© Intercept Ltd, PO Box 716, Andover, Hampshire, SP10 1YG, UK

Methods

The experiments were performed *in vitro* at 20°C on whole soleus muscle isolated from NMRI mice. The muscle were pre-stretched at rest to develop 5±2 mN, and stimulated isometrically at 1 Hz for 10 s. The last twitch was recorded and analysed. The force–velocity relationship was obtained from 18 to 22 tetani with iso-velocity releases (6), using an electromagnetic puller. The maximum velocity of unloaded shortening, V_u, was measured by the slack test method of Edman (5). After the experiments, the muscles were frozen in liquid nitrogen. The myosins were extracted at 0°C with a Guba-Straub solution containing 10 mM PPi and separated by PPi-PAA electrophoresis in a nondissociating solution. The myosin heavy chains were separated by the method of Carraro & Catani (7). The relative proportions of the isomyosins and of the myosin heavy chains were measured by densitometry of the gels. The statistical calculations were done with the SPSS package on a AT computer. A more detailed description of our methods is reported in Asmussen & Maréchal (2).

Results

Table 1 shows that the twitch is slow in type; it also shows that the maximum isometric force varies quite a lot between muscles of different animals. These variations are still present when the force is expressed per unit area (maximum

Table 1 Isometric twitch and tetanic forces of soleus muscle of mouse (strain NMRI 20°C).

Variables	Means	Standard Deviation	Minimum	Maximum
Body weight (g)	32.43	5.28	25	42
Age (days)	81.1	17.6	55	120
Muscle weight (mg)	8.45	2.69	4.3	13
Fibre length (mm)	7.41	.64	6.3	8.5
Twitch force (mN)	22.53	7.40	11	37
Twitch time to peak (ms)	65.5	22.6	40	104
Twitch time of half relaxation (ms)	158.4	80.4	40	320
Twitch to tetanus force (%)	12.9	5.0	4.8	23
Maximum isometric force (mN)	186.1	51.8	106	292
Maximum isometric stress (mN/mm^2)	168.4	37.2	100	237
Fatigue index (5)	95.7	14.7	73.7	154.1

The means are computed on 21 experiments. The fatigue index is the ratio of the maximum isometric force of the first tetanus to that of the last tetanus in an experiment (n= 18–22). The maximum isometric force and stress reported in the table are those of the first tetanus in the experiment.

isometric stress). The maximum isometric force of the last tetanus in the experiment is on the average 5% lower than that of the first one (see 'fatigue index').

Table 2 Parameters of the force–velocity relationship of soleus muscle of mouse (strain NMRI, 20°C).

Variables	Means	Standard Deviation	Minimum	Maximum
Force constant a/F_o (%)	0.11	0.05	0.03	0.22
Velocity constant b (fibre length/s)	0.45	0.08	0.32	0.61
SSH	86	62	11	203
Velocity constant ba (with a/F_o=.11) (fibre length/s)	0.47	0.10	.33	0.65
SSHa	105	83	12	308
Maximum power (mW/mm^3)	39.5	9.1	27.4	64.5
V_{max} (fibre length/s)	5.0	3.21	2.0	16
V_u (fibre length/s)	5.7	0.97	3.9	8.2
V_u/V_{max}	1.43	0.58	0.33	2.4

Means of 21 experiments. V_{max} equals b/(a/F_o). ba is the velocity constant computed by setting a/F_o = 0.11 in all experiments. SSH is the residual sum of squares when b and a/F_o are evaluated by a non linear least squares procedure; SSHa is the residual sum of squares when ba is evaluated. The estimates of the parameters of each force-velocity curve are based on 18 to 22 tetani with relases at velocities ranging from 0.1 to 3 fibre length/s. The velocity of release during the slack test (to evaluate V_u) was 267 mm/s, with step sizes ranging from 0.2 mm to 2.0 mm.

Table 2 reports the observations on the force–velocity relationships. The velocity constant, b, and the force constant, a/F_o have been computed by a non-linear least squares procedure based on 18 to 22 tetanus-and-releases. The maximum velocity of shortening, V_{max}, is calculated as the ratio $b/(a/F_o)$. The velocity of unloaded shortening, V_u (5), is 40% higher than V_{max}. We have also computed the velocity constant *ba* by the least squares method when the force constant is set at 0.11 for all experiments. The goodness of fit can be evaluated by the residual sum of squares, i.e. the variance of the data not explained by the fitted curves. SSH and SSHa are the residual sums of squares corresponding to the two estimates *b* and *ba* of the velocity constant. The two estimates *b* and *ba* are not different. As expected, SSHa is higher than SSH, but the difference is small: it is thus reasonable to accept that a/F_o is the same for all muscles, whilst *b* may vary between muscles. This conclusion is supported by the absence of correlation between a/F_o and any of the other variables, whilst *b* and *ba* show strong correlations, with the isomyosins and the myosin heavy chains (see Table 4).

The maximum power P_{max} (Table 2) is the power per unit volume; it is equal to:

$$P_{max} = S_0 * b * M^2/(a/F_0) \tag{1}$$

in which M^2 is the maximum power normalized for force and velocity (8), F_o is the maximum isometric force and S_o is the maximum isometric stress:

$$M^2 = \{1 + 2\, a/F_0 - 2\, \mathrm{SQR}[(a/F_0)^2 + a/F_0]\} * a/F_0 \tag{2}$$

P_{max} varies between muscles quite markedly (Table 2) but these variations are not correlated with the type of isomyosins or myosin heavy chains (see Table 4).

The proportions of isomyosins and myosin heavy chains are reported in Table 3. They are similar for SM_2 and MHC_1 on the one hand, and IM and MHC_{2A} on the other hand. Fig. 1 shows the data for SM_2 and MHC_1 and the line of

Table 3 Isomyosins and myosin heavy chains of soleus muscle of mouse (strain NMRI, 20°C).

Variables	Means	Standard Deviation	Minimum	Maximum
SM_2	66.0	8.9	52	84
IM	34.0	8.9	16	48
MHC_1	64.7	13.9	46	90
MHC_{2A}	33.7	14.0	8	54
MHC_{2B}	1.7	4.3	0	19

Means of 21 experiments, expressed in % of total isomyosins or total myosin heavy chains. The zone labelled MHC_{2B} may contain myosin heavy chain 2X (10) or 2d (11).

Table 4 Coefficients of correlation between the mechanical parameters and weights, isomyosins and myosin heavy chains.

Variables	Muscle weight	SM_2	MHC_1	MHC_{2A}
Twitch force	0.231	-0.231	-0.245	0.213
Time to peak	-0.691**	0.321	0.290	-0.424
Half relaxation	-0.592**	0.368	0.488	0.444
Max. isometric force	0.728**	-0.310	-0.265	0.392
Max. isometric force	0.368	0.390	0.549*	-0.487
a/F_o	0.488	0.403	0.409	-0.428
b	0.454	-0.511*	-0.641**	0.670**
ba	0.638**	-0.621*	-0.728**	0.767**
Max. power	0.327	-0.294	-0.269	0.371

Number of cases: 21. 1-tailed significance: *:0.01; **:0.001. The coefficients of correlation between the mechanical parameters and isomyosin IM are equal to those of SM_2 changed sign.

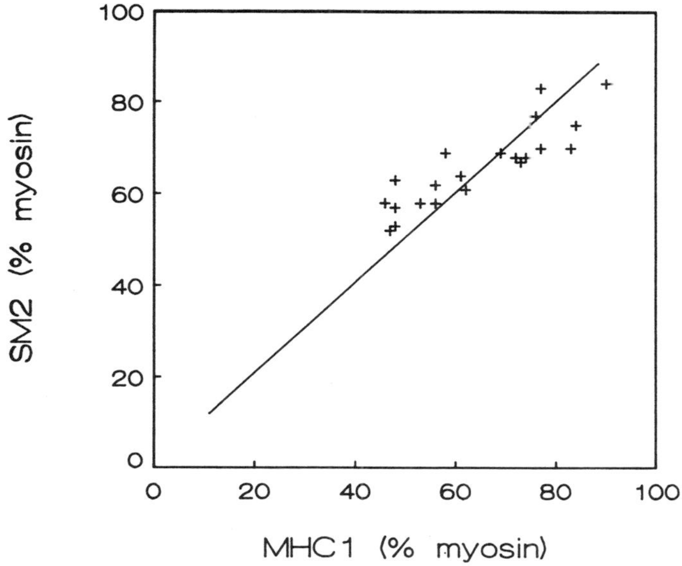

Fig. 1 Content in slow-twitch isomyosin SM_2 (in % of total isomyosins) as a function of the content in slow myosin heavy chain MHC1 (in % of total myosin heavy chain in 21 soleus muscle of mouse (strain NMRI). The slope of the line is 1 (identity line).

identity (slope = 1). This result is expected if all MHC_1 is in SM_2. It also demonstrates that our estimates of the myosin proportions are quantitative.

The correlations between some variables are reported in Table 4. The twitch time to peak and time to half relaxation depend on the muscle weight: apparently heavier animals develop shorter twitches. They do not depend on the proportions in isomyosins: presumably the shape of the twitch is entirely determined by the mechanisms of release and uptake of sarcoplasmic calcium. The force constant is influenced neither by muscle weight nor by the isomyosins. To the contrary, the velocity constants strongly depend on the isomyosins.

We further analysed the correlation between b and the slow myosin heavy chain, MHC_1, (Fig. 2). The regression line was computed on the assumption that both variables have similar coefficients of variation; this means that the slope of the regression line is the mean between the slopes of the regression lines of b on MHC_1 and of MHC_1 on b and the intercept is computed using the means of b and MHC_1. The equation of the best linear regression between b and MHC_1 is:

$$b = 0.85\ (\pm 0.12)\ \ 0.0061\ (\pm 0.0017)\ MHC_1 \qquad (3)$$
$$\text{where } n = 21 \quad r = 0.641 \quad p = 0.002$$

Twelve of the mice were male and 10 were female: except for the body and muscle weights, which were smaller in the female animals, no variable was correlated with sex.

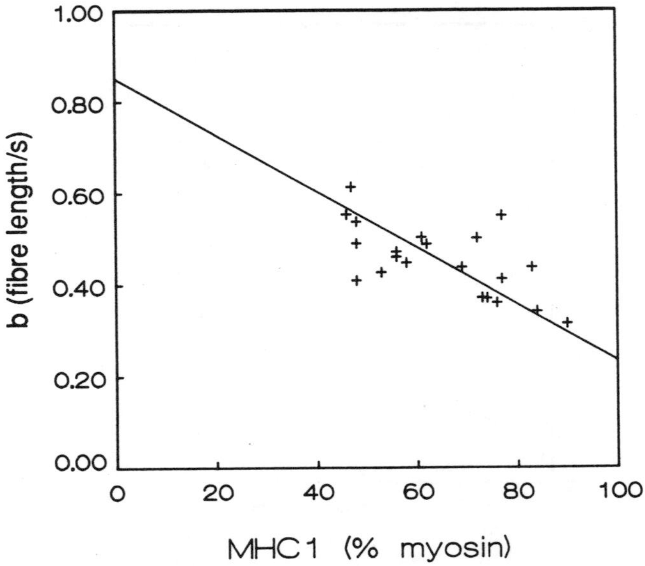

Fig. 2 Velocity constant of the Hill's force–velocity relationship as a function of the proportion in slow myosin heavy chain MHC1. Data of 21 mouse soleus muscles (strain NMRI).

Discussion

Soleus muscle is a relatively simple muscle: its fibres segregate into two varieties of isomyosins (SM_2 and IM), each isomyosin being homogeneous for its myosin heavy chain (MHC_1 and MHC_{2A} respectively). In unpublished experiments, we observed no fibres simultaneously stained by specific antibodies directed against MHC_1 or MHC_{2A} (gift of Dr F. Pons, Montpellier): most of the muscle fibres therefore have only one isomyosin. The muscle fibres being parallel, they add their forces to produce the force–velocity relationship of the whole muscle. It is reasonable to hypothesize that the fibres are subdivided into two homogeneous groups which contract at the same velocity during the tetanus and add their force. It is then justified to evaluate the parameters of the force–velocity of the two populations by an extrapolation of Equation 1. If soleus muscle had only type I fibres, or, equivalently, had only MHC_1, putting $MHC_1 = 100$ in Equation 3 gives an evaluation of the velocity constant of type I fibres. The velocity constant of the type IIA(IM) fibres is obtained if there are no type I fibres in the muscle, i.e., setting $MHC_1 = 0$ in Equation 3. The force constant is the same for all fibres, since it shows no correlation with the isomyosins. V_{max} is computed by the ratio $b/(a/F_o)$. Table 5 summarizes the mechanical properties inferred for the two varieties of fibres.

The parameters of type I fibres are close to those observed in guinea-pig soleus, which is a muscle containing only type I fibres: the velocity constant in

Table 5 Force–velocity parameters of type I and type IIA(IM) fibres of mouse soleus (strain NMRI, 20°C).

Fibre type	b (*Lf*/s)	a/F_o	V_{max} (*Lf*/s)
Type I	0.24	0.11	2.2
Type IIA(IM)	0.85	0.11	7.7

soleus guinea-pig is b = 0.13 ± 0.02 fibre length/s, and its force constant is a/F_o = 0.10 ± 0.02 (2). Type IIA(IM) fibres have the subscript 'IM' to recall that their isomyosin is hybrid in its alkali light chains. The parameters of the force–velocity of type IIA fibres homogeneous in fast light chains 1 is not known, but we know those of a fast-twitch muscle that contains approximately 20% of type IIA fibres, 75% of type IIB fibres and 5% of type I fibres in mouse C57/BL extensor digitorum longus: b = 1.8 ± 0.2 fibre length/s, a/F_o= 0.23 ± 0.03 and V_{max} = 8.2 ± 0.2 fibre length/s (9). We conclude that the substitution of one fast light chain LC_{1f} in type IIA fast fibre by one slow light chain LC_{1s} in type IIA(IM) fibres does not change the maximal velocity of shortening but reduces both b and a/F_o. In molecular terms, this could mean that the alkali light chain does not influence the maximum ATPase activity of the crossbridges, but modulates the energy transfer from the ATPase site to the force-transducing site.

References

1. Ariano, M. A., Armstrong, R. B., Edgerton, V. R., 1973, *J. Histochem. Cytochem.*, **21**, 51–55.
2. Asmussen, G., Maréchal, G., 1989, *J. Physiol. (Lond.)*, **416**,245–254.
3. Fitzsimons, R. B., Hoh, J. F. Y., 1983, *J. Physiol. (Lond.)*, **343**, 539–550.
4. Maréchal, G., Biral, D., Beckers-Bleukx, G., ColsonVan Schoor, M., 1989, *Biomed. Bioch. Acta*, **48**, 417–421.
5. Edman, K. A. P., 1979, *J. Physiol. (Lond.)*, **291**, 143–159.
6. Cecchi, G., Colomo, F., Lombardi, V., 1978, *J. Physiol.*, **285**, 257–273.
7. Carraro, U., Catani, C., 1983, *Bioch. Biophys. Res. Com.*, **116**, 793–802.
8. Woledge, R. C., Curtin, N. A. & Homsher, E., 1985, *Energetic Aspects of Muscle Contraction*. Monographs of the Physiol. Soc. 41.
9. Moens P. D., 1990, *Energétique et régénération des muscles striés de souris normales et dystrophiques*. Mémoire de Licence en Sciences Biomédicales. Université Catholique de Louvain.
10. Schiaffino, S., Ausoni, S., Gorza, L., Saggin, L., Gundersen, K., Lømo, T., 1988, *Acta Physiol. Scand.*, **134**, 575–576.
11. Termin, A., Staron, R. S., Pette, D., 1989, *Histochem.* **92**, 453–4457.

47
Maximal Isometric Forces of mdx Muscles

P.D. MOENS, G. BECKERS-BLEUKX AND G. MARÉCHAL

Department of Physiology, Catholic University of Louvain, B-1200 Brussels, Belgium

Introduction

Duchenne muscular dystrophy, an X-linked disease, is characterized by the absence of a 400 kDa protein, discovered by Hoffman *et al.* (1), called dystrophin. Zubrzycka-Gaarn *et al.* (2) have localized this protein in the sarcolemma and have suggested that dystrophin is a binding molecule that anchors elements of the internal cytoskeleton to the plasma membrane thereby providing mechanical strength for the sarcolemma.

According to this hypothesis, we would expect muscles lacking dystrophin to develop less force than normal muscles.

Bulfield *et al.* (3) have discovered a mutant of the C57BL10, the mdx mice. These mice have the same gene deletion as in the human Duchenne dystrophy and do not express dystrophin (1).

We have analysed the maximal isometric forces developed by the soleus and extensor digitorum longus (EDL) muscles from dystrophic mice (mdx) and their control strain.

Methods

Six C57 and six mdx mice matched for age (60 ± 5 days) and weight (28 ± 2 g) are killed by neck-breaking. The soleus and EDL muscles of each mouse are dissected, measured, weighted and mounted on a ergometer. Muscles are stimulated isometrically at 20°C in an oxygenated Krebs solution. We have recorded the force of the twitch (*FT*) and the maximal tetanic isometric force

Muscle and Motility Vol. 2. Proceedings of XIXth European Conference in Brussels
© Intercept Ltd, PO Box 716, Andover, Hampshire, SP10 1YG, UK

(*Fo*). The maximal tetanic isometric force was normalized giving the stress (So) (*So* = *Fo** fibres length (*Lf*)/ muscle weight (*W*m); density = 1). The length of the fibres was obtained by microscopic measurements of few superficial fibres. We found that in EDL the fibre length is 0.48% of muscle length, and in soleus it is 0.68%. This is in agreement with Brooks and Faulkner (4). Results are expressed as mean ± SEM.

Results

The length (*Lo*) and the weight (*Wm*) of soleus muscles are not different in the two strains (Fig. 1a). The maximal isometric force (*Fo*) developed by the mdx soleus (243 ± 26 mN) is higher, but not significantly different from that of controls (227 ± 11 mN). The stress (*So*) and the maximal force of the twitch (*FT*) aresimilar in mdx and C57 soleus.

The mdx EDL muscle (Fig. 1b) is shorter (11.7 ± 0.2 mm) than the C57 EDL (12.8 ± 0.2 mm), but there is no difference in the weights. The maximal isometric force of the mdx muscle (33 ± 19 mN) is significantly higher than that of the control muscle (263 ± 21 mN). However, the stress and the maximal

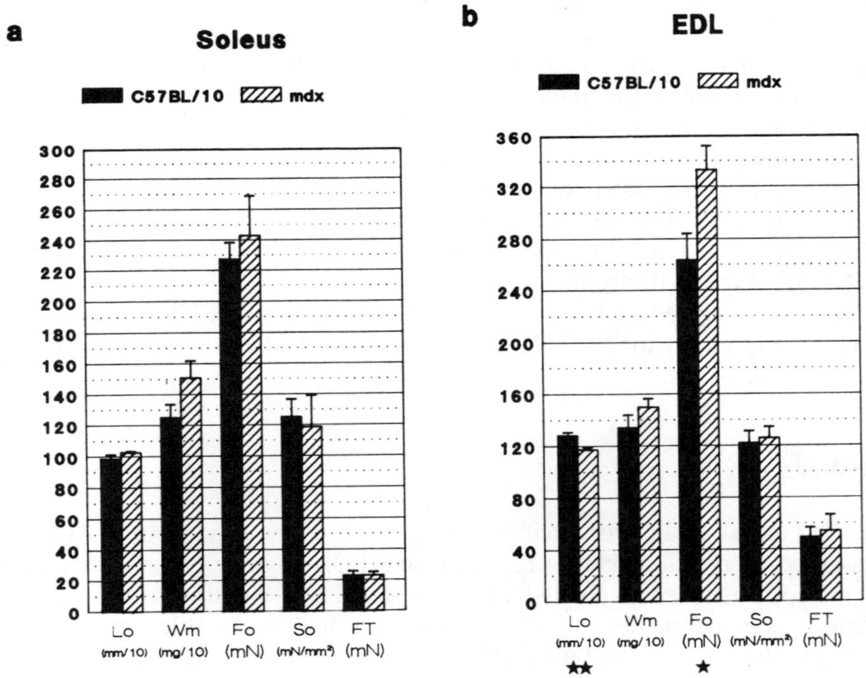

Fig. 1 Comparison between morphological and mechanical parameters of the C57BL/10 and the mdx mice. (a) Soleus muscles (C57: *n* = 6; mdx:*n* = 6). (b) EDL muscles (C57: *n* = 6; mdx: *n* = 6). The bar is the SEM. *$P<5\%$. **$P<1\%$.

force of the twitch are not different between the mdx and the C57 EDL muscles.

Discussion

Our estimates of the maximal isometric force of the soleus muscles are in agreement with those reported by Maréchal and Beckers-Bleukx (5) but they are somewhat greater than those reported by Anderson *et al.* (6) and Coulton *et al.* (7) for both mdx and C57 mice.

The maximal isometric force of the EDL muscle as well as the maximal force of the twitch for the EDL and soleus muscle correspond to the values given by Anderson *et al.* (6). However, they show that the stress of the mdx EDL muscles is significantly lower than the stress of the C57 EDL muscles. We found no differences between the two strains concerning the stress of the muscles. Such differences in the results could be due to the different control strain used. Anderson *et al.* (6) have worked with C57BL/6 mice whilst we used the C57BL/10 strain, the original strain of the mutant mdx (3).

The force developed by the mdx muscle during a twitch and a tetanus was similar to the force developed by soleus and EDL muscles of C57 mice. Thus the absence of dystrophin is not correlated with a decrease of the force developed by the muscle and could not account for the weakness of the muscles characteristic of the Duchenne muscular dystrophy.

Acknowledgements

We are grateful to Dr Partridge and Dr Morgan for donating the animals.

References

1. Hoffman, E. P., Brown, R. H., Kunkel, L. M., 1987, *Cell*, **51**, 919–928.
2. Zubrzycka-Gaarn, E. E., Bulman, D. E., Karpati, D. E., Burghes, A. H. M., Belfall, B., Klamut, H. J., Talbot, J., Hodges, R. S., Ray, P. N., Worton, R. G., 1988, *Nature*, **333**, 466–469.
3. Bulfield, G., Siller, W. G., Wight, P. A. L., Moore, K. J., 1984, *Proc. Natl. Acad. Sci. USA*, **81**, 1189–1192.
4. Brooks, S. V., Faulkner, J. A., 1988, *J. Physiol.*, **404**, 71–82.
5. Maréchal, G., Beckers-Bleukx, G., 1991, abstract in *J. Musc. Res. Cell. Mot.*, **19**, 74.
6. Anderson, J. E., Bressler, B. H., Ovalle, W. K., 1988, *J. Muscle Res. Cell. Mot.*, **9**, 499–515.
7. Coulton, G. R., Curtin, N. A., Morgan, J. E., Partridge, T. A., 1988, *Neuropathol. Appl. Neurobiol.*, **14**, 299–314.

48
Type I and Type II Fibres Work with the same Mechanical Efficiency during Bicycling

J. I. MEDBØ

Department of Physiology, National Institute of Occupational Health, PO Box 8149 dep., N-0033-8149 Oslo 1, Norway

This paper discusses the mechanical efficiency of type I and type II fibres during dynamic work. The background for the study is conflicting evidences of the mechanical efficiency for the two main fibre types in muscle. Results from *in vitro* experiments suggest that type I (slow twitch, ST) fibres work more efficiently than type II (fast twitch, FT) fibres (e.g. 1, 2). These data suggest that type I fibres are twice as efficient as type II fibres. Results from studies on exercising humans suggest that the two main fibre types work with a similar mechanical efficiency during dynamic exercise like bicycling. These data are reviewed and re-evaluated. In addition new data are presented.

Muscles function by developing force, and this force development costs energy in the form of ATP. When muscles shorten against an external force, work is done. The mechanical efficiency (e) of this work (W) is defined as the ratio between the work done and the amount of energy released to do the work (E_w), or equivalently as their time derivatives:

$$e = W/E_w \tag{1}$$

Muscle and Motility Vol. 2. Proceedings of XIXth European Conference in Brussels
© Intercept Ltd, PO Box 716, Andover, Hampshire, SP10 1YG, UK

RE-EVALUATION OF FORMER DATA ON HUMANS DURING BICYCLE

Vøllestad and colleagues have studied the recruitment pattern of different muscle fibre types during exercise (3, 4). Muscle fibres were classified as type I, type IIA, type II AB, or type II B by staining for myosin ATPase activity after different preincubations (the type II AB fibres are fibres showing physiological and histochemical properties between the type II A and II B fibres). The degree of activation of different fibre types during exercise was determined from the fall in the glycogen concentration in single muscle fibres. It is here important to note that even well trained subjects are unable to exercise at the powers used if fat is the only substrate for oxidative phosphorylation. Hence, when a fibre becomes depleted of glycogen, it has to reduce its power output or even become wholly passive, and other fibres take over. The O_2 uptake was measured and used as a measure of the rate of energy release during exercise. The braking power was kept constant during each experiment, which means that a possible fall in the efficiency would raise the O_2 uptake (see equation 1). The experiments were done at several intensities (3, 4).

The results in Vøllestad's experiments of interest here can be summarized as follows: type I fibres were active from the beginning of exercise. At the higher intensities used some type II fibres were also recruited from the onset of exercise. As the exercise went on these fibres ran short of glycogen and accordingly reduced their power output. First type II A and later type II AB and type II B fibres were recruited. Since the power was kept constant, a possibly lower efficiency of type II fibres would therefore require a larger O_2 uptake. The O_2 uptake rose only 2.5% during the first hour of exercise (4). It can be shown by simple calculations that if this higher O_2 uptake was caused by a lower mechanical efficiency for type II fibres, the recruited type II fibres could only be ~5% less efficient than the type I fibres. However, oxidation of carbohydrates gives more ATP per mole O_2 consumed than fat oxidation, and there may have been a slight shift from carbohydrate to fat as substrate for oxidation during exercise. The rise in the O_2 uptake can be fully accounted for if the ATP turnover rate was kept constant throughout the exercise, but fat oxidation accounted for 15% more of the total ATP turnover rate after one hour than at the beginning of exercise. Hence, a re-examination of Vøllestad's data (4) suggests that type II fibres work as efficiently as type I fibres during bicycling.

Introduction

The purpose of this study was to find a possible relationship between mechanical efficiency during bicycling and fibre type recruitment. This was done in two ways. Type I fibres are recruited at lower intensities than type II fibres, and a recruitment threshold for type II fibers was estimated as described below. A constant efficiency means that the O_2 uptake increases linearly with power both below and above this threshold. If type II fibres are less efficient than type I fibres, the O_2 uptake will be disproportionately high above the recruitment threshold of type II fibres. In addition the mechanical efficiency was correlated to the fraction of type I fibres in the quadriceps muscle. A positive correlation indicates a lower mechanical efficiency of type II fibres.

Subjects and Methods

Sixteen healthy young men served as subjects, and the exercise was done on a Krogh-type bicycle ergometer at 1.5 Hz pedalling rate. For each subject we established a relationship between bicycle power and steady state O_2 uptake as described in detail elsewhere (5–7). The fraction of type I fibres in m. *vastus lateralis* was determined from at least four muscle biopsies from each subject. In addition to the data discussed above Vøllestad & Blom (3) examined the recruitment of different fibre types at the onset of exercise at different intensities. This examination showed that as the exercise intensity was increased a larger and larger fraction of the whole quadriceps muscle mass was recruited. There was a one-to-one relationship between the exercise intensity relative to the maximal O_2 uptake and the fraction of the quadriceps muscle mass engaged. This means that at powers corresponding to 40%, 60% or 80% of the maximal O_2 uptake, 40%, 60% or 80% respectively of the whole quadricepts muscle was used. Type I fibres were recruited first and were the only active fibres at the lowest intensities, and all type I fibres were recruited before type II fibres. We have used this finding to indicate the power where type II fibres were first recruited. Different subjects had different proportions of type I fibres in their quadriceps muscle. Some subjects were more physically fit than others. Accordingly the estimated recruitment thresholds varied considerably (see arrows in Fig. 1).

Results

The mechanical efficiency was 0.27 (range, 0.25–0.29). For each subject the O_2 uptake increased linearly with power at all intensitites examined, which means that the efficiency was constant. The estimated recruitment threshold of type II fibres varied between 60 and 260 W for the different subjects (Fig. 1). Hence, some subjects probably used type II fibres at all powers examined, while other subjects used mainly type I fibres until quite high powers were reached.

There was no significant correlation between the mechanical efficiency and the fraction of type I fibres recruited ($r = -0.3$, $P = 0.3$).

Discussion

This study showed no effect of fibre type recruitment on the mechanical efficiency, suggesting that both fibre types were equally efficient during bicycling. This conclusion is based on the poor correlation between the two parameters and is in line with a study of Bosco and co-workers who found a correlation coefficient of 0.0 between efficiency and fibre type occurrence in

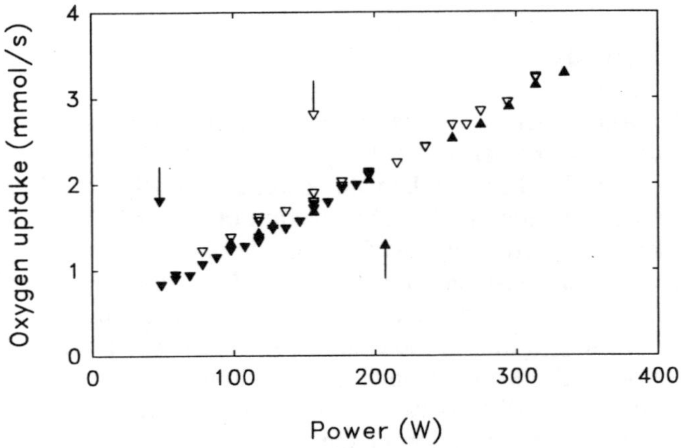

Fig. 1 Steady-state O_2 uptake versus power for three typical subjects marked with three different symbols. The arrows indicate the power where type II fibres in m. vastus lateralis are recruited for each subject. This threshold is estimated as described in Methods. Different muscles have different proportions of type II fibres, but a separate analysis suggests that the recruitment threshold for other muscles is within ±50 W of the indicated value (7).

m. vastus lateralis (8). The efficiency in this study was the same at low powers where only type I fibres were active as at high powers where both fibre types were working. This observation is in close agreement with the data of Vøllestad and co-workers (3, 4) re-evaluated in the beginning of this paper.

A separate paper discusses several methodological questions and concludes that possible shortcomings in the methods used cannot have masked a difference in the efficiency between different fibre types (7). I will here only emphasize that an insignificant correlation between efficiency and fibre type occurrence does not prove equal efficiencies of the two fibre types. However, a statistical analysis where a 95% confidence interval was calculated, showed that if type II fibres are less efficient than type I fibres, the present data suggest that the difference is at most 5%.

In conclusion, all evidence suggests that type II fibres in humans work as efficiently as type I fibres during bicycling.

Why do *in vitro* Experiments suggest that Type II Fibres are Inefficient?

Type II fibres (fast twitch, FT) have usually been assumed to work less efficiently than type I fibres (slow twitch, ST), and this conclusion is based on *in vitro* experiments on muscles from different animals. Results from Gibbs and co-workers are quite typical in this respect (1, 2). They let rat soleus (ST) and EDL (FT) muscles work isotonically *in vitro* under electrical stimulation. The efficiency varied with the load, and the peak efficiencies reported were 0.185 ± 0.010 for the soleus muscle and 0.095 ± 0.005 or only half as much for the EDL. Both values are significantly less than our values for cycling men. Here I discuss possible causes for a low and varying mechanical efficiency.

ENERGY USED IN PROCESSES NOT DOING WORK

Both in single muscles and in whole body there are processes not doing physical work but still requiring energy. If the efficiency is measured as the ratio between the work done and the total energy release, the efficiency will be too low, particularly if the non-work component (E_n) is large:

$$e = W/(E_w + E_n) \tag{2}$$

In Gibbs and coworkers' experiments the EDL muscle was originally stimulated for 1 s, giving a peak efficiency of 0.095. The experiments were repeated with only 0.5 s stimulation of this muscle, and the peak efficiency in these experiments were 0.15, much closer to the efficiency of the soleus muscle (1). Recordings of the isotonic work showed that most of the work was done in the first part of the stimulation, while in the latter part of the stimulation the force development was almost static (Fig. 3 in ref. 1). Energy was probably released at a fairly constant rate. This means that the efficiency was high in the beginning of the stimulation and very low at the end. The reported efficiences are means of the efficiency during different parts of the stimulation, and the low mechanical efficiency reported may be due to a large energy release to do static 'work'. Fast-twitch muscles shorten faster than slow-twitch muscles. If the stimulation period is the same for both muscle types, the static component may be larger for the fast-twitch muscles and thus lower the measured efficiency more in these muscles.

SPECIES DIFFERENCES

It is likely that animals, including their muscles, through ten thousands of years evolve to fit their natural environment as well as possible. Insufficient food supply is a situation all wild animals meet at intervals, and an efficient use of energy is therefore important for energy-demanding processes. Animals such as dogs, horses, many large wild mammals, and humans naturally perform long-lasting exercise of high intensity like running. Because of the long duration the energy released during exercise is quite large. Experiments on humans have shown that both fibre types are active during this kind of exercise (4), and type II fibres with a high mechanical efficiency are therefore of great importance. Most *in vitor* experiments have been carried out on muscles from animals like frog, rate, mouse, guinea pig, rabbit and cat. The type II muscles of these animals are active only during short bursts of intense activity. Because of the short duration this activity requires little energy, and a low mechanical efficiency in type II fibres may therefore be no drawback for these animals. Hence, evolution may have resulted in differences between species concerning the efficiency of type II fibres during dynamic work.

DIFFERENT STIMULATION PATTERNS

When muscles are stimulated *in vitro*, the muscles are usually stimulated for one second or more using a constant stimulation frequency. During dynamic exercise like running or bicycling the activity is cyclic, and during each cycle a muscle is usually active for only 0.1–0.3 s. The natural stimulation of the muscle during these periods is initiated with a duplet or triplet at very high frequency. There is thus a striking difference between *in vivo* and *in vitro* experiments in the way muscles are used. I am not aware of any evidence showing that this difference has consequences for the efficiency, but the possibility cannot be ruled out.

In conclusion, the cause of conflicting data on the mechanical efficiency of different muscle fibres is not solved, and further studies are needed.

References

1. Gibbs, C. L., Gibson, R., 1972, *Am. J. Physiol.*, **223**, 864–871.
2. Wendt, I. R., Gibbs, C. L., 1973, *Am. J. Physiol.*, **224**, 1081–1086.
3. Vøllestad, N. K., Blom, P., 1985, *Acta Physiol. Scand.*, **125**, 395–405.
4. Vøllestad, N. K., Vaage, O., Hermansen, L., 1984, *Acta Physiol. Scand.*, **122**, 433–441.
5. Medbø, J. I., Mohn, A.-C., Tabata, I., Bahr, R., Vaage, O., Sejersted, O. M., 1988, *J. Appl. Physiol.*, **64**, 50–60.
6. Medbø, J. I., Tabata, I., 1989, *J. Appl. Physiol.*, **67**, 1881–1886.
7. Medbø, J. I., Tabata, I., 1991, *Acta Physiol. Scand.*, submitted.
8. Bosco, C., Komi, P. V., Sinkkonen, K., 1980, *Scand. J. Sports Sci.*, **2**, 47–51.

SECTION SEVEN

Contractile Proteins

49
Nucleotide-induced Changes in Domain Structure of Myosin Subfragment 1

D. I. LEVITSKY*, N. V. KHVOROV*, V. L. SHNYROV**, A. E. BUKATINA**, N. S. VEDENKINA**, E. A. PERMYAKOV** AND B. F. POGLAZOV*

** A. N. Bach Institute of Biochemistry, USSR Academy of Sciences, Leninsky prospect 33, Moscow 117071, USSR*
*** Institute of Biological Physics, USSR Academy of Sciences, Pushchino, Moscow region 142292, USSR*

Introduction

Myosin subfragment 1 (S1) is an isolated myosin head, i.e. the segment of myosin molecule containing the sites responsible for ATPase activity and actin binding. In recent years the problem of elucidating the domain structure in the S1 molecule has attracted considerable research activity. This is a very important problem since interdomain interactions in myosin head might play an essential role in the force generation during muscle contraction. At present there are data indicating the existence of separate structural domans in S1. These are data on electron microscopy of myosin heads (1) and X-ray diffraction on crystallized S1 (2) and also data on segmental motion of S1 in solution (3). S1 fragments 25, 50 and 20 kDa obtained by cleavage of S1 heavy chain by trypsin and many other proteinases (4, 5) are also sometimes called S1 'domains'. The data on selective denaturation of 50 kDa segment suggest that the S1 molecule is composed of at least two domains with different stabilities (6, 7).

One of us has already (8) proposed the 'successive annealing' method as a way of revealing separate structural domains in a protein molecule (i.e. the parts whose structures fold and unfold co-operatively and independently from each other) by means of differential scanning microcalorimetry. Recently we applied this method for a study of S1 structure and revealed three domains in

Muscle and Motility Vol. 2. Proceedings of XIXth European Conference in Brussels
© Intercept Ltd, PO Box 716, Andover, Hampshire, SP10 1YG, UK

the S1 molecule (9, 10). It turned out that the domains do not correspond completely to the proteolytic fragments of the S1 heavy chain. For example, the most thermolabile domain was identified as a region in the N-terminal part of the 50 kDa segment (9).

The aim of the present work was to reveal the changes in S1 domain structure induced by nucleotide binding to S1. For this purpose we have studied the thermal unfolding and domain structure of S1 in the complexes of S1 with ADP, non-hydrolysable ATP analogue 5′-adenylyl imidodiphosphate (AMP-PNP), and with ADP in the presence of vanadate ion (V_i). These complexes are stable analogues of intermediate complexes S1-ADP, S1-ATP and S1-ADP-P_i, existing during the S1 steady-state ATPase reaction.

Materials and Methods

S1 from rabbit skeletal muscle myosin was prepared by digestion of myosin filaments with chymotrypsin (11). Concentration of S1 was determined spectrophotometrically using $E^{1\%}_{280\,nm} = 7.5$.

All temperature-dependent measurements were carried out in a solution of 10 mN Hepes, pH 7.3, containing 1 mM $MgCl_2$ and nucleotides in concentrations indicated below, at S1 concentrations from 0.5 to 2.0 mg/ml and a heating rate of 1 K/min.

Calorimetric measurements were carried out on a DASM-4 differential adiabatic scanning microcalorimeter (USSR) whose construction and principles of work are described elsewhere (12). The special construction of the 0.47 ml platinum-measuring cell enables artifacts caused by a thermally induced protein aggregation to a great extent to be avoided. Decomposition of the complex heat sorption curves into elementary components was done by means of the 'successive annealing' method as described earlier (8, 10).

Thermally induced aggregation of S1 and its complexes with nucleotides was detected by changes in apparent absorbance at 350 nm measured on a Cary-219 spectrophotometer (Varian). Tryptophan fluorescence of S1 was measured on a laboratory-built spectrofluorimeter with monochromatic excitation at 280.4 nm.

Results and Discussion

Fig. 1 shows calorimetric data on the thermally induced unfolding of S1 and its complexes with nucleotide-free S1 possesses a single peak with the maximum at 47.2°C (Fig. 1A). The binding of Mg-ADP to S1 by itself practically does not influence the temperature of the thermal transition but increases its sharpness (Fig. 1B). At the same time, the binding of Mg-AMP-PNP to S1 shifts the heat sorption curve for S1 towards higher temperatures (up to 53.2°C) and essentially changes its shape (Fig. 1C). The binding of Mg-ADP and vanadate ion (V_i) induces an effect similar in character but more pronounced (the maximum of the peak shifts up to 56.1°C and the transition becomes more distinct) (Fig. 1D).

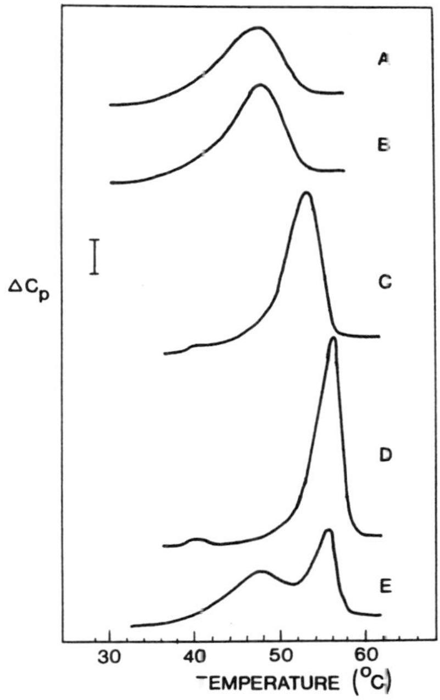

Fig. 1 Temperature dependences of excess heat capacity (Δc_p) for S1 (17.5 μM) in the absence of added nucleotides (A) and in the presence of 0.25 mM ADP (B); 0.1 mM AMP-PNP (C); 0.25 mM ADP, 0.25 mM V_i (D); 10 μM ADP, 10 μM V_i (E). Vertical bar corresponds to 500 J/K.kg. See Materials and Methods for other conditions.

Due to the high affinity of S1 to ADP and V_i such change occurs even at equimolar concentrations of S1, ADP and V_i. Fig. 1E shows the results of the experiment when the concentrations of ADP and V_i added are insufficient for complete saturation of the S1 binding sites. In this case, when only 50–60% of S1 molecules are able to form the complex with ADP and V_i, two heat sorption peaks are found, one of which is characteristic for nucleotide-free S1 and the other one for the S1-ADP-V_i complex. The complete absence of any intermediate transitions suggests that all the S1 molecules are in one of the two possible conformational states. Similar coexistence of two heat sorption peaks may be achieved as a result of ultraviolet irradiation (above 320 nm) of the S1-ADP-V_i complex, which is known to cause a rapid release of trapped Mg-ADP and V_i from S1 (13). So, the transition of the S1 molecule to a completely different conformational state, which is characteristic for the S1-ADP-V_i complex, is fully reversible.

The increase of S1 thermostability in the complexes S1-AMP-PNP and S1-ADP-V_i is confirmed by results of turbidity and tryptophan fluorescence measurements. For example, the termperatures of the half-maximal increase of turbidity are 40.5, 47.0 and 49.0°C for S1, S1-AMP-PNP and S1-ADP-V_i, respectively. In the case of S1-AMP-PNP and S1-ADP-V_i the blue shift of the

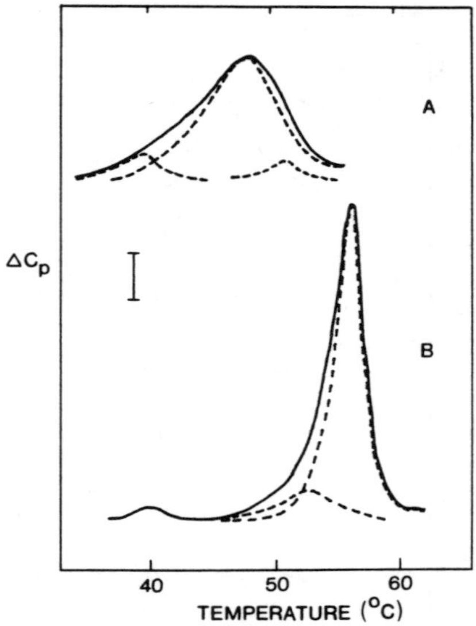

Fig. 2 Decomposition by the 'successive annealing' method of the total heat sorption curves (——) for nucleotide-free S1 (A) and S1-ADP-V_i (B) into elementary peaks (- - -) corresponding to the melting of separate domains. Vertical bar corresponds to 500 J/K.kg. See the text for details.

tryptophan fluorescence spectrum position occurs at higher temperatures than in the case of S1 and S1-ADP (data not shown).

Figure 2A shows results of a decomposition of the total heat sorption curve of S1 into elementary peaks. Each peak corresponds to the melting of a separate co-operative region (domain) in the S1 molecule. It is clearly seen that S1 contains three such domains which melt with maxima at 39.0, 47.2 and 50.8°C. S1-ADP has similar three-domain structure. On the other hand, S1 molecule possesses absolutely different domain structure in the S1-AMP-PNP and S1-ADP-V_i complexes. For example, in the case of S1-ADP-V_i the total heat sorption curve is decomposed into the components with maxima at 52.4 and 56.2°C (Fig. 2B). Quantitative analysis of these data, including the calculation of temperature dependences of free energies of stabilization for the experimentally revealed domains, was carried out. The results of this analysis show that the domain structure of S1 in the complexes of S1-AMP-PNP and S1-ADP-V_i is similar and radically differs from that of nucleotide-free S1 and S1 in the S1-ADP complex.

So, we can definitively subdivide all the S1 samples studied into two classes with absolutely different domain structures. The first class comprises nucleotide-free S1 and S1-ADP, while the second one includes S1-AMP-PNP and S1-ADP-V_i. Since S1-AMP-PNP and S1-ADP-V_i are stable analogues of the intermediates of the ATPase reaction, S1-ATP and S1-ADP-P_i, we can conclude that the domain structure of S1 in these intermediates is absolutely different from that of nucleotide-free S1 or S1-ADP.

S1 cross-linked by N,N′-*p*-phyenylenedimaleimide (*p*PDM) in the presence of Mg-ADP is also often used as a stable analogue of the S1-ADP-P_i. Surprisingly, it turned out that the domain structure of *p*PDM-cross-linked S1 almost does not differ from that of the nucleotide-free S1 (data not shown). It means that *p*PDM-cross-linked S1, in contrast to S1-AMP-PNP and S1-ADP-V_i, can not be considered as a structural analogue of the intermediate complexes S1-ATP and S1-ADP-P_i.

In conclusion it should be noted that the data obtained are direct evidences of the existence of S1 in two main conformational states during the ATP hydrolysis, one of which is realized in nucleotide-free S1 and in S1-ADP and another one is present in S1-ATP and S1-ADP-P_i. At the same time S1-ADP slightly differs from S1, and S1-ATP from S1-ADP-P_i, but such relatively small differences are not comparable with the global differences between the main conformation states.

References

1. Walker, M. L., Trinick, J. A., 1988, *J. Muscl. Res. Cell Motil.*, **9**, 359–366.
2. Winkelmann, D. A., Mekeel, H., Rayment, I., 1985, *J. Mol. Biol.*, **181**, 487–501.
3. Highsmith, S., Eden, D., 1986, *Biochemistry*, **25**, 2237–2242.
4. Applegate, D., Reisler, E., 1983, *Proc. Natl. Acad. Sci. USA*, **80**, 7109–7112.
5. Mornet, D., Ue, K., Morales, M. F., 1984, *Proc. Natl. Acad. Sci. USA*, **81**, 736–739.
6. Burke, M., Sivaramakrishnan, M., 1986, *J. Biol. Chem.*, **261**, 12330–12336.
7. Burke, M., Zaager, S., Bliss, J., 1987, *Biochemistry*, **26**, 1492–1496.
8. Shnyrov, V. L., Zhadan, G. G., Akoev, I. G., 1984, *Bioelectromagnetics*, **5**, 411–418.
9. Levitsky, D. I., Khvorov, N. V., Shnyrov, V. L., Vedenkina, N. S., Permyakov, E. A., Poglazov, B. G., 1990, *FEBS Lett.*, **264**, 176–178.
10. Shnyrov, V. L., Levitsky, D. I., Vedenkina, N. S., Nikolaeva, O. P., Khvorov, N. V., Permyakov, E. A., Poglazov, B. F., 1989, *Dokl. Acad. Nauk USSR*, **304**, 1497–1499 (in Russian).
11. Weeds, A. G., Taylor, R. S., 1975, *Nature (Lond.)*, **257**, 54–56.
12. Privalov, P. L., Potekhin, S. A., 1986, *Methods Enzymol.*, **131**, 4–51.
13. Grammer, J. C., Cremo, C. R., Yount, R. G., 1988, *Biochemistry*, **27**, 8408–8415.

50
Divalent Metal Ions Mediate the Interaction of the Regulatory Light Chain with the Heavy Chain in Skeletal Muscle Myosin

M. C. SCHAUB, P. A. J. HUBER AND A. JACOBER

Institute of Pharmacology, University of Zürich, Gloriastrasse, 32, CH-8006 Zürich, Switzerland

Introduction

All double-headed myosin species contain a pair each of two types of light chains (LC), the alkali-LC (LC1/A1 and LC3/A2) and the regulatory-LC (RLC/P-LC/LC2). In smooth muscle and in nonmuscle cells the RLC is directly involved in regulation of contraction. The homologue RLC of rabbit fast skeletal muscle myosin comprises 169 amino acid residues giving it a molecular mass of 18 950 Da (1, 2). One RLC is nonconvalently associated with each myosin head near the head-rod junction (3) and its function in skeletal muscle is far from clear at present (4, 5). It belongs to a superfamily of intracellular Ca^{2+}-binding proteins whose common feature is the presence of domains which are structurally related to the EF-hand conformation in carp parvalbumin (6, 7). It is able to bind a divalent metal ion (Me^{2+}) by its first EF-domain and it can be phosphorylated reversibly at its Ser-16. We report here on the divalent metal ion dependent reactivity characteristics of the two unique Cys-128 and Cys-157 of the RLC in intact myosin, actomyosin and papain subframent-1 (PAP-S1) which retains its RLC, and that the formation of an intrachain disulphide bond in the RLC impairs its interaction with the myosin heavy chain (HC).

Muscle and Motility Vol. 2. Proceedings of XIXth European Conference in Brussels
© Intercept Ltd, PO Box 716, Andover, Hampshire, SP10 1YG, UK

Materials and Methods

Myosin, actomyosin, papain subfragment-1 retaining its RLC (PAP-S1) and isolated RLC were prepared from rabbit fast skeletal muscles (8). Treatment of the proteins with 5,5′-dithiobis(2-nitrobenzoic acid) (DTNB) and with N-ethylmaleimide (NEM) and subsequent electrophoresis on 7.5% polyacrylamide gels in SDS (SDS-PAGE) without prior reduction by dithiothreitol (DTT) have been described in (8). UV difference spectroscopy of isolated RLC was performed as in (9) and free Ca^{2+} and Mg^{2+} ion concentrations were calculated as in (10).

Results

Treatment of myosin with DTNB in concentrations ranging from 0–2 mM leads to the formation of 5-thio-2-nitrobenzoic acid (TNB) mixed disulphides as well as to protein disulphide bonds. The protein disulphide bonds may be comprised within one peptide subunit inducing an intrachain loop, or else, may lead to crosslinking two different chains together. Both types of protein disulphide formations may be visualized in SDS-PAGE since complexes of two protein chains exhibit a larger apparent molecular mass when crosslinked together while, on the other hand, a single chain with an intrachain loop will migrate with a lower apparent molecular mass. In the latter case the intrachain loop prevents the protein chain from unfolding completely in SDS. To preserve these migration characteristics the protein had to be electrophoresed without prior reduction by DTT. When myosin was subjected to SDS after treatment

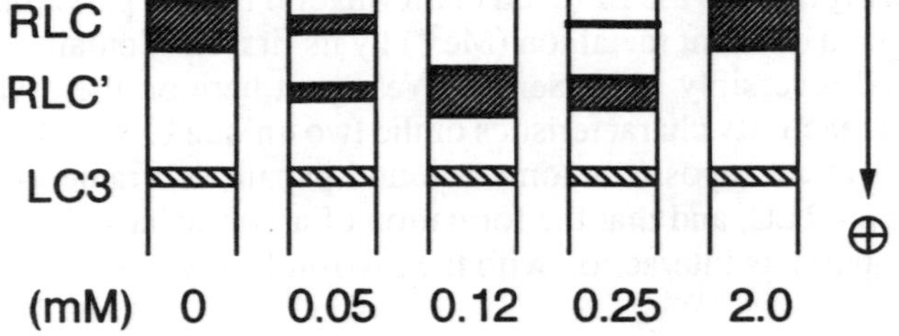

Fig. 1 Scheme of SDS-PAGE in the absence of DTT after incubation of myosin with DTNB in the presence of 5 mM EDTA at pH 7.6 followed by treatment with 5 mM NEM. DTNB incubation for 20 min at 25°C with concentrations as indicated.

Table 1 Modification of the two unique Cys-128 and Cys-157 in the RLC by DTNB treatment of myosin in the presence of EDTA, followed by radioactive incorporation of N-ethyl[1–14C]succinimide. % RLC and % RLC′ were assessed by densitometry from SDS-PAGE and radioactivity was counted in the respective protein bands. RLC′ denotes the faster migrating RLC species containing an intrachain disulphide bond.

Concentration of DTNB during incubation (μM)	% RLC protein band	% RLC′ protein band	% radioact. recovered in the RLC	% radioact. recovered in the RLC′	Free thiols per RLC left after DTNB treatment
0	100	0	100	—	2.0
40	51	49	40	6	0.9
120	2	98	3	2	0.1
400	39	61	3	0	0.1
2000	98	2	0	0	0

with DTNB, additional disulphide exchange reactions occurred, however, in this milieu. In order to prevent further rearrangements the still free thiol groups were blocked by treatment with a large excess of NEM (2–5 mM) prior to SDS-PAGE (8). By this technique it could be shown that treatment of myosin with increasing DTNB concentrations in the presence of 5 mM EDTA induces the RLC to be transformed completely into the faster migrating RLC′ species which with higher DTNB concentrations (above 200 μM) reverts back to the original position (Fig. 1). In such an experiment the thiol groups in the RLC after treatment of myosin with DTNB were probed by an additional subsequent treatment with radioactive [$^{1\text{-}14}$C]-NEM. The results in Table 1 indicate: (i) the native RLC contains two free thiols (Cys-128 and Cys-157); (ii) the faster migrating RLC′ has blocked both thiols and therefore no radioactivity is incorporated; and (iii) the two thiols of the RLC after treatment with higher DTNB concentrations are equally blocked and do not incorporate radioactivity. The latter RLC has its two thiols blocked by a TNB group each and comigrates in SDS-PAGE together with the native RLC. The RLC′ species, however, represents an RLC with an intrachain disulphide bond between its two thiols forming a loop with 28 amino acid residues and migrating with a lower apparent molecular mass. Reduction of the RLC′ with DTT prior to SDS-PAGE restores its original migration position (8).

The presence of Mg^{2+} or Ca^{2+} prevents the formation of the intrachain disulphide bond in the RLC in myosin and both thiols become blocked separately by a TNB group as shown by subsequent probing with radioactive [$^{1\text{-}14}$C]-NEM (8). Incubation of myosin with DTNB in the presence of various amounts of Ca^{2+} or Mg^{2+} in an EDTA buffer system allowed to determine the free Me^{2+} concentration causing 50% inhibition of the intrachain loop formation as assessed densitometrically from the extent of RLC′ species occurring in SDS-PAGE. The corresponding apparent affinity constants are given in Table 2. These are close to those we determined earlier in direct binding studies of Ca^{2+} and Mg^{2+} to intact myosin (10). It came as a surprise that the formation of the intrachain loop in the RLC and its prevention by Me^{2+} was also observed after treatment of actomyosin or PAP-S1 with DTNB. The apparent affinities of Ca^{2+} and Mg^{2+}

Table 2 Apparent affinity constants for binding of divalent metal ions to the RLC in intact myosin, actomyosin and papain-subfragment-1 (PAP-S1). The extent of intrachain disulphide formation in the RLC by incubation with DTNB with various concentrations of free metal ions was assessed densitometrically from the RLC′ species in SDS-PAGE (affinity constant *K* is given as reciprocal molarity).

Metal ion	Myosin	Actomyosin	PAP-S1
Calcium	8.9×10^7	1.2×10^8	2.9×10^7
Magnesium	2.5×10^6	3.6×10^6	1.6×10^6

to these two protein preparations were very similar to those found with myosin (Table 2). The tight interaction with actin in rigor thus does not affect the thiol group reactivity of the RLC in myosin. It is well known, however, that rigor interaction with actin greatly affects the reactivity of the so-called thiol-1 (Cys-707) and thiol-2 (Cys-697) groups in the HC of the myosin head portion (11). On the other hand, the thiol group reactivity of the RLC in isolated PAP-S1 exhibits the same characteristics as in myosin and actomyosin. In addition, the apparent affinities for Ca^{2+} and Mg^{2+} are almost identical to those in the intact molecular structure of myosin. The affinity for Ca^{2+} is in all three proteins 18–36 times higher than for Mg^{2+}. It has to be emphasized here the DTNB at moderate concentrations always induces an intrachain loop in isolated RLC in the presence or absence of Me^{2+} (8). The degree of phosphorylation of the RLC was found to have no influence on the thiol group reactivites.

Preliminary experiments indicated that the RLC with an intrachain disulphide bond can be removed from myosin or from actomyosin by repeated washings with low ionic strength buffer containing Ca^{2+} or Mg^{2+}. Unmodified RLC could not be removed by this procedure. Isolated RLC into which an intrachain loop had been introduced did not rebind to actomyosin that was depleted of its endogenous RLC. Native isolated RLC was found to rebind stoichiometrically.

Since DTNB induces an intrachaindisulphidebond in the RLC of intact myosin only in the absence of Me^{2+}, and since the looped RLC could the be removed from myosin, one wonders whether the looped RLC is able to bind Me^{2+} at all. UV difference spectroscopy indicated that Ca^{2+} affected the signal at 285 nm wave length both in the isolated native RLC and in the DTNB-looped RLC′ (Fig. 2). Ca^{2+} neither affected the spectral signal of ribonuclease which was used as control protein here, nor had it any effect on a solution containing free amino acids in the same concentration and proportion as in the RLC protein chain. The apparent affinity constants were deduced to be $2.9 \times 10^4 M^{-1}$ for the native RLC and $1.6 \times 10^4 M^{-1}$ for the DTNB-looped RLC′. The affinity of Ca^{2+} for the isolated native RLC is 3000 times lower than that for intact myosin (compare Table 2). This low affinity of Ca^{2+} to the isolated RLC is in the same range as found earlier by various techniques (12, 13, 14). The extent of the signal Ca^{2+} induced in the DTNB-looped RLC′ was consistently one-third of that with the native RLC.

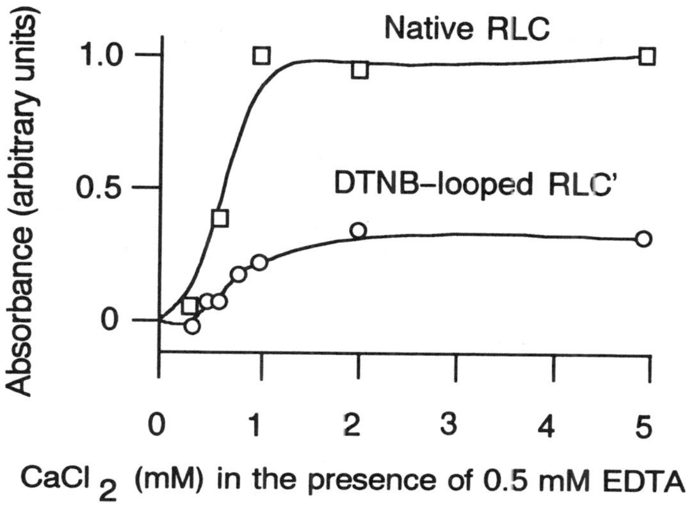

Fig. 2 UV difference spectroscopy at 285 nm of native RLC and of RLC′ containing an intrachain disulphide bond between its two unique Cys (DTNB-looped RLC′) with increasing concentrations of $CaCl_2$ in the presence of 5 mM EDTA. The arbitrary value of 1.0 corresponds to 0.023 absorbance units. Each point represents the mean of 6 determinations stemming from two different preparations.

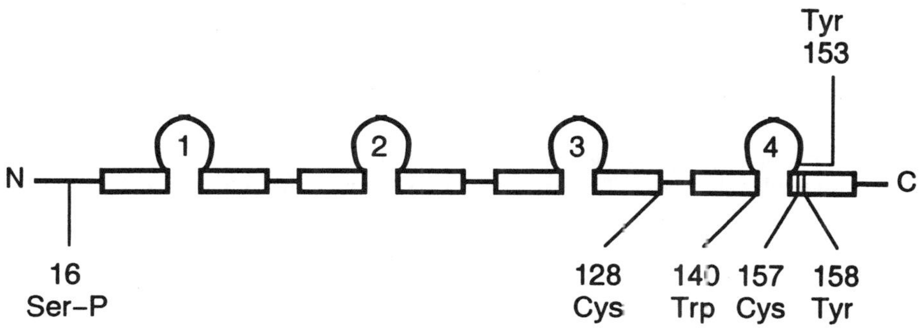

Fig. 3 Scheme of the RLC indicating the location of the two unique Cys-128 and Cys-157, the two unique Tyr-153 and Tyr-158, the only Trp-140 and the phosphorylatable Ser-16. The EF-domains are numbered; only the first one is able to bind divalent metal ions.

Discussion

The divalent metal ions (Me^{2+}) are thought to bind the first EF-domain (6) which is running from Ile-27 through Gly-58 including the two α-helices flanking the binding site (7). It is located near the N-terminus. Binding of divalent Me^{2+} to this domain in the isolated RLC does not affect the induction of an intrachain disulphide bond between the two unique Cys-128 and Cys-157 which are both located in the c-terminal quarter of the RLC (Fig. 3). However, binding of Ca^{2+} to the native RLC affects the UV signal at 285 nm while Phe does hardly so at this wavelength. The only Trp-140 and the two unique Tyr-153 and Tyr-158 are all located within the region of the two Cys near the c-terminus. The UV signal originates probably from some, or from all these three, aromatic amino acids and, moreover, its response is reduced when Ca^{2+} binds to the DTNB-looped RLC′ where the two Cys in the C-terminus are firmly crosslinked. Taken together, these findings indicate that an exchange of information takes place between the N-terminus and the C-terminus in the isolated RLC. In intact myosin the inducibility of the intrachain disulphide bond in the RLC by DTNB becomes strongly dependent on Me^{2+}. The Me^{2+} is preventing the formation of the intrachain disulphide bond. The interaction with the myosin HC thus enhances the flow of information between the two terminal portions of the RLC. This could be compatible with a compact, bent conformation of the RLC in the myosin head region with the two terminal RLC portions coming into close proximity of each other (3). Furthermore, the intrachain crosslinking of its two Cys prevents the RLC from associating properly with the myosin HC. This is in agreement with the notion that the C-terminal portion of the RLC is responsible for its binding to myosin. (10).

Acknowledgement

Supported by the Swiss National Science Foundation Grant 3.075.87

References

1. Collins, J. H., 1976, *Nature*, **259**, 699–700.
2. Matsuda, G., Maita, T., Suzuyama, Y., Setoguchi, M., Umegane, T., 1977, *J. Biochem.*, **81**, 809–811.
3. Vibert, P. J., Cohen, C., 1988, *J. Muscle Res. Cell Mot.*, **9**, 296–305.
4. England, P. J., 1984, *J. Mol. Cell. Cardiol.*, **16**, 591–594.
5. Hofmann, P. A., Metzger, J. M., Greaser, M. L., Moss, R. L., 1990, *J. Gen. Physiol.*, **95**, 477–498.
6. Kretsinger, R. H., 1980, *CRC Crit. Rev. Biochem.*, **8**, 119–174.
7. Kretsinger, R. H., Nockolds, C. E., 1973, *J. Biol. Chem.*, **248**, 3313–3326.
8. Huber, P. A. J., Brunner, U. T., Schaub, M. C., 1989, *Biochemistry*, **28**, 9116–9123.

10. Mitchell, E. J., Jakes, R., Kendrick-Jones, J., 1986, *Eur. J. Biochem.*, **161**, 25–35.

10. Watterson, J. G., Kohler, L., Schaub, M. C., 1979, *J. Biol. Chem.*, **254**, 6470–6477.

11. Watterson, J. G., Schaub, M. C., 1973, *Cold Spring Harbor Symp. Quant. Biol.* **37**, 153–156.
12. Alexis, M. N., Gratzer, W. B., 1978, *Biochemistry*, **17**, 2319–2325.
13. Werber, M. M., 1978, *Experientia*, **34**, 575–576.

51
Effect of Myosin Light Chain Phosphorylation on Myosin ATPase Activity as Examined by Native Pyrophosphate Polyacrylamide Gel Electrophoresis

H. TAKANO-OHMURO* AND K. KOHAMA**

**The Tokyo Metropolitan Institute of Medical Science, Honkomagome, 3-18-22, Bunkyo-ku, Tokyo 113, Japan*
***Department of Pharmacology, Gunma University School of Medicine, Maebashi, Gunma 371, Japan*

Introduction

It is well established that regulatory light chain (RLC) phosphorylation of vertebrate smooth muscle myosin and non-muscle myosin modulates actin-activated Mg^{2+}-ATPase activities (see ref. 1 for review).

We have shown that native pyrophosphate polyacrylamide gel electrophoresis (PPi PAGE) (2, 3) is an effective means by which to analyze the myosin phosphorylation state of smooth muscle both *in vitro* and *in vivo*:

Muscle and Motility Vol. 2. Proceedings of XIXth European Conference in Brussels
© Intercept Ltd, PO Box 716, Andover, Hampshire, SP10 1YG, UK

1. smooth muscle myosin phosphorylated at its RLC with myosin light chain kinase (MLCK) increases its mobility on PPi PAGE according to the degree of its RLC phosphorylation (4, 5).
2. Smooth muscle myosin phosphorylated at its RLC with protein kinase C (PKC) does not show this mobility increase on PPi PAGE (6).
3. Phosphorylation state of the myosin in contracting smooth muscle is analyzed by PPi PAGE without myosin purification (5).
4. Phosphorylation does not affect the mobility of striated muscle myosin on PPi PAGE (4).

These results lead us to speculate that RLC phosphorylation, which is effective in enhancing actin-activated myosin Mg^{2+}-ATPase activity, is closely associated with the increase in myosin mobility on PPi PAGE.

In this paper, we present further evidence to support this hypothesis using various kinds of myosins and kinases.

Methods

PREPARATION OF PROTEINS

Myosin from chicken brain was prepared as described by Barylko and Sobieszek (7). Myosin from human platelet was prepared according to the method of Adelstein and Conti. (8). Liver cell myosin and brush border myosins were prepared according to the method of Citi and Kendrick-Jones (9), and Ueno and Sekine (10), respectively. Smooth muscle myosins from adult and embryonic chicken gizzards were prepared by the Ebashi method (11) with minor modifications (4). MLCK, calmodulin and PKC were prepared as described previously (4, 6). Insulin receptor kinase was prepared as described previously (12).

PHOSPHORYLATION OF MYOSINS

Myosins were phosphorylated with MLCK, PKC or insulin receptor kinase as described previously (4, 6, 12). A control experiment without the kinase was carried out at the same time.

GEL ELECTROPHORESIS

Native PPi PAGE was carried out as previously reported (4) without dissociating myosin subunits. Phosphorylation degree of RLC in each myosin band was demonstrated by excising myosin bands from PPi PAGE gels followed by urea-glycerol PAGE as previously described (4). In some experiments the excised bands were subjected to SDS PAGE, and phosphorylation of RLC was detected by autoradiography as previously described (4).

Results

ANALYSIS OF NON-MUSCLE MYOSINS PHOSPHORYLATED WITH MLCK BY PPi PAGE

Myosins from non-muscle cells, i.e. brain, platelet, liver, and brush border, were phosphorylated by MLCK at their RLC. Then they were subiected to PPi PAGE. We found that the phosphorylated myosins migrated faster than did each corresponding control unphosphorylated myosin. The results conformed with those from the experiment on smooth muscle myosin. As shown in Fig.1, bands of control non-muscle myosin correspond to GM, those of phosphorylated non-muscle myosin to GUP1 or GMP2.

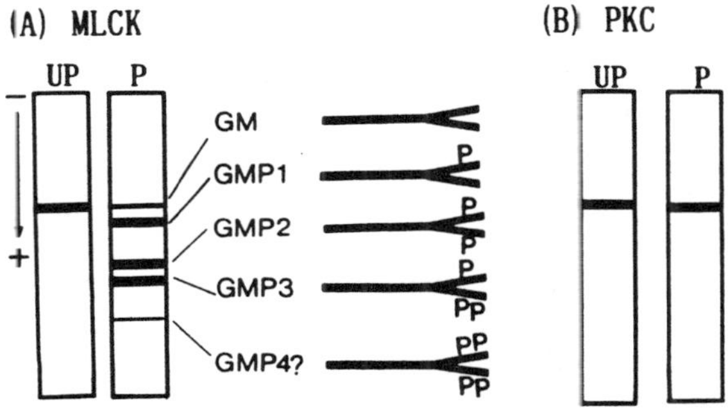

Fig. 1 Schema of native pyrophosphate polyacrylamide gel electrophoresis (PPi PAGE) analysis of gizzard smooth muscle myosin light chain kinase (MLCK) or protein kinase C (PKC).

(A) Gizzard myosin phosphorylated at its regulatory light chain (RLC) in various degrees with MLCK is separated into five isolated bands, tentatively called GM, GMP1, GMP2, GMP3 and GMP4 on PPI PAGE. The phosphorylation degree of RLC in each band was examined by urea-glycerol PAGE as decribed in Methods. Their mobilities on PPi PAGE and degree of RLC phosphorylation are: myosin with 2 unphosphorylated RLC (GM) < myosin with 1 unphosphorylated and 1 mono-phophorylated RLC (GMP1) < myosin with 2 mono-phosphorylated RLC (GMP2) < myosin with 1 mono-phosphorylated and 1 di-phosphorylated RLC (GMP3) < myosin with 2 di-phosphorylated RLC (GMP4). (B) Gizzard myosin phosphorylated at its RLC with PKC does not show any mobility change.

It must be noted that actin-activated Mg^{2+}-ATPase activity of these myosins was enhanced by phosphorylation with MLCK (ref. 7 for brain myosin, ref. 8 for platelet myosin, ref. 9 for brush border myosin, ref. 10 for liver cell myosin). Thus the increase in the mobility of non-muscle myosins is compatible with our hyphothesis as stated in the introduction.

ANALYSIS OF EMBRYONIC GIZZARD MYOSIN PHOSPHORYLATED WITH PKC OR MLCK BY PPi PAGE

Phosphorylation of RLC with PKC activates actin-activated myosin Mg^{2+}-ATPase activity of 12-day-old embryonic gizzard myosin (EGM) (13), but it does not activate that of adult gizzard myosin (14, 15).

Twelve-day-old EGM phosphorylated with PKC at its RLC was analyzed by PPi PAGE. The mobility of 12-day-old EGM increased but that of adult gizzard did not, while both myosins phosphorylated with MLCK showed increased mobility. Phosphorylation of both myosins with MLCK activated the ATPase activity (1, 13).

These results strongly suggest that phosphorylation which enhances actin-activated myosin Mg^{2+}-ATPase activity is associated with the increase in myosin mobility on PPi PAGE.

ANALYSIS OF GIZZARD SMOOTH MUSCLE MYOSIN PHOSPHORYLATED WITH INSULIN RECEPTOR KINASE BY PPi PAGE

Insulin receptor kinase could phosphorylate gizzard myosin at its RLC in the presence of poly [L-lysine] as described previously (12). When gizzard myosin phosphorylated with insulin receptor kinase was analyzed by PPi PAGE, the phosphorylated myosin showed the same mobility as the unphosphorylated one. As we previously reported, phosphorylation with insulin receptor kinase did not affect actin-activated Mg^{2+}-ATPase activity of gizzard myosin. This result also supports our hypothesis.

Discussion

The mobility changes of myosins phosphorylated with MLCK, PKC or insulin receptor kinase, as well as the effects of the phosphorylation on actin-activated Mg^{2+}-ATPase activities are summarized in Table 1. These findings strongly support our hypothesis that phosphorylation, which is effective in enhancing actin-activated myosin Mg^{2+}-ATPase activity, is associated with the increase in myosin mobility on PPi PAGE.

Striated muscle myosins and actin-activated Mg^{2+}-ATPases, the activities of which are not modulated by phosphorylation, show no mobility change on PPi PAGE. In contrast, myosins, the ATPase activities of which are increased by phosphorylation (i.e. smooth muscle myosins and non-muscle myosins phosphorylated with MLCK, and 12-day-old embryonic gizzard myosin (emb (12d)) phosphorylated with PKC) show an increase in the mobility on PPi PAGE. Phosphorylation with PKC or insulin receptor kinase shows no enhancement on the ATPase activity of gizzard smooth muscle myosin nor mobility change of the myosin on PPi PAGE. In smooth muscle myosin sequentially phosphorylated with MLCK followed by PKC (MLCK → PKC), the fast-moving band due to phosphorylation with MLCK reverts to the slowly migrat-

Table 1 Effect of regulatory light chain phosphorylation on the mobility of myosin in PPi PAGE and on actin-activated myosin Mg^{2+}-ATPase activity.

Myosins / Kinases	Striated muscle		Smooth muscle			Non-muscle			
			Gizzard		Aorta		Brush		
	Pectoralis	Heart	Ad.	Emb (12d)		Platelet	border	Brain	Liver
MLCK	→	→	↑	↑	↑	↑	↑	↑	↑
(ATPase)	(⇒)	(⇒)	(⇑)	(⇑)	(⇑)	(⇑)	(⇑)	(⇑)	(⇑)
PKC			→	↑					
(ATPase)			(⇒)	(⇑)					
MLCK → PKC			↓						
(ATPase)			(⇓)						
PKC → MLCK			→						
(ATPase)			(⇒)						
Insulin receptor			→						
(ATPase)			(⇒)						

Black arrows indicate the mobility increase (↓), no change (→), or decrease (↑) on PPi PAGE. White arrows indicate the increase (⇑), no change (⇒), or decrease (⇓) of actin-activated myosin Mg^{2+}-ATPase activity (ATPase).

ing unphosphorylated band The ATPase activity of phosphorylated gizzard myosin with MLCK is reduced by additional phosphorylation with PKC.

Conformational change such as 10S to 6S was not observed between unphosphorylated gizzard myosin and the phosphorylated myosin with MLCK in the electrode buffer for PPi PAGE (4). It must be noted that the mobility increase could be detected in the case of H-meromyosin (4). Hence, it is possible that the mobility increase in phosphorylated myosin reflects some conformational change in the myosin head region.

As described previously, light chain composition changes during the development of gizzard smooth muscle. Twelve-day-old EGM contains a unique 23 kDa alkali light chain (L_{23}) (16, 17, 18) and an embryonic RLC (19) in addition to light chains expressed in adult gizzard myosin. Eighteen-day-old EGM contains adult type RLC and alkali light chain, and L_{23}. This myosin did not show increased mobility by phosphorylation with PKC on PPi PAGE, suggesting that L_{23} does not play a key role in the activation by PKC in 12-day-old EGM. The activation may possibly be caused by the presence of the embryonic RLC which is expressed in 12-day-old EGM not in 18-day-old EGM or adult gizzard myosin.

Acknowledgements

We thank Dr T. Ueno of Juntendo University and Dr K. Oishi of Meiji College of Pharmacy for providing us with liver cell myosin and protein kinase C.

This study was supported by grants from the Japanese Ministry of Education, Science and Culture and Uehara Memorial Foundation.

References

1. Kendrick-Jones, J., Scholey, J.M., 1981, J. *Muscl. Res. Cell Motil.*, **2**, 347–72.
2. d'Albis, A., Pantaloni, C., Bechet, J.-J., 1979, *Eur. J. Biochem.*, **99**, 261–72.
3. Hoh, J. F. Y., McGrath, R. A., White, R. I., 1976, *Biochem. J.*, **99**, 87–95.
4. Takano-Ohmuro, H., Kohama, K., 1986, *J. Biochem. (Tokyo)*, **100**, 259–262.
5. Takano-Ohmuro, H., Kohama, K., 1986, *J. Biochem. (Tokyo)*, **100**, 1681–1684.
6. Takano-Ohmuro, H, Kohama, K., 1987, *J. Biochem. (Tokyo)*, **102**, 971–974.
7. Barylko, B., Sobieszek, A. 1983, *EMBO J.*, **2**, 369–374.
8. Adelstein, R.C., Conti, M A., 1975, *Nature*, **256**, 597–598.
9. Citi, S., Kendrick-Jones, J., 1986, *J. Mol. Biol.*, **188**, 369–382.
10. Ueno, T., Sekine, T., 1987, *Biochem. Int.*, **15**, 1205–1212
11. Ebashi, S., 1976, *J. Biochem. (Tokyo)*, **79**, 229–231
12. Takano-Ohmuro, H., Kohama, K., Kathuria, S., Fujita-Yamaguchi, Y., 1989, *Proc. Japan Acad.*, **65**, 199–210.
13. de Laneroll, P., Nishikawa, M., 1988, *J. Biol. Chem.*, **263**, 9071–9074.
14. Nishikawa, M., Hidaka, H., Adelstein, R. S., 1983, *J. Biol. Chem.*, **258**, 14069–14072.
15. Nishikawa, M., Sellers, J. R., Adelstein, R. S., 1984, *J. Biol. Chem.*, **259**, 8808–8814.
16. Takano-Ohmuro, H., Obinata, T., Kawashima, M, Masaki, T., Tanaka, T., 1985, *J. Cell Biol.*, **100**, 2025–2030.
17. Takano-Ohmuro, H., Obinata, T., Mikawa, T., Masaki, T., 1983, *J. Biochem.* (Tokyo), **93**, 903–908.
18. Takano-Ohmuro, H., Ogasawara, Y., Obinata, T, 1989, *J. Biochem. (Tokyo)*, **102**, 1321–1327
19. Inoue, A., Yanagisawa, M., Takano-Ohmuro, H., Masaki, H., 1989, *Eur. J. Biochem.*, **183**, 645–651.

52
Expression of 'Cardiac-specific' Myosin Heavy Chain in Rabbit Cranial Muscles

JANTINE J. BREDMAN, WIM A. WEIJS AND ANTOON F.M. MOORMAN*

Department of Anatomy and Embryology; Academic Centre for Dentistry Amsterdam (ACTA) and Faculty of Medicine, University of Amsterdam, Meibergdreef 15; 1105 AZ Amsterdam, The Netherlands*

Summary

Rabbit cranial, trunk and limb muscles were immunohistochemically analysed with antibodies specific to 'cardiac' α, slow and fast myosin heavy chain isoforms. Cranial muscles, like masticatory, extraocular and hyoid muscle, express 'cardiac' α myosin, whereas in trunk and limb muscles the 'cardiac' a myosin is not found. The expression pattern might be determined by the developmental history of the muscles concerned because all 'cardiac' α myosin heavy chain containing muscles are found in the head area (note that the heart originates from the most cranial part of the embryo).

Introduction

The myosin heavy chain (MHC) is responsible for the calcium-dependent ATPase activity that, in turn, determines the velocity of shortening of muscle fibres (1). Using ATPase histochemistry mammalian limb muscle fibres can be characterized as type I (slow oxidative), type IIA (fast oxidative) and type IIB (fast glycolytic) and each fibre contains one or at the most two of these specific isoforms of myosin heavy chain (MHC) (2). This simple subdivision cannot

Muscle and Motility Vol. 2. Proceedings of XIXth European Conference in Brussels
© Intercept Ltd, PO Box 716, Andover, Hampshire, SP10 1YG, UK

directly be applied to the masticatory muscles. Their muscle fibres often contain mixtures of several MHCs some of them being embryonic or neonatal isoforms (3,4); others contain mixtures of 'cardiac' α with slow or fast MHC (this study and (5)). Furthermore, specific MHCs, that do not occur in limb muscles, have been found in masticatory muscles. Rowlerson *et al.* (6) found a 'superfast' MHC in primate and carnivores jaw muscles. We found 'cardiac' α myosin (or a cardiac-related myosin) in rabbit masticatory muscle (this study and (5)).

The observation that human and rabbit masticatory muscle abundantly express 'cardiac' α MHC (5) raises the question of whether the extracardiac expression of α myosin is unique for jaw muscles or that it is also present in other cranially located muscles. To answer this question we have analysed various cranial, trunk and limb muscles with three monoclonal antibodies that unambiguously distinguish slow (type I), fast (type IIA and/or B) and atrial MHC.

Methods

Samples of various muscles were taken from New Zealand rabbit as indicated in Table 1. The immunohistochemistry is fully described elsewhere (5,7) and can be summarized briefly as follows: the tissue specimens were fixed and further processed for embedding in Paraplast Plus (Monoject, Ireland). Monoclonal antibodies (Mabs) were prepared from purified myosins. Mab 249–5A4 was raised against adult human atrial myosin (7), Mab 169-ID5 against chicken-heart myosin (8) and Mab 340-3B5 was raised against rabbit anterior tibialis muscle (5). Western blot analysis has shown that all antibodies are specific for MHC (5,9). Immunohistochemical analysis on embryonic, foetal, adult heart and various adult skeletal limb muscles has shown that the Mabs 249-5A4, 169-ID5 and 340-3B5 are entirely specific for 'cardiac' α, slow and fast MHC respectively; hence, they are denoted as anti-α MHC, anti-slow MHC and anti-fast MHC, respectively. The indirect unconjugated immunoperoxidase technique (PAP-technique) was applied to detect the specific binding of the different antibodies.

Results

Table 1 shows a compendious survey of the muscles tested for the presence of 'cardiac' α MHC; some representative examples are shown in Figs. 1 and 2. In all tested limb and trunk muscles, anti-α MHC showed no reaction. In some cranially localized muscles, like the masticatory, extraocular, omohyoid and in heart muscles, an altogether different picture appeared. In these muscles anti-α MHC showed a positive staining reaction, whereas in other cranially located muscles, like the trapezius and sternocleidomastoid muscles, anti-α MHC showed no reaction. In Fig. 1(A, D) it is shown that in the trapezius and anterior tibialis muscle the extrafusal fibres did not react with anti-α MHC. But the intrafusal

Table 1 Comparison of 'cardiac' α MHC expression in rabbit muscles and origin of muscles connective tissue in birds.

Selected muscles	Anti-α MHC in rabbit muscles	Neural crest derived connective tissue
Extraocular muscle (III, IV, VI)		
six extraocular muscles	+	+
Trigeminal muscle (V)		
masseter[1], temporal[1], medial pterygoid[1], digastric[2], mylohyoid	+	+
Facial muscle (VII)		
buccinator[4]	+	?
stylohyoid	+	+
platysma	-	-
Vagal muscle (X)		
cricothyroid	+	+
oesophagus[5]	-	?
middle pharyngeal constrictor[5]	-	?
Accessorial muscle (XI)		
sternocleidomastoid, trapezius[3]	-	-
Hypoglossal muscle (XII)/ansa cervicalis		
intrinsic tongue, geniohyoid[2]	+	+
omohyoid[5]	+	?
Spinal muscle		
rhomboid	-	-
diaphragma[4]	-	?
Limb muscle		
biceps brachii[4], psoas, anterior tibialis, gastrocnemius, digitorum longus, soleus	-	-
heart muscle[5]	+	

1 = jaw closer in birds; 2 = jaw opener in birds; 3 = cucullaris muscle in birds; 4 = not present in birds; 5 = not investigated by Noden (13,14).

muscle spindle fibres showed a positive reaction (5, 10). Two types of fibres were found in the muscles not reacting with anti-α MHC. One type containing slow MHC and another type containing fast MHC (IIA and/or IIB). In Fig. 2(A, D) the masseter and extraocular muscles are shown as an example of positively reacting muscles. Five fibre types were found in these muscles: three types containing respectively exclusively slow, fast (IIA and/or IIB) or 'cardiac' α MHC, a type containing slow and 'cardiac' α MHC and a type containing fast (IIA and/or IIB) and 'cardiac' α MHC (see Table 2).

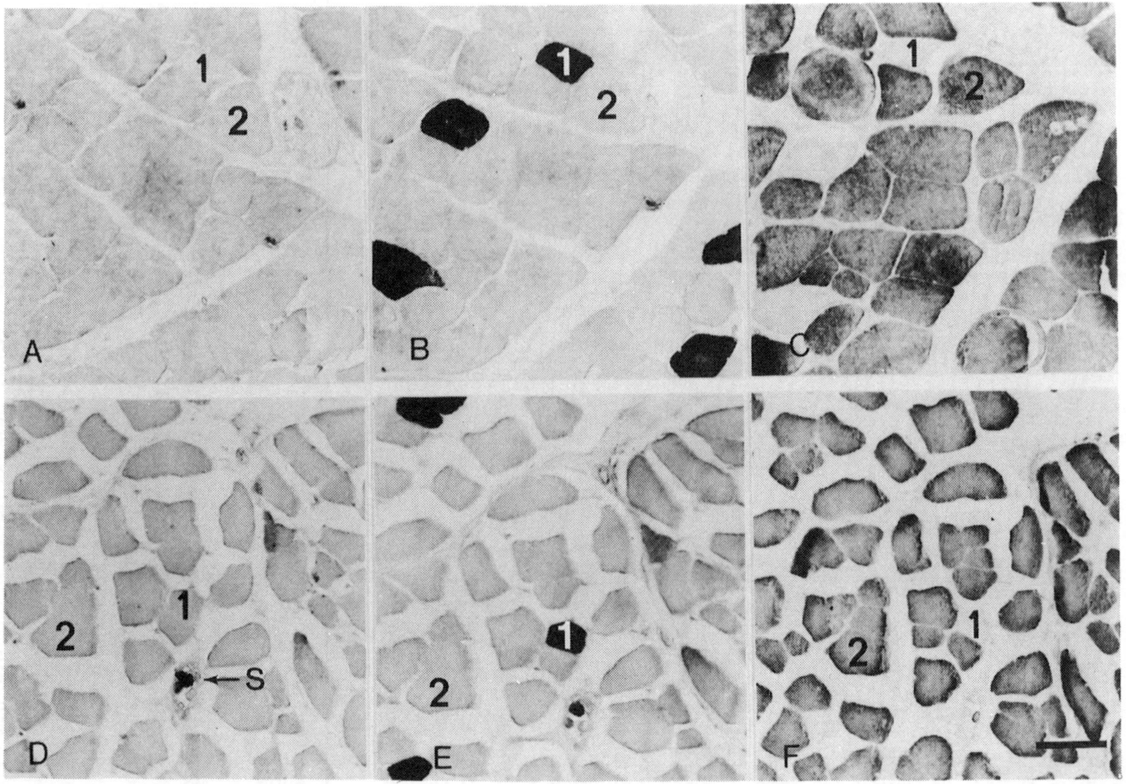

Fig. 1 Immunohistochemical analysis of serial sections of adult rabbit trapezius (A–C) and anterior tibialis (D–F) muscle. The sections were stained with anti-α MHC (A, D), anti-slow MHC (B, E) and anti-fast MHC (C, F). Fibres containing slow MHC (1), fibres containing fast MHC (2) and a muscle spindle (s) are indicated. The scale bar represents 66 μm.

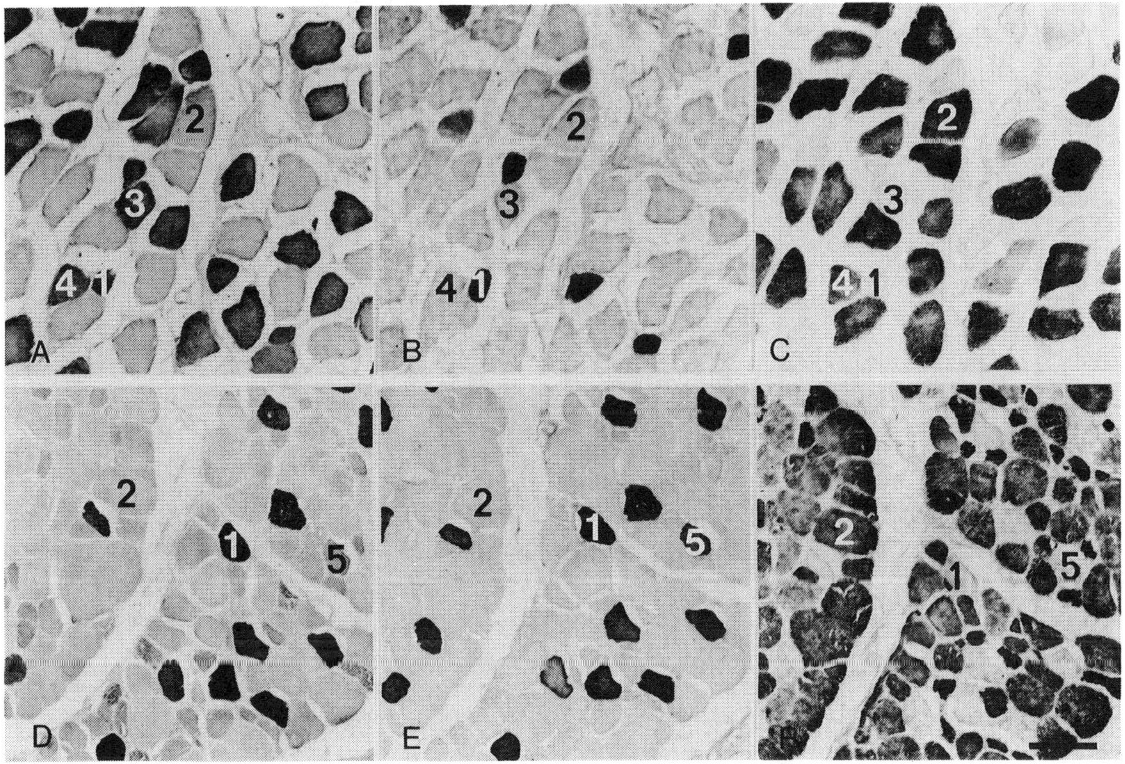

Fig. 2 Immunohistochemical analysis of serial sections of adult rabbit masseter (A–C) and extraocular (D–F) muscles. The sections were stained with anti-α MHC (A, D), anti-slow MHC (B, E) and anti-fast MHC (C, F). Fibres containing slow and 'cardiac' α MHC (1), fast MHC (2), 'cardiac' α MHC (3), 'cardiac' α and fast MHC (4) and fibres containing slow MHC (5) are indicated. The scale bar represents 63 μm.

Table 2 Fibre types in adult rabbit muscles.

Antibodies	Fibre types				
	1	2	3	4	5
ATPase*	I	II	I	IIC	I'
Anti-α MHC	+	-	+	+	-
Anti-slow MHC	+	-	-	-	+
Anti-fast MHC	-	+	-	+	-

* ATPase determined fibre types in masseter and anterior tibialis muscles, I = slow, II = fast, IIC = intermediate, I' not present in the tested parts of the masseter muscle but predominantly present in tested anterior tibialis muscle.

Discussion

The results reported in this study show that 'cardiac' α MHC is not unique for masticatory muscles. Extraocular, intrinsic and extrinsic tongue, facial and hyoid muscles reveal the expression of 'cardiac' α MHC also.

The unexpected occurrence of 'cardiac' α MHC in cranially located muscles deserves some comment. First, the cranially located muscles have a variety of functions, nevertheless they all contain 'cardiac' α MHC implying that the presence of 'cardiac' α-containing fibres cannot easily be explained on functional grounds. Secondly, it might be that the expression of 'cardiac' α MHC is related to innervation. This possibility seems unlikely since muscles innervated by cranial nerves (for example masseter) and muscles innervated by cervical nerves (geniohyoid, omohyoid) express 'cardiac' α MHC. Thirdly, most cranial muscles (possibly including the extraocular and tongue muscles (11)) are of branchiomeric (somitomeres) origin, however, the geniohyoid muscle arises from somitic myotomal cells (11, 12). Nevertheless, the geniohyoid muscle expresses 'cardiac' α MHC. So the occurrence of 'cardiac' α MHC cannot simply be correlated with the embryonic origin of the muscle fibres. Noden (13, 14), made an inventory of head and neck muscles in birds known to have the connective tissue derived from neural crest cells. Interestingly, we found a striking correspondence between these head and neck muscles in birds and the rabbit muscles that express 'cardiac' α MHC (see Table 1). (Provided this comparison between two distantly related species is allowed.) A role of the connective tissue cells in inducing the expression of specific MHCs in skeletal muscle cannot be excluded. It is known that connective tissue cells influence the pattern of differentiation and growth of the subsequently arriving myoblasts (15).

Finally, our results confirm and extend our previous observations of the extracardiac expression of so-called 'cardiac' α MHC. The expression of 'cardiac' α MHC seems to be restricted to cranially located muscles. It is interestingly that even the heart originates from the most cranial part of an embryo. 'Cranial muscles' have to cope with highly specialized functions. It might be speculated that the precise regulations would require the expression of an array of MHC isoforms (7 MHC isoforms have been described to occur in cranially located muscles as yet). This can be tested by comparing the molecular, EMG and contraction speed properties of representative examples of cranial and limb muscles.

Acknowledgements

We like to thank Cars Gravemeyer for the photographic work and Hans Korfage and Peter Brugman for their expert technical assistance.

References

1. Reiser, P. J., Moss, R. L., Giulian, G. G. & Greaser, M. L., 1985. *J. Biol. Chem.*, **260**, 9077–9080
2. Danieli-Betto, D. D., Zerbato, E., Betto, R., 1986. *Biochem. biophys. res. Comm.*, **138**, 981–987.
3. d'Albis, A., Janmot, C., Bechet, J. J., 1986. *Eur. J. Biochem.*, **156**, 291–296.
4. Butler-Browne, G. S., Eriksson, P. O., Laurent, C., Thornell, L. E., 1988. *Muscle Nerve*, **11**, 610–620.
5. Bredman, J. J., Weijs, W. A. Moorman, A. F. M., Submitted to *Histochem. J.*
6. Rowlerson, A., Mascarello, F., Veggetti, A. Carpenè, E., 1983. *J. Muscle Res. Cell Motil.*, **4**, 443–472.
7. Wessels, A., Vermeulen, J. L. M., Moorman, A. F. M., Becker, A. E., 1988. *Proc. XVIII Eur. Conf. Muscle and Mot.*. Padova Unipress, 311–316.
8. Groot de, I. J. M., Lamers, W. H., Moorman, A. F. M., 1989. *Anat. Rec.*, **224**, 365–373.
9. Wessels, A., Vermeulen, J. L. M., Virágh, Sz., Kalman, F., Lamers, W. H., Moorman, A. F. M., 1990a. *Anat. Rec.*, in press.
10. Wessels, A., Soffers, C. A. S., Vermeulen, J. L. U., Mijnders, T. A., Bredman, J. J., Moorman, A. F. M., 1990, In: Maréchal, G. and Carraro, U. (eds), Muscle and Motility, Vol. 2, pp. 71–77. Intercept Ltd, Andover.
11. Hamilton, W. J., Mossman, H. W., 1978. Macmillan Press Ltd, London.
12. Noden, D. M., 1984. *Anat. Rec.*, **208**, 1–13.
13. Noden, D. M., 1983a. *Am. J. Anat.*, **168**, 257–276.
14. Noden, D. M., 1986. *Devel. Biol.*, **116**, 347–356.
15. Noden, D. M., 1983b. *Devel. Biol.*, **96**, 144–165.

53
Implication of the C-terminal Region of Actin in the Polymerization Process

A. HOUMEIDA, Y. BENYAMIN AND C. ROUSTAN

UPR 8402 (CNRS)-U 249 (INSERM)-EPHE, Montpellier Université I CRBM (CNRS), BP 5051-34033 Montpellier Cedex, France

Introduction

Actin is essential for motility and cytoskeletal organization. Its polymorphism is controlled *in vivo* by the interaction with a number of associated proteins such as profilin, gelsolin and Gc protein (1). The polymerization of actin is known to be induced by the presence of cations, particularly Mg^{2+} *in vivo* and KCl. During the process of polymerization, each monomer would undergo large conformational changes. The monomer of actin in the filament, probably involves four regions of contact (2, 19). This identity of some of these regions can be deduced from indirect results. For example, the C-terminal extremity, whose environment is very varied (3) during polymerization, would be near one of the interfaces. Similarly, the loss of reactivity of Lys 61, 68, 113 and 283 can be explained if these residues were located in contact area (4). Lys 335, whose reactivity is strikingly enhanced, would be also in a zone of large conformational changes during polymerization process (4). This last result is in accordance with previous work which showed a decrease in antibody reactivity specific for discrete epitopes in the 300–355 sequence of F-actin by the presence of Mg^{2+} (5). Further, this region corresponded to one of the interface between actin and gelsolin, which is known to interact with the barbed end of actin filament (6). From crystallographic data describing the DNAse-I-actin complex, the C-terminal sequence 300–375 spans from subdomain III to subdomain I where C and N-terminal extremities are in close proximity to each other (7).

We have tested the effect of several antibodies, each specific to a given segment in the amino acid sequence of actin, on the polymerization of actin and

Muscle and Motility Vol. 2. Proceedings of XIXth European Conference in Brussels
© Intercept Ltd, PO Box 716, Andover, Hampshire, SP10 1YG, UK

its interaction with cations. The results are discussed in connection with the location of some associated proteins in the C-terminal sequence of actin.

Methods

Actin and Gc protein were prepared as described in (8, 9). Actin was labelled at Lys 374 by IAEDANS or at Lys 373 by NBDCl (10, 11). Antibodies to actin and the specific antibody populations were purified by affinity chromatography, as previously described (12, 5). Their specificity was determined by immunoblotting (13) or ELISA (14) using various actin fragments. The anti-(10K) antibody was specific for three epitopes in the 300–355 sequence (around Met 305, 325 and 355). Antibody populations (monospecific) were directed to 1–7, 300–310, 305–326, and dansylated Cys 374 (Fig. 1). The ELISA technic was used to monitor the interaction between coated actin and antibody population in the presence or absence of cations (14). Polymerization of actin and Gc protein interaction were followed by fluorometry by using NBD actin (13). The depolymerization effect of Gc protein on F-actin was monitored by ultracentrifugation.

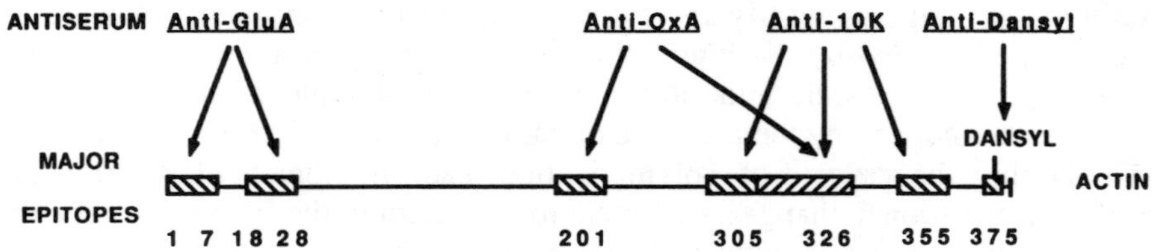

Fig. 1 Specificity of anti-actin antisera. Anti-GluA antiserum induced by F-actin treated by glutaraldehyde; anti-OxA antiserum induced by performic oxidized actin; anti-(10K); antiserum induced by the fragment 285-375 of actin. Barred boxes represent major antigenic epitopes.

Results and Discussion

The analysis of the effect of antibody populations specific for different parts of actin sequence (Table 1) on the polymerization process showed that only some populations impeded the actin–actin interaction. The study was focused on the C-terminal sequence of actin which was located in subdomain III and I of the monomer as judged by the X-ray crystallography model (7). Antibodies directed to dansylated Cys 374 as well as, those to 1-7 sequence, the two sequences being in close spatial proximity (7, 14), do not affect polymerization. In contrast, anti-(10K) inhibited strongly actin polymerization (Table 1). The anti-(305–326) affected also this process but to a less extent. In all cases, we observed a rise in the critical concentration of actin. Furthermore, purified antibodies specific to the sequence around Met 305, when extrapolated to high

Table 1 The effect of antibody populations specific for different parts of the actin sequence.

Antibody	Inhibition of actin polymerization	Inhibition of antibody interaction induced by Mg^{++} on F-actin (effect %)	Inhibition of antibody interaction by Gc-protein
Anti-(1–7 sequence)	–	0%	–
Anti-(10K)	+	25%	–
Anti-(300–326 sequence)	+	50%	–
Anti-(305–326 sequence)	+	0%	–
Anti-(dansyl)	–	n.d.	+

antibody concentrations, produced an almost total inhibition of the actin polymerization. However, as these antibodies are as reactive on F-actin as on G-actin (6), the related epitopes would not be located in an actin–actin interface. More likely, these antibodies would impede conformational changes which are necessary to allow actin filament formation. In relation to the different conformational states of actin in the filament, we noticed that these antibodies (anti-(10 K)) whose specificity is located in the 300–355 sequence, presented a

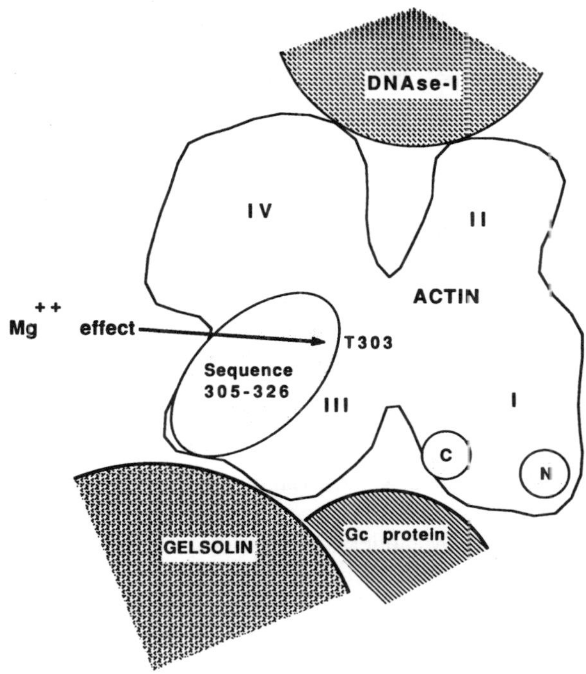

Fig. 2 Functional domains in actin monomer. I, II, III, IV, indicate the structural subdomains of actin. N and C are the N and C extremities in actin sequence.

decrease in their reactivity on Mg^{2+} binding on F-actin (Table 1). The largest effect is obtained with the subpopulation specific for the sequence around Met 305, the effect being of about 50%. So, we conclude that these different conformational at changes occurring during polymerization and Mg^{2+} binding would be concerned particularly with the subdomain III of actin (Fig. 2). This assumption is supported by the fact that this subdomain interacts with gelsolin (Fig. 2), a protein which modulates the polymorphism of actin (6). However, a protein such as Gc protein, which inhibits actin-actin interaction (9), would not interact with the subdomain III (Table 1). We have not observed, indeed, any competition towards the antibody tested, although they covered a large region at the surface of this subdomain.

If we analysed our results on the subdomain I of actin, we notice that the N-extremity is not involved in the actin interface (Table 1) (12, 15, 16). However, as evidenced by fluorescence measurements developed on actin modified in Cys 374 or Lys 373 by fluorophores, large changes in the environment occurred in the C-terminal extremity. Therefore, one interface would be located near this extremity. The location of such an interface in subdomain I, is most probably near or at the Gc protein site (Table 1). This interface can be restricted to Cys 374, as judged by the competitive effect observed between Gc protein and antibodies specific for the dansyl group (Table 1). This result is also supported by the competitive behaviour reported by others (2) between profilin and Gc protein, profilin interacting at the C-terminal extremity of actin (Fig. 2) (17). In conclusion, both subdomain III and I woud contain actin–actin interfaces and can undergo large conformational changes during actin polymerization and cation binding. They are also implicated in the interaction with some associated proteins which control the polymerization of actin.

References

1. Pollard, T. D., Cooper, J. A., 1986, *Ann. Rev. Biochem.*, **55**, 987–1035.
2. Goldsmith-Clermont, P. S. V., von Alstyne, E. L., Day, J. R., Emerson, D. L., Nel, A. E., Lazarchick, J., Galbraith, R. M., 1986, *Biochemistry*, **25**, 6467–6472.
3. Kouyama, T. and Mihashi, K., 1981, *Eur. J. Biochem.*, **114**, 33–38.
4. Lu, R. C., Szilagyi, L. 1981, *Biochemistry*, **20**, 5914–5919.
5. Mejean, C., Hue, H. K., Pons, F., Roustan, C., Benyamin, Y., 1988, *Biophys. Biochem. Res. Commun.*, **152**, 368–375.
6. Boyer M., Feinberg J., Hue H. K., Capony J. P., Benyamin Y., Roustan C. 1987, *Biochem. J.* **248**, 359–364.
7. Kabsch, N., Pai, E. F., Mannherz, H. G., Suck, D., Holmes, K. C., 1989, *J. Musc. Res. Cell Motility*, **11**, 68.
8. Spudich, J. A., Watt, S., 1971, *J. Biol. Chem.*, **246**, 4866–4871.
9. Coni, M., Constans, J., Viau, M., Olomouki, A., 1988, *Biochim. Biophys. Acta.*, **759**, 137–145.
10. Detmers, P., Weber, A., Elzinga, M., Stephens, R.E., 1981, *J. Biol. Chem.*, **256**, 99–105.

11. Takashi, R., 1979, *Biochemistry*, **18**, 5164–5169.
12. Benyamin, Y., Roustan, C., Boyer, M, 1986, *J. Immun. Meth.*, **86**, 21–29.
13. Roustan, C., Boyer, M., Fattoum, A., Jeanneau, R., Benyamin, Y., Roger, M., Pradel, L-A., 1982, *Eur. J. Biochem.*, **129**, 149–155.
14. Mejean, C., Boyer, M., Labbé, J. P., Marlier, L., Benyamin, Y., Roustan, C., 1987, *Biochem. J.*, **244**, 571–577.
15. Benyamin, Y., Roustan, C., Boyer, M., 1983, *FEBS Lett.*, **160**, 41–45.
16. Dasgupta, G., White, J., Phillips, M., Bulinski, J.C., Reisler, E., 1990, *Biochemistry*, **29**, 3319–3324.
17. Malm, B., 1984, *FEBS Lett.*, **173**, 395–402.
18. Boyer, M., Roustan, C., Benyamin, Y., 1985, *Biosci. Rept.*, **5**, 39–46.
19. Hambly, B. D., Barden, J. A., Miki, M., Dos Remedios, C.G., 1986, *Bioessays*, **4**, 124–138.

54
Smooth Muscle Caldesmon: Interaction with S.100 and Related Smooth Muscle Proteins

K. PRITCHARD AND S. B. MARSTON

Cardiac Medicine Department, National Heart and Lung Institute, Dovehouse Street, London SW3 6LY, UK

Introduction

Smooth muscle actomyosin is dually regulated by Ca^{2+}. Calmodulin is the Ca^{2+}-receptor of the thick filament-regulatory system based on myosin phosphorylation. The ability of smooth muscle thin filaments to activate myosin MgATPase is also Ca^{2+}-dependent (1, 18). However, the identity of the Ca^{2+} receptor of the thin filament- based regulatory system is uncertain (2, 3, 4).

The major proteins identified in smooth muscle native thin filaments are actin, tropomyosin and caldesmon (1, 18); caldesmon is the key regulatory component (5). In systems reconstituted from purified smooth muscle proteins, caldesmon is a potent, but not Ca^{2+}-dependent, inhibitor of the activation of myosin MgATPase by actin-tropomyosin (1, 6). In native thin filaments, a Ca^{2+}-sensitizing protein presumably reverses caldesmon-inhibition in the presence of Ca^{2+}.

Calmodulin binding to caldesmon is Ca^{2+}-dependent (3, 7) and, under certain conditions, calmodulin can confer Ca^{2+}-sensitivity on the activation of

Muscle and Motility Vol. 2. Proceedings of XIXth European Conference in Brussels
© Intercept Ltd, PO Box 716, Andover, Hampshire, SP10 1YG, UK

myosin MgATPase by purified smooth muscle F-actin, tropomyosin and caldesmon (1, 8). Under near-physiological conditions (37°C and buffer containing 120 mM KC1) Ca.calmodulin binds to caldesmon which remains bound to F-actin-tropomyosin, forming a non-inhibited quaternary complex which activates the MgATPase of myosin as much as actin-tropomyosin alone (3, 8). An intrafilament mechanism also operates in the Ca^{2+}-sensitive native thin filaments, where caldesmon remains bound in the presence of Ca^{2+}. However, the reconstituted system using calmodulin, unlike the native thin filaments, loses its Ca^{2+}-sensitivity at 25°C in buffer containing 70 mM KC1 because a non-inhibited quaternary complex is not formed and inhibition by caldesmon is then only reversed as a result of the rather ineffective competition by Ca.calmodulin against F-actin-tropomyosin for binding to caldesmon. We suspect that a protein other than calmodulin may be involved in Ca^{2+}-sensitizing native thin filaments (2, 4). The reversal of caldesmon-inhibition is not a property specific to calmodulin, since skeletal muscle troponin C can be equally effective.

Skripnikova and Gusev (9) recently proposed that S.100 protein might regulate caldesmon, since in their hands brain S.100 could reverse inhibition by gizzard caldesmon in a buffer of low ionic strength at room temperature. We have therefore investigated the regulation of caldesmon by brain S.100 and by related low-molecular-weight Ca^{2+}-binding proteins from aorta smooth muscle.

Methods

Caldesmon, F-actin and tropomyosin were prepared from sheep aortas using methods described previously (3, 8). Skeletal muscle actin was prepared from acetone powder (8) and myosin was from rabbit skeletal muscle (8). Bovine brain S.100 (mainly S.100b) was obtained by a method derived from that of Masure and co-workers (10). We purified low-molecular-weight Ca^{2+}-binding proteins by Ca^{2+}-dependent phenyl sepharose chromatography of a low-ionic strength extract from sheep aortas, followed by anion-exchange chomatography on Mono-Q FPLC. Myosin MgATPase activity was assayed as the rate of P_i release as in (8).

Results

BRAIN S.100 INTERACTION WITH SMOOTH MUSCLE CALDESMON

In the presence of 0.1 mM $CaC1_2$, S.100 from brain reversed caldesmon's inhibition of the smooth muscle actin-tropomyosin-activated MgATPase of skeletal muscle myosin. In 1mM EGTA, S.100 had no effect. The reversal of caldesmon's inhibition was dependent on the concentration of S.100. Ca^{2+}-sensitivity at any given concentration of S.100 varied markedly with temperature and ionic strength (Fig. 1).

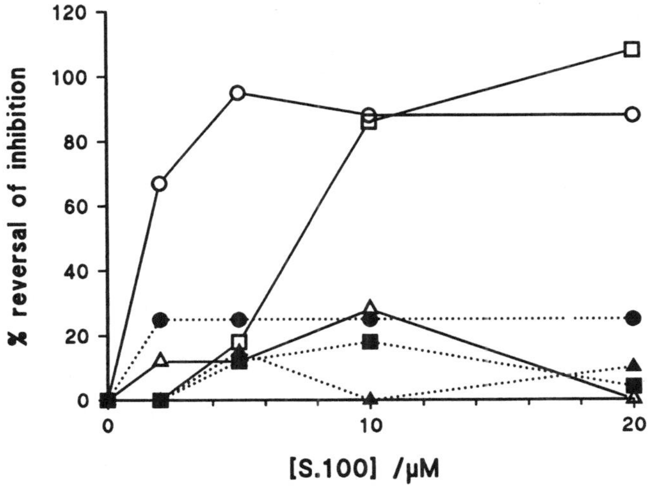

Fig. 1 Brain S.100 reverses caldesmon's inhibition of smooth muscle actin-activated myosin MgATPase. O, 37°C/120mM KC1; □, 37°C/70mM KC1; △, 25°C/ 70mM KC1. Open symbols, 0.1mM $CaC1_2$; Solid symbols, 1mM EGTA. Assay tubes contained 0.25 mg/ml skeletal muscle myosin, 24 µM vascular smooth muscle actin, 7 µM v.s.m. tropomyosin, 3 µM v.s.m. caldesmon.

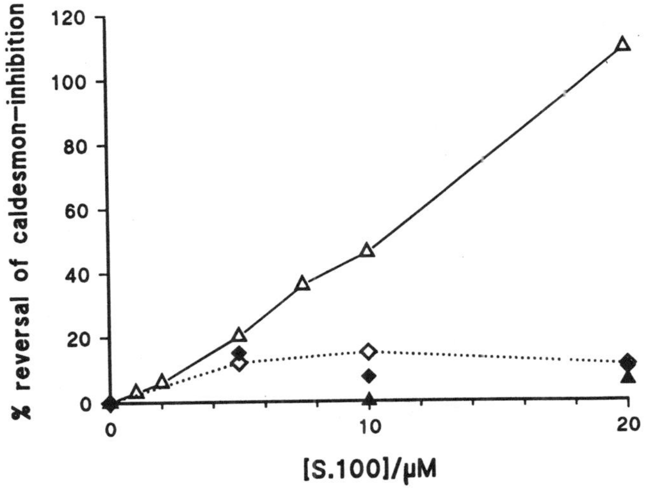

Fig. 2 At 25°C, brain S.100 reverses caldesmon's inhibition of skeletal muscle actin, but not of smooth muscle actin. △, skeletal muscle actin; ◇, smooth muscle actin. Open symbols, 0.1 mM $CaC1_2$; Solid symbols, 1 mM EGTA. Assay tubes contained 24 µM actin (either vascular smooth muscle actin or skeletal muscle actin); 0.25 mg/ml skeletal muscle myosin; 7 µM v.s.m. tropomyosin; 3 µM v.s.m. caldesmon.

INFLUENCE OF THE ACTIN ISOFORM ON Ca^{2+}-SENSITIZATION BY S.100

When skeletal muscle actin was used to reconstitute a thin filament system, S.100 reversed inhibition by caldesmon at 25°C in the presence of 70 mM KC1, whereas it did not do so when smooth muscle actin was used (Fig. 2).

EFFECT OF HEAT ON S.100 PROTEIN

A sample of S.100 protein was heated to 90°C for 3 min. The heated S.100 was tested for its ability to Ca^{2+}-sensitize a thin-filament system reconstituted from 5 μM aorta caldesmon, 7 μM aorta tropomyosin, and 25 μM aorta F-actin, at 37°C in buffer containing 120 mM KC1. At 10 μM, the heated S.100 protein reversed caldesmon's inhibition of the actin-activated ATPase by 100%, as did the non-heated control S.100 (not illustrated); neither sample of S.100 had any effect on MgATPase activity in the presence of EGTA.

SMOOTH MUSCLE S.100-LIKE PROTEINS AND THEIR INTERACTION WITH CALDESMON

We obtained three proteins which appeared pure on SDS/polyacrylamide gels.

(*a*) A 17kDa M_r protein (yield approx. 16mg/kg wet weight). This protein eluted from anion exchange (Mono-Q) at the same salt concentration as brain calmodulin. In our reconstituted thin filament system it reversed caldesmon's inhibition in a manner indistinguishable from brain calmodulin (Fig. 3).

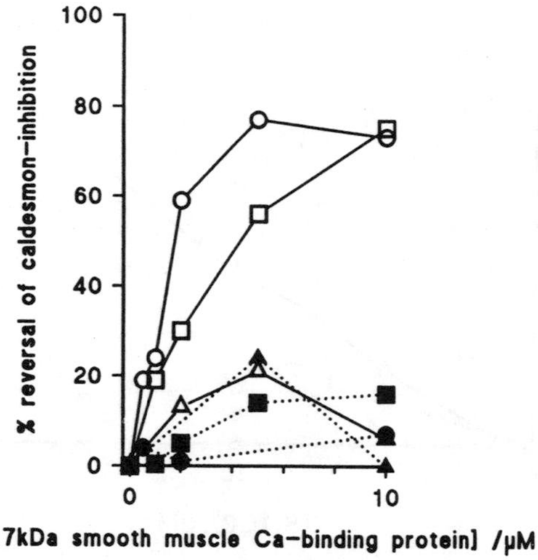

Fig. 3 Reversal of caldesmon's inhibition by a 17 kDa Ca-binding protein from sheep aorta vascular smooth muscle. O, 37°C/120 mM KC1; □, 37°C/70mM KC1; △, 25°C/ 70 mM KC1. Open symbols, 0.1 mM $CaC1_2$; Solid symbols, 1 mM EGTA. Assay tubes contained 0.25 mg/ml skeletal muscle myosin, 24 μM vascular smooth muscle actin, 7 μM v.s.m. tropomyosin and 2.4 μM v.s.m. caldesmon.

(*b*) An 11kDa M_r protein (SDS/PAGE) obtained at a yield of approximately 48mg/kg. This eluted from Mono-Q at very low salt concentration. Neither of two independent preparations of this protein, at up to 20 μM, had any effect on inhibition by 3μM caldesmon, in the presence of either 0.1mM $CaCl_2$ or 1.0mM EGTA, at 37°C in buffer containing 120mM or 70mM KC1, or at 25°C with 70mM KC1 (not illustrated).

(*c*) An approximately 11 kDa protein which was much less abundant (about 4mg/kg wet weight) and which eluted from Mono-Q at about the same salt concentration as brain S.100. In one experiment we were unable to reverse the inhibition due to 5 μM caldesmon, using this protein at up to 10 μM at 37°C/120mM KC1.

Discussion

BRAIN S.100 INTERACTS WITH SMOOTH MUSCLE CALDESMON

Brain S.100, like brain calmodulin (or skeletal muscle troponin C) reversed the inhibition by caldesmon in a reconstituted smooth muscle thin filament system. It seems unusual for S.100 and calmodulin to interact with the same target protein, and have the same effects (11). Experimental conditions of temperature and ionic strength had a large influence on the Ca^{2+}-sensitivity of the system reconstituted using brain S.100, with the same pattern of effects as for brain calmodulin (3, 8). It seems likely that the underlying mechanism is the same with brain S.100 as with brain calmodulin (3). S.100, like calmodulin, appears to be heat-stable and in this respect both differ from the Ca^{2+}-sensitizing protein(s) which we previously detected in aorta extracts (2, 4).

THIN FILAMENTS RECONSTITUTED WITH SKELETAL MUSCLE ACTIN

Brain S.100 Ca^{2+}-sensitized a thin filament system reconstituted using skeletal muscle F-actin, under conditions (25°C/70mM KC1) where no Ca^{2+}-sensitivity was observed when smooth muscle F-actin was used. We previously reported this phenomenon with brain calmodulin (8). Binding experiments with calmodulin (8) demonstrated that skeletal muscle actin had a greater tendency than smooth muscle actin to form a non-inhibited quaternary complex. The same explanation may apply with brain S.100. Since skeletal muscle actin and caldesmon probably do not interact in intact muscle cells, this finding is of doubtful physiological relevance and does not support a role for S.100 rather than calmodulin in Ca^{2+}-sensitizing physiological systems involving caldesmon and smooth muscle actin.

COULD CALDESMON INTERACT WITH S.100 IN INTACT SMOOTH MUSCLE?

If the intracellular content of either calmodulin or of S.100 is greater than that of caldesmon, then the possibility exists that one of these proteins might regulate actin–myosin interactions controlled by caldesmon. The caldesmon content of smooth muscle may be about 20 μM (12, 13) and the total calmodulin content is perhaps 20–30 μM (14). Little is known about the S.100 content of smooth muscle; published estimates of *ca.* 1 μM could be underestimates owing to limited cross-reacitivity of antibodies to brain S.100 against a smooth muscle S.100 (9). Early reports (14, 15, 17, 18) suggested that thin-filament-linked regulation of smooth muscle actomyosin might be mediated by a troponin-C-like protein other than calmodulin, but of similar molecular weight. We suspect that the 17 kDa protein we isolated from aorta is actually calmodulin, and we did not find that this protein had any special properties in our system.

The major 11 kDa protein which eluted from anion exchange at low salt concentration resembles in this respect, and in its abundance, the 'SMCaBP-11' recently discovered in gizzard by Mani and Kay (16). If this is the same protein, then as a dimer its intracellular concentration is at least 6 μM (assuming 100% yield and 40% smooth muscle in aorta). Mani and Kay reported significant differences between SMCaBP-11 and brain S.100, and in our experiments this similar protein lacked any ability to regulate smooth muscle actin via caldesmon.

The other approximately 11 kDa protein, which we obtained in low yield, is possibly smooth muscle S.100. It was perhaps obtained in an inactive condition, although the calmodulin-like fraction from the same preparations had full activity, and the activity of brain S.100 was robust. Brain S.100 has mainly β-subunits, but smooth muscle S.100 may, like striated muscle S.100, be mainly α-subunits; isoform specificity has been reported for other S.100 interactions (11).

Our results with these S.100-like smooth muscle proteins provide no evidence to suggest that such a protein controls caldesmon in intact smooth muscles. However, caldesmon of the low Mw class ('CD_I' or 'CD_{77}') is abundant in various type of secretory and motile cells some of which may also have abundant S.100; if so, the S.100–caldesmon interaction might confer Ca^{2+}-dependence on actin filament function in certain non muscle cells.

Acknowledgements

This work was generously funded by the British Heart Foundation.

References

1. Marston, S. B., Smith, C. W. J., 1984, *J. Muscle Research*, **5**, 559–575.
2. Pritchard, K. P. , Marston, S. B., 1989. In: Carraro, U. (ed.), *Sarcomeric and Non-Sarcomeric Muscles: Basic and Applied Research Prospects for the 90s*, pp. 649–654. Unipress, Padova.
3. Pritchard, K. P., Marston, S. B., 1989, *Biochem. J.*, **257**, 839–843.
4. Pritchard, K., Marston, S., 1988, *Trans. Biochem. Soc.*, **16**, 355–356.
5. Marston, S. B., Redwood, C. S., Lehman, W., 1988, *Biochem. Biophys. Res. Comm.*, **155**, 197–203.
6. Smith, C. W. J., Marston, S. B., 1985, *FEBS Lett.*, **184**, 115–119.

7. Sobue, K., Muramoto, Y., Fujita, M., Kakiuchi, S., 1981, *Proc. Natl. Acad. Sci. USA*, **78**, 5652–5655.
8. Smith, C. W. J., Pritchard, K., Marston, S., 1987, *J. Biol. Chem.*, **262**, 116–122.
9. Skripnikova, E. V., Gusev, N. B., 1989, *FEBS Lett.*, **257**, 380–382.
10. Masure, H. R., Head, J. F., Tice, H. M., 1984, *Biochem. J.*, **218**, 691–696.
11. Kligman, D., Hilt, D. C., 1988, *Trends in Biochem. Sci.*, **13**, 437–443.
12. Ngai, P. K., Walsh, M. P., 1984, *J. Biol. Chem.*, **259**, 13656–13659.
13. Ngai, P. K., Walsh, M. P., 1985, *Biochem. J.*, **230**, 517–522.
14. Grand, R. J. A., Perry, S. V., Weeks, R. A., 1979, *Biochem. J.*, **177**, 521–529.
15. Ebashi, S., 1980, *Proc. Roy. Soc.* B, **207**, 259–286.
16. Mani, R. S., Kay, C. M., 1990, *Biochemistry*, **29**, 1398–1404.
17. Grand, R. J. A., Perry, S. V., Weeks, R. A., 1977. In: Casteels, R., Godfraind, T., Ruegg, J. C. (eds), *Excitation-contraction Coupling in Smooth Muscles*, pp. 335–341. Celsevier/North Holland, Amsterdam.
18. Marston, S. B., Smith, C. W. J., 1985, *J. Muscle Research*, **6**, 669–708.

55
Crosslinking Between Smooth Muscle Actin and Myosin by Caldesmon

S. MARSTON*, C. REDWOOD* AND P. BENNETT**

**Department of Cardiac Medicine, National Heart and Lung Institute, Dovehouse Street, London SW3 6LY, UK*
***MRC Cell Biophysics Unit, 26–29 Drury Lane, London WC2B 5RC, UK*

Introduction

Caldesmon is a protein of 87 000 molecular weight which is a component of the thin filaments of all smooth muscles (1, 2, 3). A lower molecular weight isoform is also present in thin filament systems of non muscle cells (4, 5). Recently both biochemical and structural approaches have indicated that the intact thin filament contains caldesmon at a stoichiometry of 1 per 14 actins (two tropomyosins) (6, 7).

Purified caldesmon binds to actin and actin-tropomyosin at an affinity ranging from $>10^8$ M^{-1} at low ionic strength to 10^7 M^{-1} in 100 mM KCl (8). Caldesmon is an elongated molecule (3 nm × 75 nm) which binds along the axis of the actin filament such that it is in contact with 14 actins and two tropomyosins (6, 9).

Over the last six years caldesmon has been implicated as a central component of the Ca^{2+}-dependent regulatory system of smooth muscle thin filaments in concert with myosin regulation by phosphorylation and its mechanism has been elucidated (8, 10, 11). Caldesmon is a powerful inhibitor of actin-tropomyposin activation of myosin ATPase; caldesmon also binds to calcium-

Muscle and Motility Vol. 2. Proceedings of XIXth European Conference in Brussels
© Intercept Ltd, PO Box 716, Andover, Hampshire, SP10 1YG, UK

binding proteins (see Chapter 54 in this volume) in such a way that caldesmon inhibition becomes Ca^{2+}-regulated in native thin filaments. The primary effect that caldesmon has is to decrease V_{max} of actin activation of myosin ATPase without a significant increase in the affinity of the actin filament for the myosin filament (12); indeed, in some studies, caldesmon has reduced the affinity of actin for myosin subfragments (13, 14). This activity seems to be confined to the C-terminal region of caldesmon since chymotryptic fragments of caldesmon (amino acids 483–758 (C-terminus)) bind to actin and have inhibitory activity (15) whilst a genetically engineered caldesmon containing amino acids 1–578 is inactive (16).

Interest in caldesmon function has been greatly increased by the observation that it binds, not only to actin but also to myosin. This property has been observed for isolated caldesmon (17), actin-caldesmon complex (18) and intact native smooth muscle thin filaments (19). This interaction therefore gives caldesmon the potential of being a crosslinking protein between actin and myosin filaments entirely separate from the actin–myosin interaction which is responsible for force generation. The speculation has been made that such a crosslinking could contribute to force maintenance in smooth muscles (20). In this report we have examined caldesmon binding to myosin and its crosslinking abilities using whole and genetically engineered partial caldesmon molecules and whole aorta myosin and myosin rods.

Methods

Aorta caldesmon, actin, tropomyosin, thin filaments and myosin were prepared by our standard methods (8, 12, 21). Myosin rod was prepared by papain digestion of aorta myosin and ethanol precipitation (22). The preparation of whole caldesmon and the mutant containing amino acids 1–578 by expression of the cDNA in *E. coli* is described by Redwood *et al.*(16)

Caldesmon binding to myosin or rod was assayed by sedimentation; 2–3 μM caldesmon and up to 10 μM myosin were mixed and centrifuged to sediment myosin and myosin–caldesmon complex. Portions of the supernatant were separated on SDS gels and the caldesmon band was quantitatively scanned. Less than 10% of caldesmon sediments in the absence of myosin and up to 95% sediments in its presence. The percentage of bound caldesmon was defined as the percentage of caldesmon missing from the supernatant when myosin was present.

For electron microscopy observation myosin and rod at about 1.5 mg/ml were dialysed against 0.3M KCl, pH 6.0 to promote the formation of side polar filaments (23). Samples were diluted into low ionic strength buffer and loaded onto thin carbon grids and negatively stained with uranyl acetate. In some cases a low electron-dose procedure was employed in order to preserve fragile structural detail.

Results and discussion

THE BINDING OF CALDESMON TO AORTA MYOSIN

Binding curves for caldesmon binding to aorta myosin have been obtained under a variety of conditions. At low ionic strength the binding is virtually stoichiometric indicating an affinity of better than 10^6 M^{-1} but the binding is strongly dependent on salt concentration (Fig. 1A) decreasing to around 10^5 M^{-1} at an ionic strength of 0.1 mM. Binding affinities are expressed in terms of myosin molecule concentration (M wt 450 000).

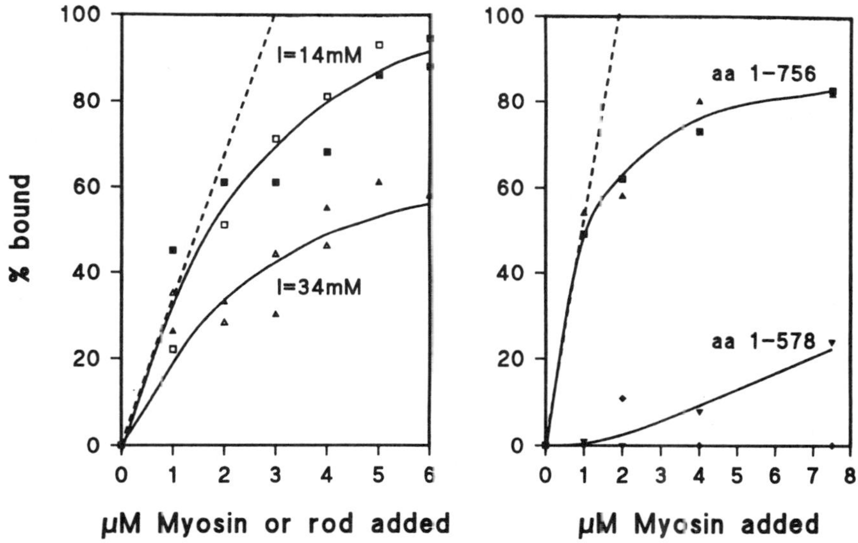

Fig. 1 Caldesmon binding to aorta myosin and rod.
Left: The effect of ionic strength and comparison of myosin and rod binding. Conditions: 3 µM caldesmon in 5 mM PIPES, 2.5 mM $MgCl_2$, 1 mM DTT, pH 7.0 plus KCl to adjust ionic strength. ■, myosin, I = 14 mM; □, rod, I = 14 mM; ▲, myosin, I = 34 mM; △, rod, I = 34 mM; dotted line represents stoichiometric binding.

Right: Binding of various types of caldesmon and myosin. Conditions as above, 2 µM caldesmon, I = 14 mM. ■, native chicken gizzard caldesmon; ▲, caldesmon expressed in *E. coli*; ▼, amino acids 1–578 of caldesmon expressed in *E. coli*; ■, native caldesmon with skeletal muscle myosin.

LOCATION OF THE CALDESMON-BINDING SITE ON MYOSIN

Ikebe & Reardon (17) reported that the S2 portion of myosin was responsible for caldesmon binding on the basis of studies using a caldesmon affinity column and a number of workers have observed that caldesmon promotes a tight binding between actin-tropomyosin and smooth muscle HMM, but not S1 (13, 18). To test this prediction we prepared the rod portion of aorta myosin.

This contains all the alpha helical tail of myosin (i.e. S2 and 'LMM') but lacks the S1 portion. Very pure rod could be obtained by ethanol precipitation and the preparation had the advantage that it was much less viscous in low ionic strength solutions than myosin and was thus easier to handle. We assumed a molecular weight of 240 000 for the rod (24). Fig. 1A shows that aorta myosin rod did indeed bind to caldesmon and the binding was indistinguishable from whole aorta myosin, thus confirming the location of the binding site in the S2 portion of myosin (amino acids 849–1279 (25)). It is noteworthy that the ability of myosin to bind caldesmon has so far only been observed in smooth muscle myosin. Skeletal muscle myosin does not bind caldesmon at all (Fig. 1B).

LOCATION OF MYOSIN BINDING SITE ON CALDESMON

Two groups have reported that caldesmon fragments containing the N-terminal 200 or so amino acids have an ability to bind myosin whilst the rest of the molecule does not (14, 26). In apparent contradiction, our studies of the interaction of caldesmon containing native thin filaments with HMM showed that there was a tight interaction independent of the ATPase site which had a stoichiometry of one per actin which indicated that one caldesmon could interact with many myosins either directly or via a cooperative interaction involving actin (19). Such behaviour is hard to reconcile with a single discrete binding domain. We tested the myosin binding ability of caldesmon expressed in *E. coli* and found it indistinguishable from native caldesmon (Fig. 1B) and yet a mutant containing amino acids 1–578 and including the N-terminus bound extremely weakly to myosin (Fig. 1B). Bearing in mind that the previous studies did not provide a quantitative estimate of binding affinity we would conclude that the N-terminus of caldesmon may be necessary for myosin binding but it is certainly not sufficient; indeed it is possible that end to end interactions are involved as well as intramolecular interactions.

STRUCTURE OF CALDESMON–MYOSIN COMPLEXES

We examined the caldesmon–myosin rod complex by electron microscopy. Aorta myosin rod can form short side polar filaments in the same way as intact myosin (23) moreover, since the myosin heads are absent, the filament backbone can be clearly observed (Fig. 2, top). The rod filaments had a clean outline with occasional 'fraying' due to the S2 portions sticking out from the filament. In the presence of saturating caldesmon this fraying effect was dramatically increased with filamentous 'whiskers' curving out from the filaments (Fig. 2 middle). The backbone structure of the caldesmon rod complex is similar to pure rod suggesting that there is no disruption of LMM packing and that caldesmon is bound at the head end of the myosin rod (S2) which is peeling away from the filament. The length and thickness of the whiskers suggested that they were complexes of caldesmon and S2 but it is uncertain if each whisker contained one or more molecules of each.

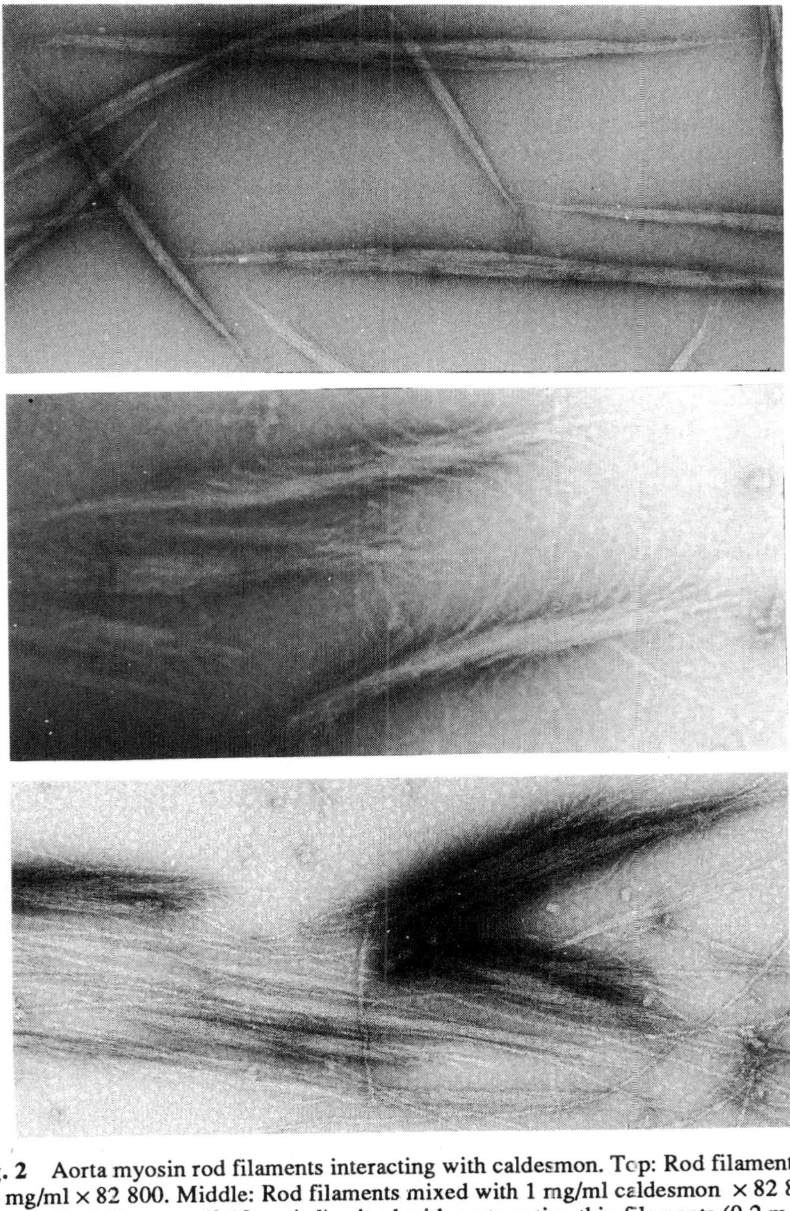

Fig. 2 Aorta myosin rod filaments interacting with caldesmon. Top: Rod filaments at 0.4 mg/ml × 82 800. Middle: Rod filaments mixed with 1 mg/ml caldesmon × 82 800. Bottom: Rod filaments (0.18 mg/ml) mixed with aorta native thin filaments (0.2 mg/ml) × 82 800.

CROSSLINKING OF ACTIN AND MYOSIN BY CALDESMON

We observed crosslinking of caldesmon containing sheep aorta thin filaments to both myosin filaments in ATP (relaxing conditions) and rod filaments. The latter complexes were particularly clear since there were no myosin heads present (Fig. 2, bottom). In both cases, thick and thin filaments were formed into small parallel aggregates which were never observed with myosin + pure actin and therefore must be due to the crosslinking ability of caldesmon. Thick and thin filaments seemed to form a loose structure in which the thin filament was rarely touching the surface of the thick filament. Occasionally we could see material crossing the gap which may represent the S2 of myosin contacting the thin filaments.

Conclusion

Caldesmon binds strongly to both F-actin and the S2 portion of smooth muscle myosin. This means that caldesmon crosslinks native thin filaments and myosin filaments by a mechanism quite independent of the contractile interaction.

Acknowledgements

This work has been supported by grants from the British Heart Foundation, Wellcome Trust and Medical Research Council.

References

1. Bryan, J., Imai, M., Lee, R., Moor, J., Cooke, G., Lin, S., 1989, *J. Biol. Chem.*, **264**, 13873–13879.
2. Marston, S. B., Lehman, W., 1985, *Biochem. J.*, **231**, 517–522.
3. Furst, D. O., Cross, R. A., DeMey, R. A., Small, J. V., 1986, *EMBO J.*, **5**, 251–257.
4. Bretscher, A., Lynch, ,W., 1985, *J. Cell Biol.*, **100**, 1656–1663.
5. Ban, T., Ishimura, K., Fujita, H., Sobue, K., Kakiuchi, S., 1984, *Acta Histochem. Cytochem.*, **17**, 331–338.
6. Lehman, W., Craig, R., Lui, J., Moody, C., 1989, *J. Muscl. Res.*, **10**, 101–112.
7. Marston, S. B., 1990, *Biochem. J.* in press.
8. Smith, C. W. J., Pritchard, K., Marston, S., 1987, *J. Biol. Chem.*, **262**, 116–122.
9. Moody, C. J., Lehman, W., Craig, R., 1990, *J. Muscl. Res.*, **11**, 176–185.
10. Pritchard, K. P., Marston, S. B., 1989, *Biochem. J.*, **257**, 839–843.
11. Marston, S. B., Redwood, C. S., Lehman, W., 1988, *Biochem. Biophys. Res. Comm.*, **155**, 197–203.
12. Marston, S. B., 1988, *FEBS Lett.*, **238**, 147–151.
13. Hemric, M. E., Chalovich, J.M., 1988, *J. Biol. Chem.*, **263**, 1878–1885.
14. Velaz, L., Ingraham, R. H., Chalovich, J. M., 1990, *J. Biol. Chem.*, **265**, 2231–2237.
15. Szpacenko, A., Dabrowska, R., 1986, *FEBS Lett.*, **202**, 182–186.

16. Redwood, C. R., Bryan, J., Cross, R.A., Kendrick-Jones, J., Marston, S. B., 1990, *FEBS Lett.*, in press.
17. Ikebe, M., Reardon, S., 1988, *J. Biol. Chem.*, **263**, 3055–3058.
18. Lash, J. A., Sellers, J. R., Hathaway, D. R., 1986, *J. Biol. Chem.*, **261**, 16155–16160.
19. Marston, S. B., 1989, *Biochem. J.*, **259**, 303–306.
20. Marston, S. B., 1989, *J. Muscl. Res.*, **10**, 97–100.
21. Marston, S. B., Smith, C. W. J., 1984, *J. Muscl. Res.*, **5**, 559–575.
22. Margossian, S. S., Lowey, S., 1982, *Methods Enzymol.*, **85**, 55–71.
23. Craig, R., Megerman, J., 1977, *J. Cell Biol.*, **75**, 990–996.
24. Small, J. V., Sobieszek, A., 1983. In: Stephens, N. L. (ed.), *The Biochemistry of smooth muscle*, pp. 85–140. Chemical Rubber Company Press, Boca Raton, Florida.
25. Yanagisawa, M., Hamada, Y., Katsuragawa, Y., Imamura, M., Mikawa, T., Masaki, T., 1987, *J. Mol. Biol.*, **198**, 143–157.

SECTION EIGHT
New Methods

56
Demonstration of Myosin Heavy Chain mRNA by using Degenerated Oligonucleotide PCR Primer

W. F. H. M. MOMMAERTS AND H. P. VOSBERG

Max Planck Institute for Medical Research, Heidelberg, Germany.

Muscle is subject to neurogenic and hormonal influences, and the myocardium is likewise affected by various functionally induced controls. Alterations in these regulatory relationships occur in development, in normal or pathological adaptation, in ageing and as a result of experimental perturbations. It would be timely to analyse such dependencies as due to selective isogene expressions, but one should keep in mind that these gene expressions occur in the midst of complex biological events, which may not be ignored. Even at the level of changes induced in selective gene expression, the situation is usually complex in that not one but several gene families are separately controlled, as if collectively striving toward a new functional harmony. Within this multitude, we limit ourselves to one problem: the alternate expression of myosin (MHC) isoforms. Such changes have thus far mostly been assessed by the analytical detection of isoproteins: S vs. F in muscle, alpha vs. beta in the heart. While keeping our above admonition in mind, we applied ourselves to the separate messenger RNAs, so as to learn about the selective transcription and translation processes which dominate the problematics between the gene on the one, the protein end-product on the other hand.

Against this background, then, we attempt to establish a system of techniques for the separate determination of iso-messengers side by side. Our approach is based upon the DNA-polymerase chain reaction, PCR.

When the relevant sequences are known, it is a relatively simple matter to design and synthesize the forward and reverse primers by which to amplify, in the PCR, any domain of a mRNA and its cDNA. To maximize the isoform specificity, we want such domains to include the non-translated sequences (here called non-*trans*) between the stop signal and the poly-A or poly-T. We have this preference because the nontrans domains carry the greatest diversity and difference incidence in any comparison between isoforms and in species comparisons as well. As to the forward primer, we have constructed a general

Muscle and Motility Vol. 2. Proceedings of XIXth European Conference in Brussels
© Intercept Ltd, PO Box 716, Andover, Hampshire, SP10 1YG, UK

MHC primer, reacting equally with alpha and beta isogenes, and in species comparisons as well.

Turning to the reverse primers, we first intended to use a 20-mer oligo-T as the universal primer, leaving it to the forward primer to limit the product to the MHC subfamily in question. However, this technique does not yield discrete bands: the product molecules are length-heterogeneous, because of the stochastic placement of the T(20)primers anywhere along the poly-A. While a PCR product accumulates, it would be difficult to purify, and its sequencing would be in the forward direction only. When all the relevant sequences are known, there is of course no difficulty, one simply constructs primers to fit. But a new approach is needed when that knowledge is absent. We have developed a solution in two steps.

The first is the introduction of the principle of anchor primers. While a complete primer is an oligonucleotide (e.g. 20-mer) of completely known identity, complementary to a template of interest, an anchor primer is a structure of the form T 20-n Nn in which the oligo-portion has affinity to a part of the poly-A while the remainder is specifically complementary to the penultimate sequence of the non-*trans* just at the poly-A or -T: the anchor. This anchor attaches the primer precisely at the junction of the non-*trans* and the tail (Fig. 1). This assures that all product molecules are polymerase-synthesized and that all are in register. Thus, also, they are all of the same lengths and form sharp bands. And, they can be purified, and sequenced. A four-base anchor can

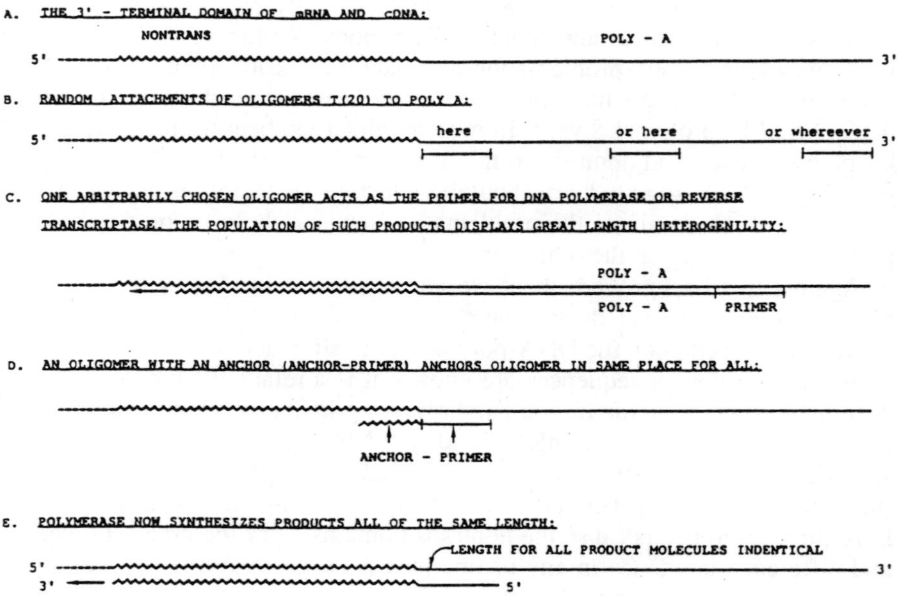

Fig. 1 Complete primers are those of which the sequences are completely known, and which have maximal specificity. Anchor primers have a lesser number of known bases such as 8, 6, 4, or 3 out of 20 or 28. All of these yield PCR products showing sharp bands.

suffice. However, accidental similarities may cause cross reactivity: the polymerase is tolerant.

Application of the anchor principle still requires sequence knowledge to chose the right anchor. This could be a gamble, but, as the last step in our build-up, we have a more systematic solution: the principle of degenerate anchors. In these, a T-tail and an anchor are again in tandem but the anchor, at each base position, permutates among the four bases A, G, C and T. Advantage was taken of the limitation that in the contemporary practice of solid state synthesis it would be difficult to obtain the substrate-bound starting base other than as a single A, G, C, or T. Thus we prepare four separate primers, each having one lone base, which is to hybridize on the template at position (-4), whereas the other positions are degenerate (Fig. 2). This shows the distinction of the two human cardiac isoforms.

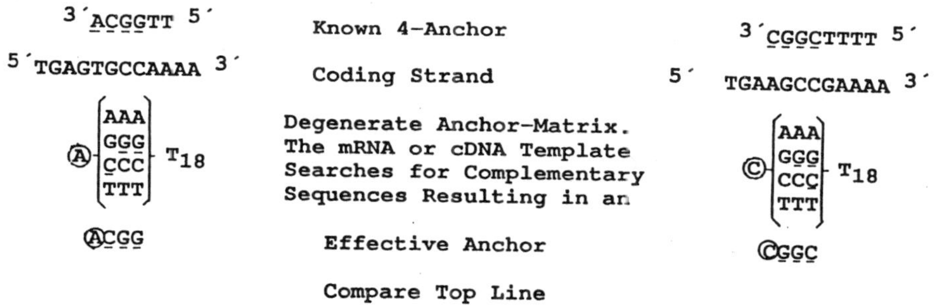

Fig. 2 Principle of the action of degenerate anchor primers. The matrix is a shorthand representation of the possible nucleotide compositon at positions 0 to -3, and the lone nucleotide at position -4 from the junction to the T-tails. The template (e.g. cDNA) finds its optimal affinity among the possible sequences. The lone nucleotide A corresponds to Myosin α, C to β. The lone nucleotides G and T are not illustrated.

Care is taken in the design that the differential choices can be identified. We have settled upon two criteria. The first is the length of the PCR product; second, is the use of restriction analysis for the human cardiac isogenes we have these restrictions: Hind III for the α- and MboI for the β-isogene.

References

Innis, M. A., Gelfand, D. H., Sninsky, J. J., White, T. J. , 1990, *PCR-Protocols.* Academic Press, London.

57
The Use of Microdialysis for the *in vivo* Study of Skeletal Muscle Glucose Metabolism

J. HENRIKSSON, H. ROSDAHL, T. FUCHI, Y. OSHIDA AND U. UNGERSTEDT

Departments of Physiology III and Pharmacology, Karolinska Institutet, Box 5626, S-114 86, Stockholm, Sweden

Introduction

In this project a relatively new technique, microdialysis, has been used in studies of skeletal muscle glucose metabolism. Microdialysis combines the advantage of the *in vitro* approach with that of *in vivo* studies. Thus, the method enables continuous control of the extracellular environment of muscle with minimal interference with the normal physiological state. In addition, microdialysis renders possible the study of effects of substances which, if administered into the blood, would have complex effects involving a number of different organs. This technique therefore promises to be an important tool in the arsenal of muscle metabolic techniques.

Principle

The idea of tissue dialysis was first applied by Delgado *et al.* (1) and then by Ungerstedt and Pycock (2), who introduced the use of hollow fibres perfused

Muscle and Motility Vol. 2. Proceedings of XIXth European Conference in Brussels
© Intercept Ltd, PO Box 716, Andover, Hampshire, SP10 1YG, UK

continuously with physiological fluid. To date the technique has been used mainly in brain research for the analysis of such compounds as catecholamines, amino acids and peptides, but it has in recent years also been applied to other organs, such as the adipose tissue (3). The microdialysis probe can be said to constitute an 'artificial blood vessel', inserted into the tissue, where diffusion of chemical substances will occur in the direction of the lowest concentration. Constituents of the muscle extracellular space can be collected for analysis and foreign compounds may be added to the extracellular space via the dialysis probe and their effects can be monitored.

Methods

Male Wistar rats weighing 255–300 g were anaesthetized with sodium pentobarbital and placed on a heating pad connected to a temperature controller. A catheter was inserted towards the heart via the left jugular vein and the skin over both calves were cut open. The muscle fascia just above the Achilles tendon was punctured with a fine needle and microdialysis probes (Carnegie Medicin, Stockholm, Sweden) were gently inserted superficially into the medial part of the gastrocnemius muscle. In some experiments probes were also placed in the inguinal adipose tissue or deep in the quadriceps femoris muscle. Microdialysis probes (4) consisted of a tubular membrane (polycarbonate, outer diameter 0.5 mm; length 10 mm; molecular weight cutoff 5000–20 000 daltons); glued at one end to a steel cannula with the other end sealed with glue. The perfusate is carried to the distal end via a fine cannula inside the probe, flows back between the membrane and the inner cannula in a retrograde direction and is collected at the outlet of the probe. Diffusion through the membrane is limited by the total membrane length (10 mm), given its constant diameter, as well as by the characteristics of the tissue surrounding the probe. The perfusion is controlled by a micro-injection pump with three different syringes which may contain perfusion fluids of different compositions. It is possible rapidly to change the perfusion fluid by a liquid switch. From the liquid switch, a single tube connects the perfusion medium to the probe. As a rule, the outflow tubing from the probes was cut to a length of 10 cm. This limits the total dead space from tissue to outlet to 5 µl. The samples were in the present study collected in 300 µl polyethylene tubes; we also commonly collect samples in small glass vials in a refrigerated microfraction collector (Carnegie Medicin, Stockholm, Sweden) or (for very low perfusion flow rates) under mineral oil in teflon oil wells (5). In the present experiments the perfusion medium was Krebs-Henseleit bicarbonate buffer (pH 7.4) (6). Unless otherwise stated, the perfusion rate was 1 µl /min.

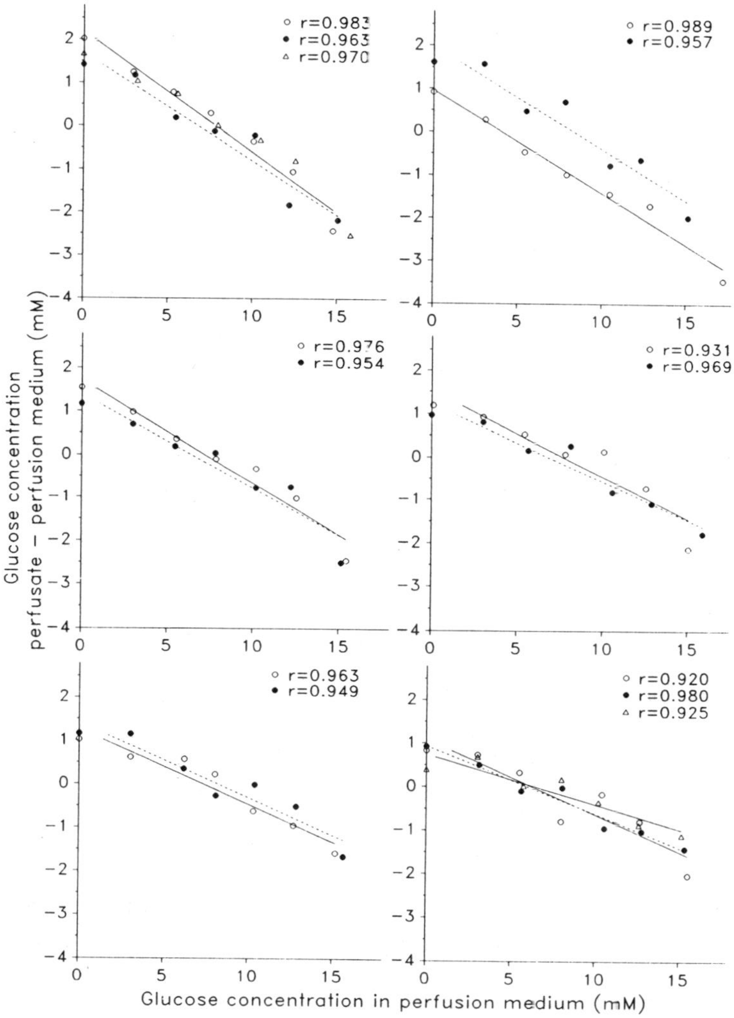

Fig. 1 Results of 14 different *in vivo* recovery experiments. Probes were placed in the gastrocnemius muscle of the rat and perfused at 1 μl/min with media with different glucose concentrations (x-axis). The y-axis shows the net change in the glucose concentration of the perfusion medium after dialysis has occurred (glucose concentration in the collected perfusate minus the glucose concentration in the infused perfusion medium). The *in vivo* relative recovery (i.e. the fraction of an absolute change in the diffusion gradient across the dialysis membrane that is recovered as a net change in the glucose concentration of the perfusate collected at the probe outlet) is calculated as the slope of the line ($\Delta y/\Delta x$). The x-value at $y = 0$ (i.e. no net flux of glucose) is taken to represent the tissue extracellular glucose concentration. This value, 6.9 ± 1.1 mM (mean ± SD), was not significantly different from the plasma glucose concentration of 7.3 ± 0.6 mM ($n = 5$).

Results and Discussion

RECOVERY

The term 'recovery' (alternatively, relative recovery or extraction fraction) is defined as the ratio between the concentration of a substance at the dialysate outlet and the concentration of the same substance in the solution outside the probe. Absolute recovery is the total amount of a substance collected in the outflow per unit time. The recovery can be determined *in vitro* by placing the perfused dialysis probe in a solution of known solute concentration. In a typical experiment the *in vitro* recoveries for glucose at different perfusion flow rates were 46.1 ± 2.1% at 1 µl/min (mean ± SD, n = 6; 8.22 mM external glucose concentration, room temperature), 70.72 ± 7% at 0.5 µl/min, 27.61 ± 7% at 2 µl/min, 15.41 ± 3% at 4 µl/min, 8.7 ± 1.0% at 8 µl/min and 6.41 ± 1% at 12 µl/min. Consequently, the absolute recovery increased with the proportions of 1 (1 = 0.5µl/min):1.30: 1.56: 1.74: 1.96: 2.17 with the increasing perfusion flow rates. These *in vitro* recovery data are useful in describing the diffusion characteristics of the probe itself, but are of limited value in predicting the probe behaviour *in vivo*. The reason is that the relative and absolute recovery values *in vivo* is generally more dependent on tissue related-factors than on the characteristics of the probe membrane *per se*. An example of this will be given here.

At steady state the relative recovery of glucose *in vivo* may be calculated as the glucose concentration at the dialysate outlet divided by the blood glucose concentration. This can be done since in the basal state the extracellular glucose concentration in adipose tissue (7) and muscle (Fig. 1) is very close to the blood glucose concentration. In a series of experiments ($n = 8$) at 1 µl/min we found the recovery to be significantly higher in adipose tissue than in muscle (37.65 ± 9% vs. 17.1 ± 1.8%, $P<0.01$, mean SEM). When the perfusion flow rate was lowered to 0.5 µl/min the recovery increased, but, somewhat unexpectedly, the increase was much greater in skeletal muscle than in adipose tissue. This nearly abolished the difference in in vivo recoveries between the two organs (adipose tissue: 51.3 ± 4.1%; muscle: 50.0 ± 3.1%, n.s.). There may be multiple reasons for the different dialysis performance in the two tissues, but the finding illustrates the need for caution in interpreting organ-specific microdialysis results. This may be particularly true regarding tissues of small laboratory animals, where the microdialysis probe may disturb the spatial concentration profile of a substance within a much larger fraction of a specific organ than in a larger animal. An approach to experimentally overcome the problem in determining the undisturbed extracellular concentration of a compound was devised by Lönnroth *et al.* (7). We have applied this technique to determine muscle extracellular glucose concentration, the results are shown in Fig. 1.

EFFECT OF INSULIN INFUSION

The changes in the glucose concentration at the dialysate outlet and in the blood were followed after intravenous injection of bolus doses of insulin (60–1200 mU/kg body wt). Glucose rapidly equilibrates between different fluid compartments and therefore a close similarity in the time course of insulin-induced change between muscle, adipose tissue and blood was expected. Interestingly, at a perfusion flow rate of 1 μl/min, the relative decrease in the muscle dialysate glucose concentration (to 30.0 ± 3.5% of basal values at the maximal effect, mean ± SEM, $n = 8$) was more pronounced ($P<0.01$) than the change in the blood glucose concentration (to 44.3 ± 2.9% of basal values). This tended to be true also for the decrease in the glucose concentration of the adipose tissue dialysate (to 35.4 ± 6.6% of initial values). This has an important methodological implication; it confirms that we are performing tissue dialysis in a space clearly different from the blood and therefore indicates that damage to cells and blood vessels when inserting the dialysis probe has no detrimental effect on the results in this respect. The apparently higher insulin responsiveness in muscle than in adipose tissue at this perfusion flow rate may, however, be secondary to the much lower relative recovery. This is supported by the fact that the difference between muscle and adipose tissue was reversed when the flow rate was decreased to 0.5 μl/min, at which flow rate, as shown above, the relative recoveries were nearly equal in the two tissues. (Maximal effects: blood glucose concentration decreased to 40.7 ± 5.6% of initial values, mean SEM, $n = 5$. The corresponding values for the muscle and adipose tissue dialysates were 36.2 ± 6.5% and 23.5 ± 6.2%, respectively; ANOVA: muscle vs. adipose tissue, $P<0.01$; adipose tissue vs. blood, $P<0.01$; muscle vs. blood, $P<0.1$.) The difference between muscle and adipose tissue noted at 1 μl/min disappeared when the local tissue glucose drainage was minimized by including 2.5 mM glucose in the perfusion solution. It may therefore be that, during hypoglycaemia, the muscle tissue is depleted of glucose by the microdialysis probes at too high perfusion flow rates.

EFFECTS OF ADENOSINE ON INSULIN-STIMULATED MUSCLE GLYCOLYSIS

When insulin was infused i.v. at five different rates, 4.4–333 mU $\times$ kg^{-1} $\times$ min^{-1}, the lactate concentration in the collected muscle perfusates increased dosedependently (up to fourfold). This measure of muscle insulin sensitivity with respect to glucose metabolism was significantly increased and decreased, respectively, when an adenosine receptor antagonist (8-phenyltheophylline, 30 μM) or adenosine (5 μM) was included in the perfusion medium (Fig. 2). These findings illustrate that the microdialysis technique is sensitive enough to pick up local differences in muscle release of a compound although the activator, in this case insulin, is administered via the blood. This underlines the unique possibilities offered by microdialysis for the *in vivo* study of metabolic regulation in skeletal muscle.

Acknowledgements

The studies were supported by the Swedish Medical Research Council (Project B8714X07917), the Karolinska Institute, the Magnus Bergvall foundation, and the Research Council of the Swedish Sports Federation.

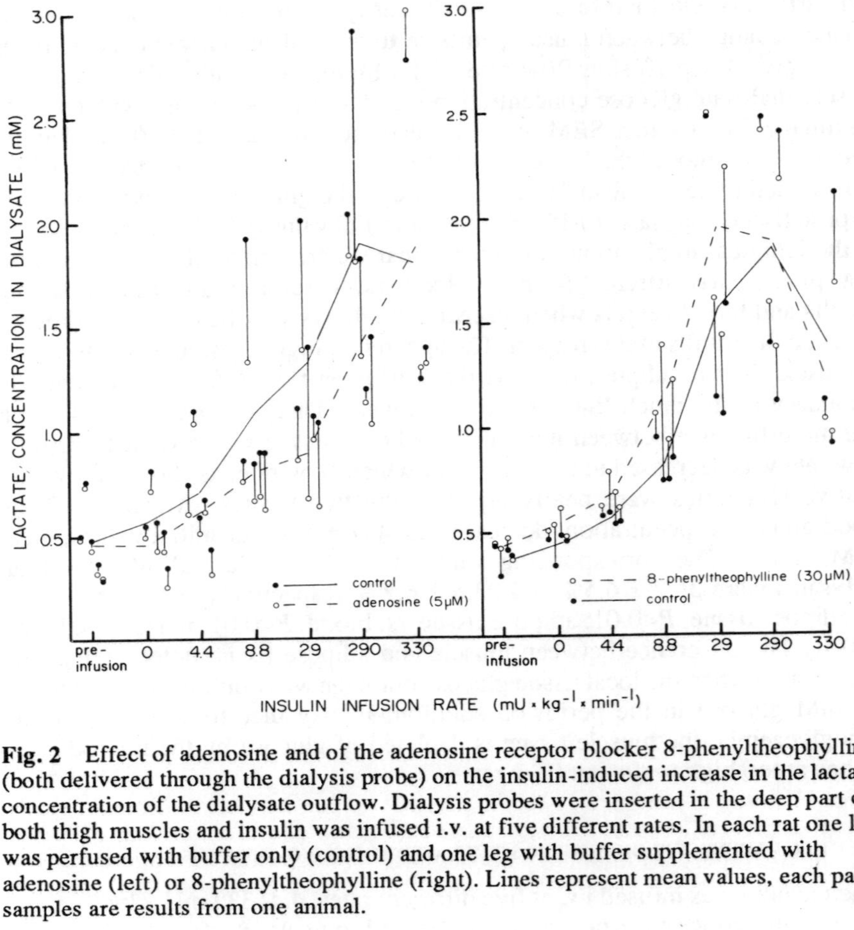

Fig. 2 Effect of adenosine and of the adenosine receptor blocker 8-phenyltheophylline (both delivered through the dialysis probe) on the insulin-induced increase in the lactate concentration of the dialysate outflow. Dialysis probes were inserted in the deep part of both thigh muscles and insulin was infused i.v. at five different rates. In each rat one leg was perfused with buffer only (control) and one leg with buffer supplemented with adenosine (left) or 8-phenyltheophylline (right). Lines represent mean values, each pair of samples are results from one animal.

References

1. Delgado, J. M. R., DeFeudis, F. V., Roth, R. H., Ryugo, D. K., Mitruka, B. K., 1972, *Arch. Int. Pharmacodyn.*, **198**, 9–21.
2. Ungerstedt, U., Pycock, C., 1974, *Bull. Schweiz. Akad. Med. Wiss.*, **1278**, 1–13.
3. Arnér, P., Bolinder, J., Eliasson, A., Lundin, A., Ungerstedt, U., 1988, *Am. J. Physiol.*, **255**, E737–E742.
4. Tossman, U., Ungerstedt, U., 1986, *Acta Physiol Scand.*, **128**, 9–14.
5. Lowry, O. H., Passonneau, J. V., 1972, *A Flexible System of Enzymatic Analysis*. Academic Press, New York.
6. Krebs, H. A., Henseleit, K., 1932, *Hoppe-Seyler Z. Physiol. Chem.*, **210**, 33–66.
7. Lönnroth, P., Jansson, P. A., Smith, U., 1987, *Am. J. Physiol.* **253**, E228–E231.

58
A New Method Combining Histo- and Immunohistochemistry for Muscle Tissue

B. ZAWADOWSKA AND M. MÜNTENER

Institute of Anatomy, University of Zürich-Irchel, Winterthurerstrasse 190, 8057 Zürich, Switzerland

Introduction

Today muscle fibre typing is mostly performed by the histochemical reaction for myofibrillar ATPase (mATPase) activity (1, 2, 20). On this basis type IA, IB, IIA, IIB, and IIC muscle fibres can be identified (3, 4). Type I and II fibres can also be distinguished immunohistochemically using antibodies against myosin heavy chains (5–8). Myofibrillar ATPase activity largely correlates with the content of parvalbumin (PV) (9, 10; personal observation). The soluble, Ca^{2+}-binding protein PV is thought to be involved in the relaxation process of muscle fibres (11, 12, 13). However, immunolocalization of PV requires fixed and paraffin-embedded tissue while the histochemical reaction for mATPase is performed on unfixed frozen tissue. For this reason histo- and immunohistochemistry can only be performed on consecutive tissue blocks (fixed paraffin-embedded and unfixed frozen, respectively) and the correlation of the findings is very time consuming and usually requiring a great amount of material. The aim of our study was to establish a method allowing histo- and immunohistochemical reactions on consecutive sections. This method would also be very useful when only a scarce amount of tissue is available, as for instance in clinical investigations.

Muscle and Motility Vol. 2. Proceedings of XIXth European Conference in Brussels
© Intercept Ltd, PO Box 716, Andover, Hampshire, SP10 1YG, UK

Materials and Methods

Male rats (strain Zur:Siv) supplied by the Institute of Laboratory Animal Science, University of Zürich were used. In preliminary experiments rats were anaesthetized with 100 mg/kg body weight Trapanal (Byk-Gulden) i.p. and perfused for 5 min via the abdominal aorta with a fixative containing 2.5% paraformaldehyde, 0.1% glutaraldehyde, and 0.05% picric acid in 0.1 M cacodylate buffer with 3.5% saccharose and 4% HAES (Fresenius), pH 7.4, followed by perfusion with 3.5% saccharose in the same buffer for 10 min. The sternomastoideus muscles were excised, washed in cacodylate buffer, dipped in talcum powder mixed with tissue-tek, and frozen in isopentane cooled by liquid nitrogen (-160°C).

In further experiments the animals were anaesthetized with 0.25 ml/kg body weight Innovar-Vet (Pitman-Moore) i.m. combined with 2.5 mg Valium (Roche) and 7.5 mg Nembutal (Abbott) i.p. Three fast muscles — extensor digitorum longus (EDL), sternomastoideus, and sternohyoideus —were excised, divided transversally and longitudinally with a razor blade, and fixed by immersion. Different fixatives were tested:

1. the fixative described above for perfusion;
2. 2 and 4% paraformaldehyde in 0.1 M cacodylate buffer with 4% HAES and without HAES pH 7.4;
3. 2 and 4% paraformaldehyde in 0.2 M phosphate buffer pH 7.4;
4. Bouin's solution;
5. Bouin's solution diluted 1:1 in distilled water.

All these fixatives were applied for 5, 10, 15, 20, 30, 45, 60, and 120 min, respectively. Subsequently the samples were washed in 0.1% cacodylate buffer (pH 7.4) for 1 h. After wiping away excess of buffer from the surface, the samples were dipped in talcum powder mixed with tissue-tek and frozen in isopentane cooled by liquid nitrogen. Consecutive transversal sections (10–12 μm) were cut in a cryostat at -20°C and mounted on slides coated with chrome-gelatine.

HISTOCHEMISTRY

The reaction for mATPase activity was carried out according to refs. 2, 14, 15) with certain modifications. Consecutive sections without air drying and without additional fixation by formaldehyde were treated with pHs ranging from 10.0 to 10.5 for alkaline and 3.9 to 5.0 for acid preincubation, respectively. Preincubation times ranged from 1 to 10 min.

Consecutive sections were incubated (also without additional fixation) for cytochrome *c* oxidase according to Wong-Riley for 1 to 2 h until various degrees of staining intensities in the different fibre types were clearly visible.

IMMUNOHISTOCHEMISTRY

For the immunolocalization of PV, monoclonal antibodies (antibody 235, ref. 17) were used. Consecutive sections were mounted on slides coated with chrome-gelatine and dried for 10 min at room temperature. Visualization of PV was performed according to the following techniques:

1. indirect immunoperoxidase
2. avidin-biotin-peroxidase complex (ABC) (18),
3. single and double peroxidase anti-peroxidase (PAP)- and avidin-biotin-peroxidase complex (ABC) (19).

The primary antibody was used in dilutions from 1:500 to 1:10000. Incubations were tested for 24, 48, and 72 h at 4°C and for 20, 30, 60, 90, 120, and 180 min at room temperature and 37°C.

In immunoperoxidase technique sections were treated with 10% foetal calf serum (Flow Laboratories) or 5% bovine serum albumine (Fluka) to reduce the non-specific background staining. As the second layer, peroxidase labelled rabbit anti-mouse IgG (Bio-Makor) was used. For the other techniques 10% normal horse serum (Vector) was applied to prevent nonspecific staining. Biotinylated anti-mouse IgG (Vector) was used as a link antibody. In ABC technique avidinbiotin-peroxidase complex (Vector) was employed as a third layer. In the combined techniques, biotinylated anti-mouse IgG was followed by mouse clono-PAP (Dakopatts) in single PAP-ABC; these two steps were repeated in double PAP-ABC technique. In all techniques the immune complex was visualized using diaminobenzidine/HCl/H_20_2 (Polyscience) as substrate. For controls, sections were either incubated with normal mouse serum in place of PV 235 or incubation with the primary antibody was omitted.

Results

The criterion for the optimal combination of the different parameters as fixative buffers, fixation times, preincubation and incubation times, pH values, temperature, and dilutions of antibodies, was the best distinguishability between the three main fibre types (I, IIA, IIB). The best results were obtained after fixation of the tissue (by perfusion or immersion) with 4% paraformaldehyde in phosphate buffer for 15 min (Fig 1). For histochemistry the best pH value for pre-incubation and incubation was 10.4 and 9.5, respectively. After acid pre-incubation the reversal staining pattern of the type I fibres was absent. Alkaline pre-incubation, however, enabled the distinction of type I, IIA, and IIB fibres. For immunohistochemistry single and double PAP-ABC techniques following incubation with the primary antibody diluted 1:10000 at 37°C for 2 h gave the best contrast between the three fibre types (Fig. 1). The results of the histo- and immunohistochemical reactions are summarized in Table 1.

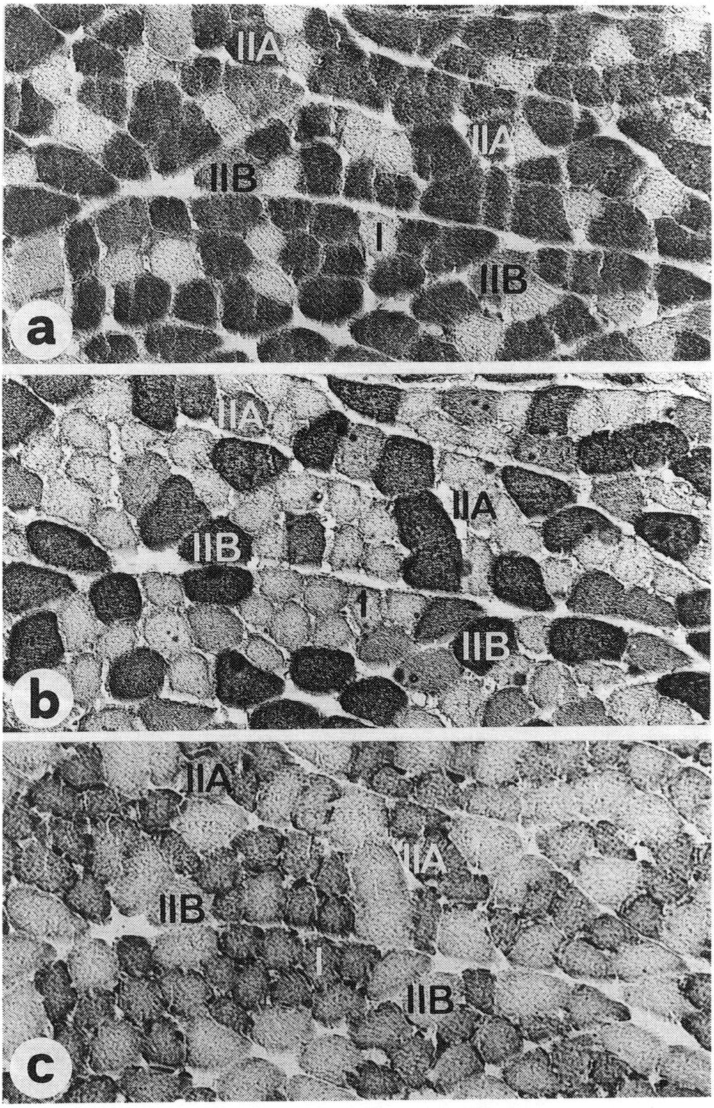

Fig. 1 Consecutive transversal sections of the EDL muscle processed for: (a) mATPase activity (incubation at pH 9.5 after pre-incubation at pH 10.4); (b) immunoreactivity of PV (dilution 1:10000); and (c) cytochrome *c* oxidase activity. Muscle fibre types are indicated. × 58.

Discussion

The present study shows for the first time mATPase activity, PV content, and cytochrome *c* oxidase activity on consecutive sections. This combination of enzyme histochemistry and immunohistochemistry renders results which are

Table 1 Staining intensity of the three main fibre types processed for thedemonstration of mATPase, cytochrome *c* oxidase, and PV. Because of similar results the techniques of indirect immunoperoxidase and ABC as well as single and double PAP+ABC have been taken together.

Fibre types	mATPase Pre-incubation pH 10.4	10.5	4.3	PV Indirect immuno-perosxydase or ABC	Single or double PAP + ABC	Cytochrome *c* oxidase
I	0	0	0	0	0	+++
IIA	+++	++	++	0 to ++	0 to +++	++ to +++
IIB	+ to ++	0 to +	0 to +	++	++	0

0, absent staining; +, week staining, ++; intermediate staining; +++ strong staining.

difficult to obtain by performing the two techniques on different consecutive blocks of tissue. In three muscles examined, positive immunostaining was observed only in IIA and IIB fibres which is in agreement with earlier findings demonstrating the association of PV with fast-contracting muscle fibres (9) and postulating the role of this protein as a mediator in the relaxation process (11, 12, 13).

In view of our results we suggest this method which combines the demonstration of mATPase activity and parvalbumin immunoreactivity as applicable to basic and applied research of muscle tissue.

Acknowledgements

We thank Ms M. Carrel, I. Drescher, S. Varol, and H. Weber for excellent technical assistance. The PV antibody 235 was a gift of Dr M. R. Celio (Fribourg, Switzerland), to whom we are also thankful for helpful discussion. We are indebted to Dr. B. Kaissling for providing us with perfused sternomastoideus muscles. This work was supported by the EMBO Stiftung Zürich, the Hartmann Muller-Stiftung, Zürich, the Roche Research Foundation, Basel, and the Stiftung für wissenschaftliche Forschung an der Universitat Zürich, Switzerland.

References

1. Brooke, M. H., Kaiser, K .K., 1969, *J. Histochem. Cytochem.*, **17**, 431–432.
2. Guth, L., Samaha, F. J., 1970, *Exp. Neurol.*, **28**, 365–367.
3. Brooke, M. H., Kaiser, K. K., 1970, *Arch. Neurol.*, **23**, 369–379.
4. Karpati, G. A., Eisen, A., Carpenter, S., 1975, *J. Histochem. Cytochem.*, **23**, 89–91.
5. Billeter, R., Weber, H., Lutz, H., Howald, H., Eppenberger, H.M., Jenny, E., 1980, *Histochemistry*, **65**, 249–259.

6. Gauthier, G. F., Lowey, S., 1977, *J. Cell Biol.*, **74**, 760–779.
7. Moore, S. E., Orest, H., Walsh, F. S., 1984, *J. Neuroimmunol.*, **7**, 137–149.
8. Pierobon-Bormioli, S., Sartore, S., Dalla Libera, L., Vitadello, M., Schiaffino, S., 1981, *J. Histochem. Cytochem.*, **29**, 1179–1188.
9. Celio, M. R., Heizmann, C. W., 1982, *Nature*, **297**, 504–506.
10. Heizmann, C. W., Celio, M. R., 1987, *Meth. Enzymol.*, **139**, 552–570.
11. Gerday, C., Gillis, J. M., 1976, *J. Physiol. (Lond.)*, **258**, 96–97.
12. Gillis, J. M., Gerday, C., 1977. In Wasserman, R.H. *et al.Calcium Binding Proteins and Calcium Function* pp. 193–196, Elsevier, Amsterdam.
13. Pechère, J. F., Derancourt, J., Haiech, J., 1977, *FEBS Lett.*, **75**, 111–114.
14. Müntener, M., 1979, *Histochemistry*, **62**, 299–304.
15. Müntener, M., 1982, *Exp. Neurol.*, **77**, 668–678.
16. Wong-Riley, M., 1979, *Brain Res.*, **171**, 11–28.
17. Celio, M. R., Baier, W., Schärer, L., de Viragh, P. A., Gerday, C. H., 1988, *Cell Calcium*, **9**, 81–86.
18. Hsu, S., Raine, L., Fanger, H., 1981, *J. Histochem. Cytochem.*, **29**, 577–580.
19. Davidoff, M., Schulze, W., 1990, *Histochemistry*, **93**, 531–536.
20. Padykula, H. A., Herman, E., 1955, *J. Histochem. Cytochem.*, **3**, 170–195.

Index

Abbreviations: MHC, myosin heavy chain; MLC, myosin light chain